Conference Proceedings of the Society for Experimental Mechanics Series

Series Editor

Kristin B. Zimmerman, Ph.D.
Society for Experimental Mechanics, Inc.,
Bethel, CT, USA

More information about this series at http://www.springer.com/series/8922

Andreas Linderholt • Matthew S. Allen • Randall L. Mayes • Daniel Rixen
Editors

Dynamics of Coupled Structures, Volume 4

Proceedings of the 36th IMAC, A Conference and Exposition on Structural Dynamics 2018

Springer

Editors
Andreas Linderholt
Department of Mechanical Engineering
Linnaeus University
Växjö, Sweden

Matthew S. Allen
Department of Engineering Physics
University of Wisconsin–Madison
Madison, Wisconsin, USA

Randall L. Mayes
Structural Dynamics Department
Sandia National Laboratory
Albuquerque, New Mexico, USA

Daniel Rixen
Lehrstuhl für Angewandte Mechanik
TU München
Garching, Bayern, Germany

ISSN 2191-5644 ISSN 2191-5652 (electronic)
Conference Proceedings of the Society for Experimental Mechanics Series
ISBN 978-3-030-09045-6 ISBN 978-3-319-74654-8 (eBook)
https://doi.org/10.1007/978-3-319-74654-8

Printed on acid-free paper

This Springer imprint is published by the registered company Springer International Publishing AG part of Springer Nature.
The registered company address is: Gewerbestrasse 11, 6330 Cham, Switzerland

Preface

Dynamics of Coupled Structures represents one of nine volumes of technical papers presented at the 36th IMAC, A Conference and Exposition on Structural Dynamics, organized by the Society for Experimental Mechanics, and held in Orlando, Florida, February 12–15, 2018. The full proceedings also include volumes on *Nonlinear Dynamics*; *Dynamics of Civil Structures*; *Model Validation and Uncertainty Quantification*; *Special Topics in Structural Dynamics*; *Structural Health Monitoring, Photogrammetry and DIC*; *Rotating Machinery, Vibro-Acoustics and Laser Vibrometry*; *Sensors and Instrumentation, Aircraft/Aerospace and Energy Harvesting*; and *Topics in Modal Analysis and Testing*.

Each collection presents early findings from experimental and computational investigations on an important area within Structural Dynamics. Coupled structures or substructuring is one of these areas.

Substructuring is a general paradigm in engineering dynamics where a complicated system is analyzed by considering the dynamic interactions between subcomponents. In numerical simulations, substructuring allows one to reduce the complexity of parts of the system in order to construct a computationally efficient model of the assembled system. A subcomponent model can also be derived experimentally, allowing one to predict the dynamic behavior of an assembly by combining experimentally and/or analytically derived models. This can be advantageous for subcomponents that are expensive or difficult to model analytically. Substructuring can also be used to couple numerical simulation with real-time testing of components. Such approaches are known as hardware-in-the-loop or hybrid testing.

Whether experimental or numerical, all substructuring approaches have a common basis, namely the equilibrium of the substructures under the action of the applied and interface forces and the compatibility of displacements at the interfaces of the subcomponents. Experimental substructuring requires special care in the way the measurements are obtained and processed in order to assure that measurement inaccuracies and noise do not invalidate the results. In numerical approaches, the fundamental quest is the efficient computation of reduced order models describing the substructure's dynamic motion. For hardware-in-the-loop applications difficulties include the fast computation of the numerical components and the proper sensing and actuation of the hardware component. Recent advances in experimental techniques, sensor/actuator technologies, novel numerical methods, and parallel computing have rekindled interest in substructuring in recent years leading to new insights and improved experimental and analytical techniques.

The organizers would like to thank the authors, presenters, session organizers, and session chairs for their participation in this track.

Växjä, Sweden — Andreas Linderholt
Madison, WI, USA — Matthew S. Allen
Albuquerque, NM, USA — Randall L. Mayes
München, Germany — Daniel Rixen

Contents

Chapter 1
Modeling an Electrodynamic Shaker Using Experimental Substructuring

Benjamin Moldenhauer, Matt Allen, Washington J. DeLima, and Eric Dodgen

Abstract When a vibration test is performed, the dynamic modes of a shaker or shaker table may be excited, changing the response of the parts depending on their location on the shaker and possibly invalidating the test. However, shaker systems are difficult to model because of joints with unknown properties, unknown stiffness and damping effects of the magnetic field and because drawings of the internal components are rarely available. This work explores the use of experimental/analytical substructuring to create a test based model of a shaker. A modal test was performed and a modal model of a shaker was obtained with an adapter plate attached. This was coupled to a finite element model of the adapter plate, and the Transmission Simulator Method was used to assemble the two and to remove one copy of the adapter plate. The model thus obtained can then be used to predict the variation in the environment across the top of the shaker as different components are connected.

Keywords Experimental Substructuring · Component Mode Synthesis · Transmission Simulator · Shaker Modeling · Modal Testing

1.1 Introduction

A random vibration test is commonly used to qualify systems that will be submitted to similar vibration loads during their life. It is important to understand the interaction between the DUT (device under test) and the test equipment (shaker, shaker-DUT adapter and control system) during a closed-loop vibration test to establish the test robustness.

A predictive model can be an integral piece of successful environmental testing by providing valuable insight into how the systems may respond during test and therefore help to achieve challenging test requirements, prevent accidental over-test of sensitive units and costly retest. Three approaches can be used to create a predictive model: experimental based model, analytical based model and experimental-numerical model. The experimental model can accurately represent the system since it captures all the information necessary to represent the actual system, but its creation can be time consuming, creating delays and adding cost. On the other hand, a detailed analytical model (e.g. FEM model) that represents all the components (shaker, shaker-adapter, fixture and DUT) involved in the test and their interactions can be computationally expensive and may not be very practical to build due to the multitude of unknowns present in the test system. One of the most difficult component to model numerically is the shaker due to the intricate geometry and unknown material properties of its internal components, the stiffness and damping from the joints where the adapter plate attaches to the shaker, and the magnetic field stiffness and damping properties [1]. An alternative is the creation of a hybrid experimental-analytical model, in which the accumulative effects of the complex behaviors due to the shaker components may be captured experimentally without specifically identifying them individually, while allowing for the other less complex components to be modeled analytically and coupled to the experimental model of the shaker to create a model of the assembled system.

This paper presents a method for modeling a vibration shaker test system using the transmission simulator method (TSM) [2, 3] of experimental component mode synthesis (CMS) substructuring. (In early works the constraint methodology used in the TSM is referred to as modal constraints for fixture and subsystem (MCFS).) The TSM addresses many of the issues that limit the effectiveness of classical structuring techniques. For example, whereas traditional methods require three translational and rotational measurements and forces at each connection point between substructures, the transmission simulator (TS) simply needs as many translational measurements as are necessary to capture the modes to be used in CMS.

B. Moldenhauer (✉) · M. Allen
University of Wisconsin – Madison, Madison, WI, USA
e-mail: bmoldenhauer@wisc.edu; msallen@engr.wisc.edu

W. J. DeLima · E. Dodgen
Honeywell - Kansas City, Kansas City, MO, USA

A. Linderholt et al. (eds.), *Dynamics of Coupled Structures, Volume 4*, Conference Proceedings of the Society for Experimental Mechanics Series, https://doi.org/10.1007/978-3-319-74654-8_1

Also, the damping and stiffness associated with the joints connecting the TS are accounted for within experimental data instead of roughly approximated or neglected. Furthermore, the presence of the TS mass loads the interface, enriching the modal basis of the identified model. Additional benefits of, and insights into, the TSM are discussed in [2]. Through the TSM, modal substructuring is employed to create a hybrid experimental/FEM model of the shaker that will allow for predictions of how the top surface will behave when potential test articles are attached.

In the following sections, a summary of the theory behind the TSM is presented, followed by a description of the various subsystems used to create the shaker model. The finite element models (FEM) and experimental procedures required to collect and process the necessary data to perform the TSM substructuring are then defined. This is followed by a discussion of the results yielded by the compiled structure. The paper concludes with remarks as to the feasibility of modeling shakers with the TSM and possible future work.

1.2 Theory

The TSM technique combines at least three subsystems: an experimental subsystem, a finite element subsystem, and a transmission simulator that is present in both. The method requires the modal parameters of each subsystem, which are gathered experimentally and analytically. The experimental subsystem consists of some tangible component and a TS, which is usually a single part machined from steel or aluminum so that it is easily modeled in finite elements [2]. A modal test is performed on this assembly using standard techniques in order to obtain its modal natural frequencies, damping ratios and mass-normalized mode shapes. Note that in common modal tests one often does not seek to estimate the scaling of the mode shapes, so some attention is given to this point later. The finite element subsystem contains a FEM of some component of interest and the TS. The mode shapes and natural frequencies of this model, as well as a FEM of the isolated TS, are computed with FEA software. These subsystems are then combined to estimate the modes of the assembly.

A visual depiction of the TSM in the current case of the shaker is given in Fig. 1.1. Simply put, the method removes the plate from an experimentally measured modal model of the shaker and plate, and replaces it with the plate and block. This is more effective than merely testing the top of the shaker and coupling it to the plate and block FEM due to the fact that the interface effects between the plate and the shaker, such as damping and stiffness, would be neglected. In the TSM, these are accounted for in the experimental data and remain after removing the plate.

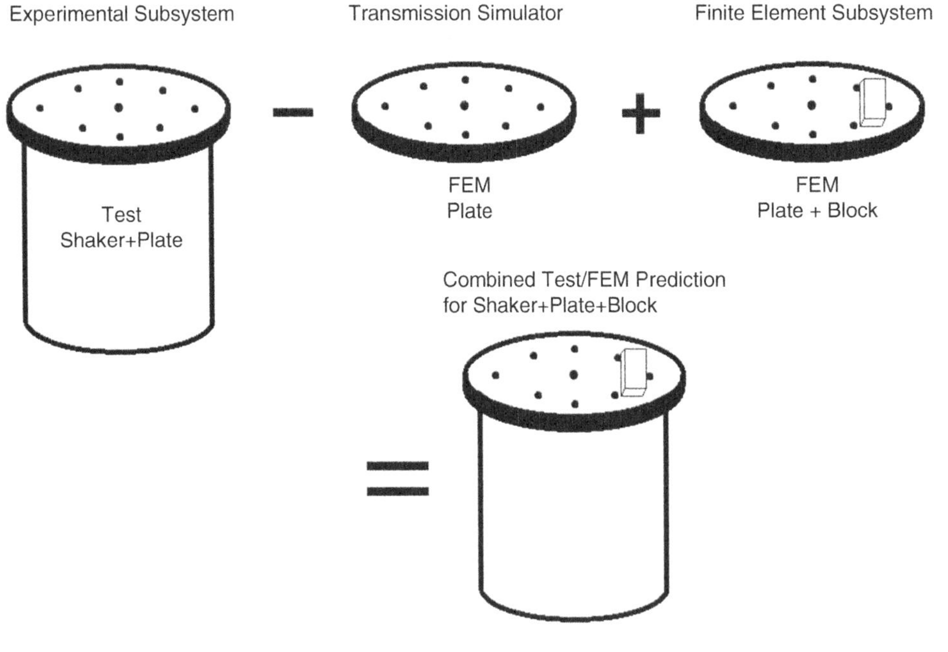

Fig. 1.1 TSM substructuring schematic

The uncoupled, undamped equations of motion of the three subsystems in modal coordinates are given below in Eq. (1.1), in which $\boldsymbol{\omega}$ is frequency, $\mathbf{I}$ is an appropriately dimensioned identity matrix, $\mathbf{q}$ are vectors of generalized modal coordinates, $\boldsymbol{\omega}_{\mathbf{n}}$ are diagonal matrices of natural frequencies of each subsystem, $\boldsymbol{\phi}^{\mathbf{T}}$ are the transpose of each mode shape matrix of each subsystem, and $\mathbf{f}$ are vectors of forces applied to each subsystem.

$$-\omega^2\begin{bmatrix}\mathbf{I} & \mathbf{0} & \mathbf{0}\\ \mathbf{0} & \mathbf{I} & \mathbf{0}\\ \mathbf{0} & \mathbf{0} & -\mathbf{I}\end{bmatrix}\begin{Bmatrix}\mathbf{q}_{\mathbf{FE}}\\ \mathbf{q}_{\mathbf{EXP}}\\ \mathbf{q}_{\mathbf{TS}}\end{Bmatrix}+\begin{bmatrix}\boldsymbol{\omega}^2_{\mathbf{n,FE}} & \mathbf{0} & \mathbf{0}\\ \mathbf{0} & \boldsymbol{\omega}^2_{\mathbf{n,EXP}} & \mathbf{0}\\ \mathbf{0} & \mathbf{0} & -\boldsymbol{\omega}^2_{\mathbf{n,TS}}\end{bmatrix}\begin{Bmatrix}\mathbf{q}_{\mathbf{FE}}\\ \mathbf{q}_{\mathbf{EXP}}\\ \mathbf{q}_{\mathbf{TS}}\end{Bmatrix}=\begin{bmatrix}\boldsymbol{\phi}^{\mathbf{T}}_{\mathbf{FE}} & \mathbf{0} & \mathbf{0}\\ \mathbf{0} & \boldsymbol{\phi}^{\mathbf{T}}_{\mathbf{EXP}} & \mathbf{0}\\ \mathbf{0} & \mathbf{0} & -\boldsymbol{\phi}^{\mathbf{T}}_{\mathbf{TS}}\end{bmatrix}\begin{Bmatrix}\mathbf{f}_{\mathbf{FE}}\\ \mathbf{f}_{\mathbf{EXP}}\\ \mathbf{f}_{\mathbf{TS}}\end{Bmatrix} \tag{1.1}$$

The top two rows are positive, signifying that the finite element and experimental subsystems are to be coupled together. The last row is negative to subtract and uncouple one copy of the TS from the built-up structure. To solve for the modal parameters of the desired structure, two constraint equations are required. These are based on equating the motion at the subsystem interfaces and are given in modal form in Eqs. (1.2 and 1.3), in which $\boldsymbol{\phi}^{\dagger}$ denotes the Moore–Penrose pseudoinverse of the mode shape matrix.

$$\boldsymbol{\phi}^{\dagger}_{\mathbf{TS}}\boldsymbol{\phi}_{\mathbf{FE}}\overline{\mathbf{q}}_{\mathbf{FE}}-\boldsymbol{\phi}^{\dagger}_{\mathbf{TS}}\boldsymbol{\phi}_{\mathbf{EXP}}\overline{\mathbf{q}}_{\mathbf{EXP}}=\mathbf{0} \tag{1.2}$$

$$\boldsymbol{\phi}^{\dagger}_{\mathbf{TS}}\boldsymbol{\phi}_{\mathbf{EXP}}\overline{\mathbf{q}}_{\mathbf{EXP}}-\boldsymbol{\phi}^{\dagger}_{\mathbf{TS}}\boldsymbol{\phi}_{\mathbf{TS}}\overline{\mathbf{q}}_{\mathbf{TS}}=\mathbf{0} \tag{1.3}$$

Applying the constraint equations yields a least squares problem that may be solved for the modal properties of the hybrid experimental/FEM model containing the experimental base, the TS interface, and the FEM component. A more rigorous and complete mathematical derivation can be found in [2].

1.3 Subsystem Definitions

1.3.1 Transmission Simulator

The TS subsystem is a simple aluminum plate, as shown in Fig. 1.2. The plate is 13 inches in diameter, 1 inch thick, has a mass of 6 kg, and attaches to the shaker with eighteen 3/8-inch steel bolts torqued to 30 Nm. The plate was machined at UW – Madison to provide a more suitable configuration and size for the test article attachment points.

1.3.2 Experimental Subsystem

In this work, the experimental subsystem is an LDS V830 electromagnetic shaker, which is located at the University of Wisconsin – Madison. The shaker, as shown in Fig. 1.3, consists of a top armature measuring 13 inches in diameter, features 21 attachment points and, according to the manufacturer specifications, has a usable frequency span of 0–3000 Hz and can produce a maximum sine force of 9.6 kN (2200 lb).

1.3.3 Finite Element Subsystem

For the purposes of this proof of concept project, a very simple structure was chosen for the finite element subsystem in the form of a steel block. While a physical representation of this subsystem is not necessary to complete the intended shaker model, one was manufactured to generate experimental truth data to compare against the substructuring results. The block, as shown in Fig. 1.4, is composed of generic stock steel, measures 2” × 2” × 4”, has a mass of 2.1 kg, and was positioned between points 18 and 25 on the aluminum adapter plate.

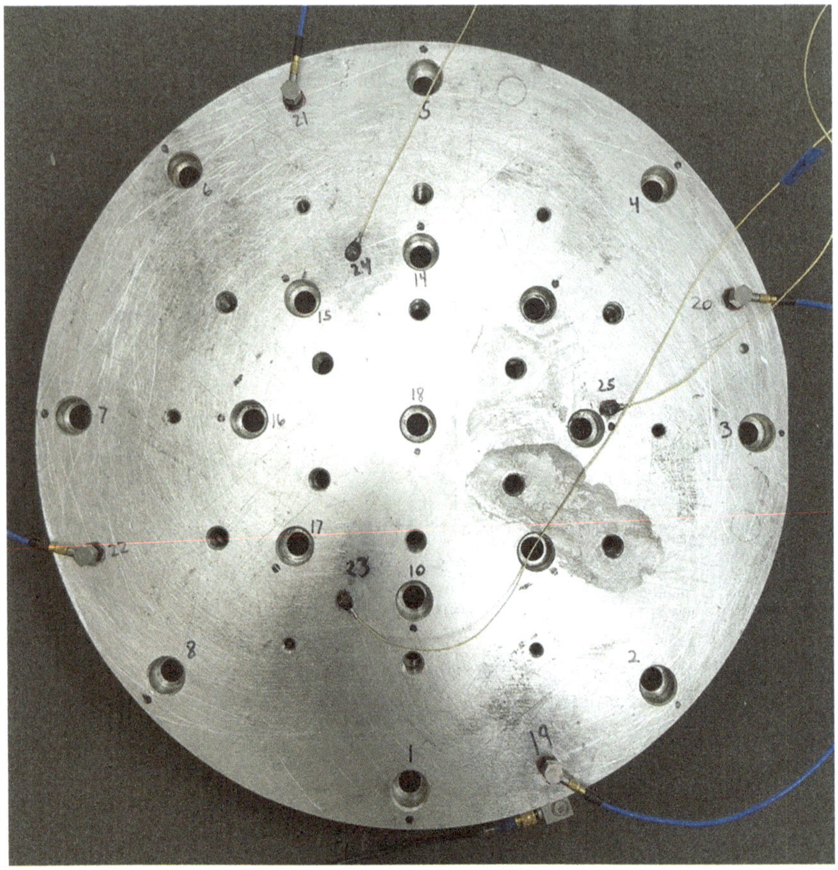

Fig. 1.2 The transmission simulator, an aluminum adapter plate

Fig. 1.3 Shaker system used to evaluate the proposed substructuring approach

1.4 Finite Element Analysis

For the substructuring procedure, finite element models (FEM) of the adapter plate and the adapter plate and block must be created. In the current work, this was done in Abaqus using shell elements, resulting in the very dense mesh shown in Fig. 1.5 for the TS FEM. A modal solution was then performed in order to determine the natural frequencies and mode shapes of the models. It is of note that typical FEA solvers will produce rigid body modes that, while mutually orthogonal,

Fig. 1.4 Steel Block test article

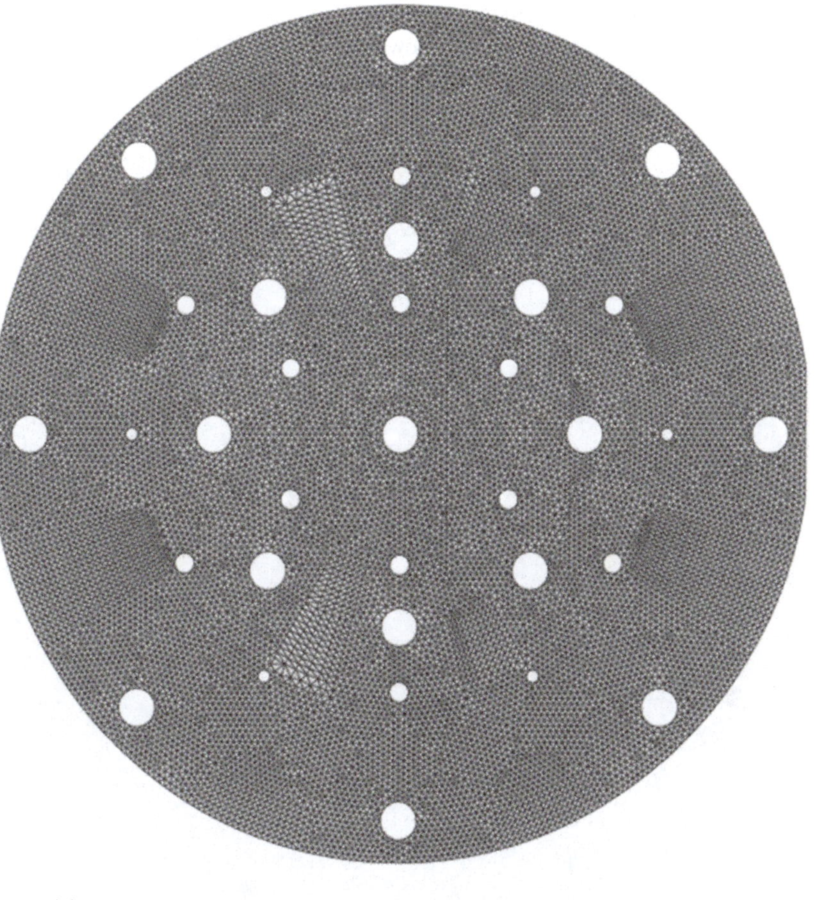

Fig. 1.5 Plate finite element model

are at an arbitrary rotation relative to the six unit directions. This may introduce difficulties later when selecting basis modes for the substructuring constraint equations. Thus, according to the recommended procedure [4], in this work the rigid body modes computed from the FEM were corrected so that each was purely about one of the six unit directions. This was done by first determining the structure principle body coordinates and principle moments of inertia. The rigid body modes are then calculated by distributing the mass, for translational, or moment of inertia, for rotational, over each DOF relative to the orientation of its local coordinate system to the principle body coordinates.

1.5 Experimental Procedure

Experimental modal data was gathered in the form of a roving hammer test. This was done to ensure that the frequency band of interest would be sufficiently excited and to provide a large array of potential drive points. The locations of each accelerometer used and where each hammer hit was applied are shown below in Fig. 1.6. The hammer strikes, given by the orange markers, were done in the vertical direction into the page at each point, and radially inward toward the center at the points along the outer edge of the plate. The accelerometers were positioned according to engineering judgement and were largely placed midway between two hammer locations in order to assume a drive point value that is the average of the values on either side. The accelerometer data was recorded in Data Physics SignalCalc® 830 software using the Data Physics Abacus® hardware interface. A frequency span of 0–5000 Hz was set to fully capture the operable range of the shaker.

The modal test was performed on four test cases: the plate isolated on foam, the plate and the block bolted together and resting on foam, the plate attached to the shaker, and the plate with block attached to the shaker. The tests on foam simulate free conditions and provide data from which to validate their respective FEMs, if necessary. Data from the shaker and plate test is used as the experimental subsystem data for the TSM. The shaker, plate, and block test results are taken as truth data from which the substructuring results may be verified.

An overlay of the results from the foam test cases are shown in Fig. 1.7. The FRFs shown are the average of all hammer locations over all accelerometers. Plots of the experimental mode shapes for the plate and plate & block are given in

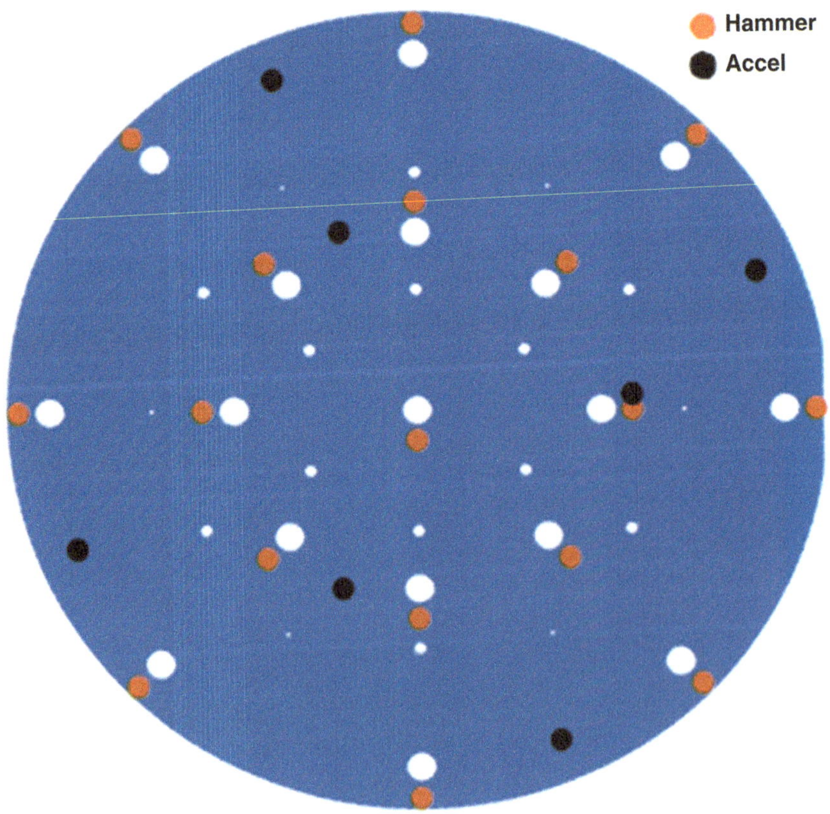

Fig. 1.6 Hammer hit and accelerometer locations

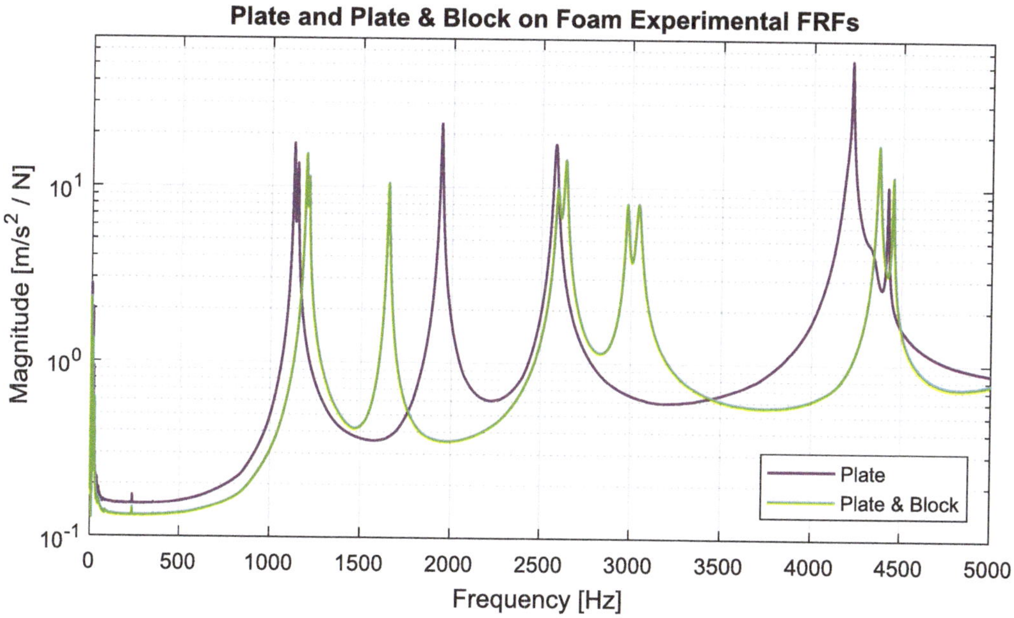

Fig. 1.7 Modal test results for the plate and plate & block on foam

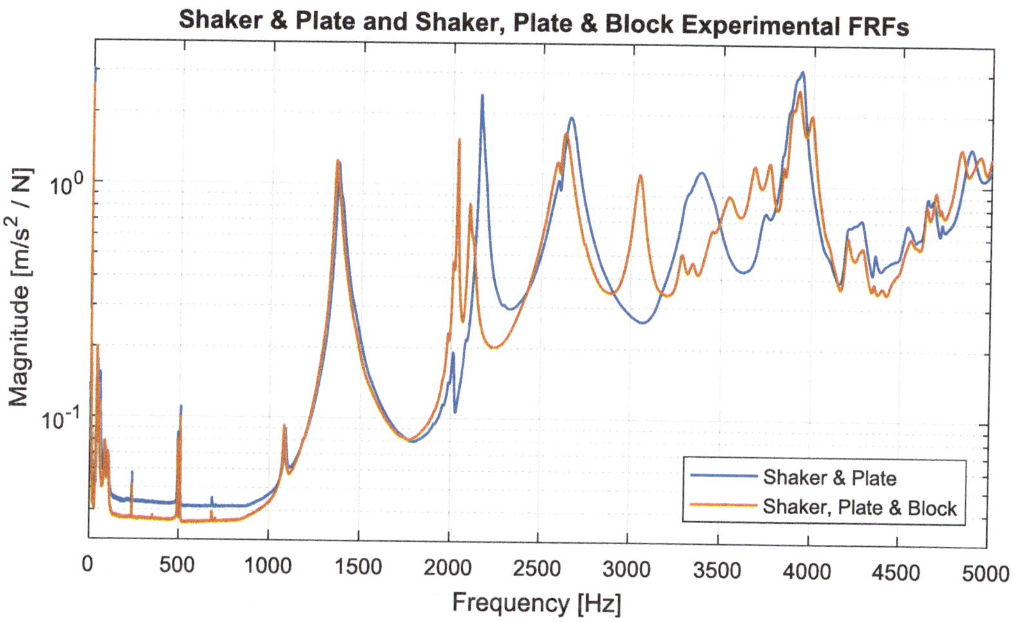

Fig. 1.8 Modal test results for the Shaker test cases

Appendixes A and B, respectively. The modes near 2000 Hz and 4000 Hz, commonly called drum modes, are noticeably shifted in frequency. These display a heightened sensitivity to the presence of the block due to their greater excitation in the vertical translation direction, which now carries more inertial mass. The other modes, commonly called potato chip modes, are relatively unaffected by the presence of the block because these modes are more akin to rotations of the block. However, these rotations are likely the cause of the additional modes observed in the plate with block case near 3000 Hz.

As observed in Fig. 1.8, the on-shaker testing yields results that are drastically different than those previously discussed. The experimental mode shapes, given in Appendixes C and D, show that most of the assembly modes are unlike the modes of a free plate. This suggests that the plate has a limited effect on the dynamics of the assembly as a whole, and that the top of the shaker is the dominant component when the two are assembled.

Curve fitting for the experimental data was done using the Algorithm of Mode Isolation (AMI) [5]. This program successively fits and subtracts modes from the experimental data until all significant peaks are accounted for. As an output, AMI supplies the natural frequencies, damping ratios, and state space modal residues for each mode. For the TSM, the mode shapes must be mass normalized. To extract these from the residues, first they are used to approximate the classical mode (i.e. real mode) residues and then the square root of the value at the drive point is used to scale the remaining values. This process is understood by observing the typical modal transfer function equation given below in Eq. (1.4), in which $\boldsymbol{\omega}$ is the frequency, $\boldsymbol{\omega_n}$ is the natural frequency at mode $\mathbf{n}$, $\boldsymbol{\xi_n}$ is the damping ratio for mode $\mathbf{n}$, $\boldsymbol{\phi_n}$ is the mass normalized mode shape, and $\boldsymbol{\phi_{n,dp}}$ is the value of the mode shape at the drive point. The product $\boldsymbol{\phi_n \phi_{n,dp}}$ is the classical mode residue and is a matrix with rows corresponding to each hammer location and columns for each mode. The value for the drive point location is then equivalent to the square of the drive point value of the mass normalized mode shape, as in Eq. (1.5). This can be used to mass normalize the residue matrix as shown in Eq. (1.6).

$$\mathbf{H}(\boldsymbol{\omega}) = \sum_{\mathbf{n}=1}^{\#\mathbf{modes}} \frac{\boldsymbol{\phi_n \phi_{n,dp}}}{-\boldsymbol{\omega}^2 + \mathbf{i}2\boldsymbol{\omega\xi_n\omega_n} + \boldsymbol{\omega_n^2}} \tag{1.4}$$

$$\sqrt{\mathbf{residue_{n,dp}}} = \boldsymbol{\phi_{n,dp}} \tag{1.5}$$

$$\boldsymbol{\phi_n} = \frac{\mathbf{residue_n}}{\sqrt{\mathbf{residue_{n,dp}}}} \tag{1.6}$$

It is of note that each mode shape can be normalized relative to a different drive point. This can be advantages when certain drive points are poorly excited by a mode, in that a more suitable location can be chosen. For this work, the drive point with the largest value was taken to normalize each respective mode.

1.6 Substructuring Results

With the experimental and FEM data collected, the TSM may be implemented as given in Eqs. (1.1, 1.2, and 1.3). To define the constraint equations, a set of independent TS FEM mode shapes must be chosen as a suitable basis. The shapes must not be linearly dependent, as this would lead to ill conditioning when computing the pseudo-inverse in Eqs. (1.2 and 1.3), (i.e. the amplitude of each of those basis vectors cannot be uniquely determined from the spatial information) and this would produce inaccurate results for the built-up structure. Several checks and criterion for determining a viable mode selection are given in [6].

The resultant solution provides the natural frequencies and mode shapes of the hybrid structure predicted by the TSM. As the first point of comparison between these and the truth data, their cross-modal assurance criterion (MAC) was computed. The MAC is a simple method for assessing the similarity of two mode shapes. A value of 1 indicates that the modes have the same shape, while a value of 0 is representative of completely independent shapes. When comparing large sets of modes, a matrix of values between 0 and 1 results showing each modes relation to all others. The cross-MAC between the experimental and predicted modes of the shaker, plate & block is given in Fig. 1.9. As shown by the color bar, a darker color represents magnitudes closer to 1. The string of large values on the diagonal from mode 1 to 15 is indicative of good agreement between the experimental and predicted modes. However, the nonzero off diagonal terms represent some degree of likeness between most of the modes. Also, the predicted set of modes show very little similarity to test modes 16 through 23, which indicates a sizable group of high frequency modes that the TSM model was unable to predict.

While it was certainly encouraging to see such good agreement between the true modes of the shaker with the block and those predicted by substructuring, further investigation revealed that the block actually changed the mode shapes of the shaker very little. Figure 1.10 shows the cross-MAC between the experimentally measured modes of the shaker & plate and those of the shaker, plate & block. The resulting MAC matrix is nearly identical to that above. This reveals that the TSM substructuring does not need to, and is not significantly altering the mode shapes of the shaker & plate due to the addition of the block. This also reveals that the off-diagonal terms in the MAC are likely due to a set of measurement DOF that are not adequate to distinguish all of these modes. Furthermore, since the group of experimental shaker, plate & block modes between 16 and 23 is not found in the shaker & plate set, it may be surmised that this set of shapes is caused by the presence of the block.

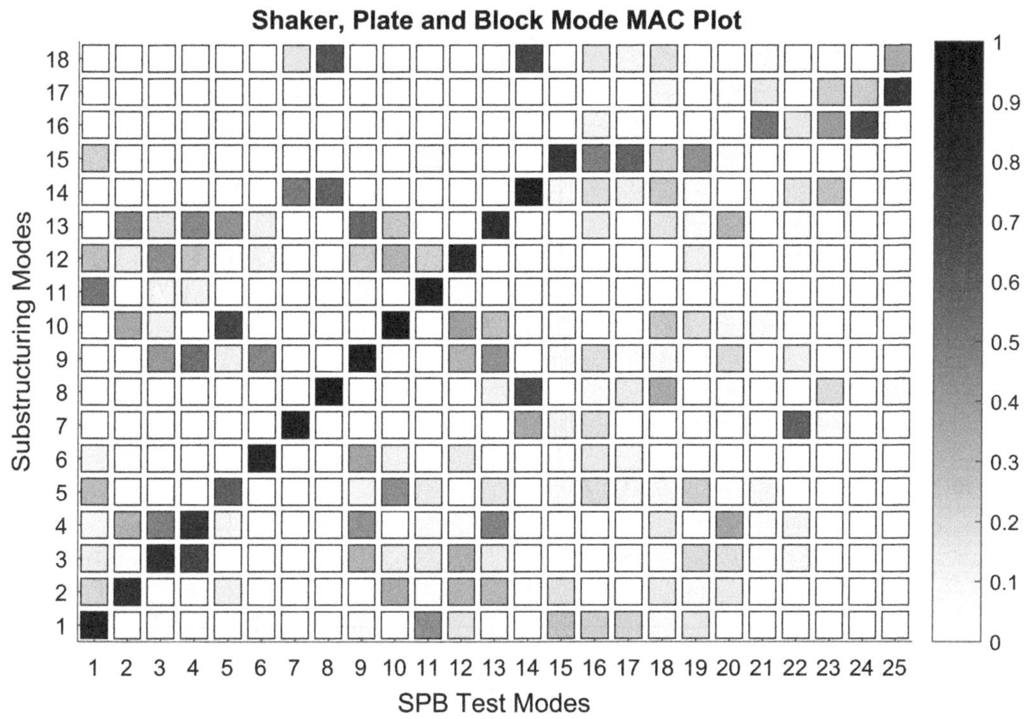

Fig. 1.9 Cross-MAC between the predicted and truth SPB mode shapes

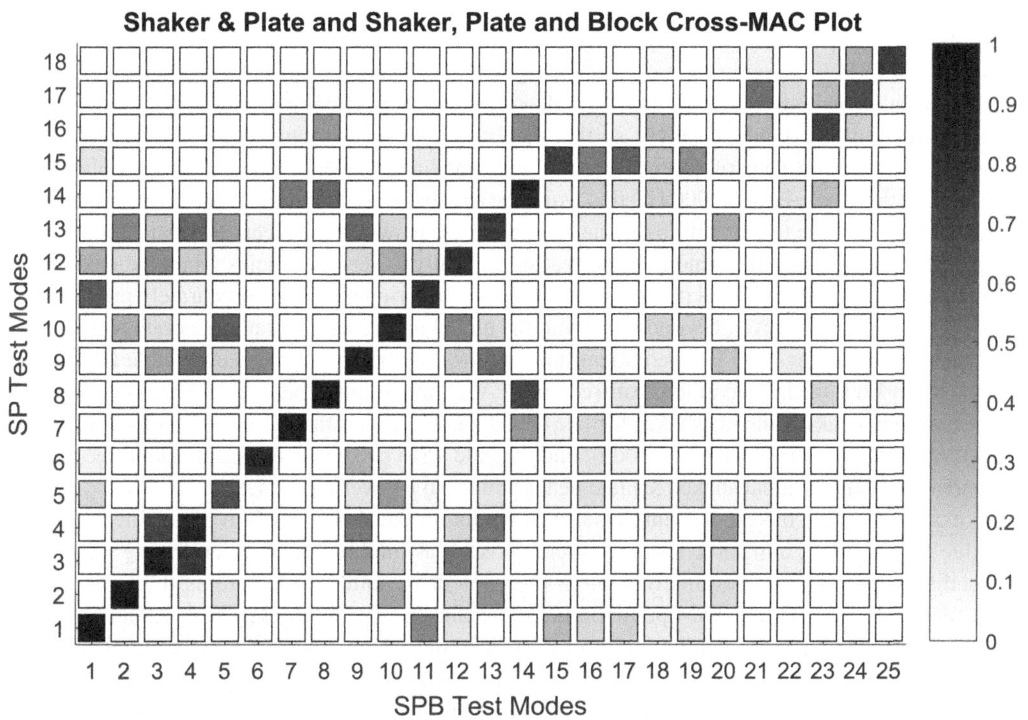

Fig. 1.10 Cross-MAC between the SP and SPB experimental mode shapes

As a method for easily comparing the shaker model to the truth data, the substructuring results were used to generate FRFs representative of the predicted parameters. It should be noted that, as a consequence of implementing FEMs without damping data in the substructuring process, the TSM model does not predict damping ratios. Thus, to construct the FRFs, the experimental damping ratios from the shaker & plate curve fits were superimposed on the predicted mode shapes and natural frequencies. This is an assumption that the damping does not significantly change due to the presence of the block.

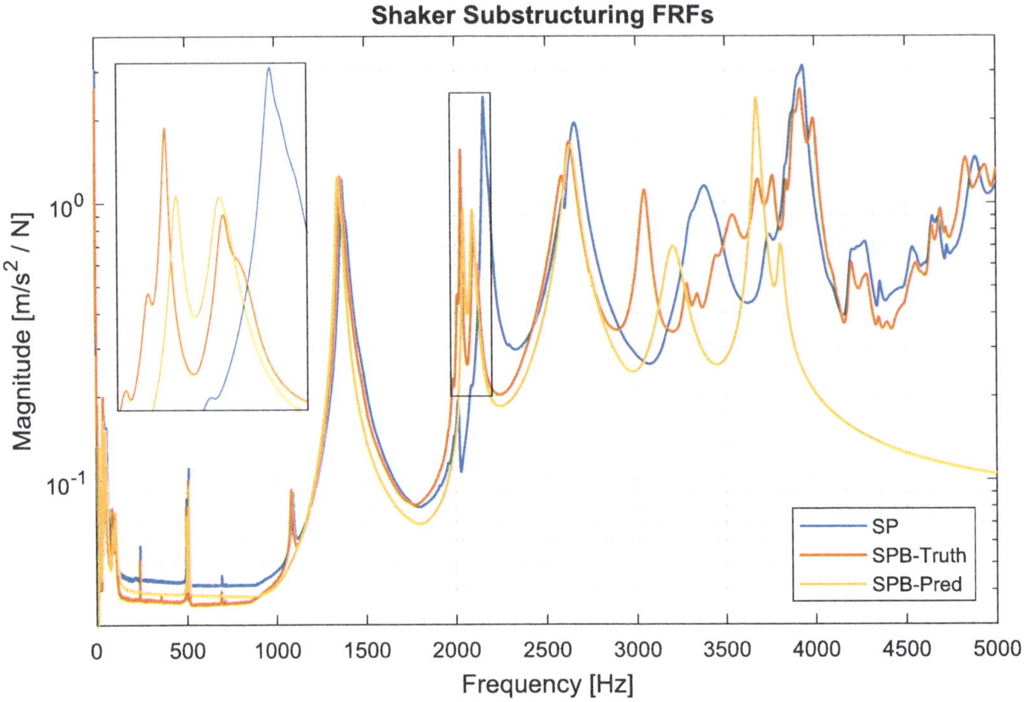

Fig. 1.11 Comparison of experimental and predicted FRFs

The results of the FRF generation and the experimental data is given in Fig. 1.11. The yellow curve, denoted 'SPB-Pred,' depicts the results of the TSM substructuring with the corresponding measurements from the truth test given by the red curve and denoted 'SPB-Truth'. The blue curve, labeled 'SP', denotes the experimental measurements from the shaker and plate, which can be compared with the predicted FRFs so that the error in the substructuring predictions can be compared with the error that one would obtain if substructuring were not performed and the shaker model was simply assumed to remain unchanged after adding the block. Below 2000 Hz, this would be a realistic assumption, since the only variance is a very slight decrease in natural frequency due to the additional mass of the block. However, between 2000 and 2500 Hz, the single large peak seen in the shaker & plate data splits into two peaks at significantly lower frequencies. The predicted model successfully emulates this change, as seen in the zoomed plot window. The substructuring model also accurately predicts the slight change in the next peak near 2600 Hz. However, beyond this point, the model no longer maintains accurate natural frequencies. This was expected, as test modes up to 4000 Hz were identified, and from previous experience, reliable CMS results should not be anticipated above 50–66% of the maximum test frequency. Also, in the range between 3000 and 4000 Hz, the truth data exhibits a number of small peaks while the shaker & plate data does not. This is the set of modes observed in the cross-MACs that display no similarity to the shaker & plate modes, and thus the TSM predicted shape set. The three peaks present in the predicted curve are simply the original shaker & plate peaks shifted to a lower frequency.

Finally, as an example of how the experimentally derived model for the shaker might be used, the response of the system was calculated due to a uniform, unit force in the Z direction, such as might be used when testing a component. This was accomplished by distributing the force evenly over the Z direction drive points while zeroing the others. This was repeated for the shaker & plate and the truth and TSM predictions for the shaker, plate & block. The resultant response as a function of frequency is given in Fig. 1.12, and can be interpreted as denoting which resonances are excited when the shaker is in operation and how much amplification is observed at each. The results show that the response is amplified by almost two orders of magnitude near 2000 Hz, which corresponds to the first resonance of the shaker. This resonance shifts in frequency and the two close modes that comprise the peak separate when the block is attached. As was seen in the comparison of the FRFs above, the TSM model is able to predict the truth data to a reasonable degree up to about 3–4 kHz, although with some variance in the peak magnitudes.

Additionally, the shape of the response at the peaks may be calculated and plotted. This provides an indication as to if certain areas of the test surface experience over or under testing when the system is at a resonance. The shapes shown in Fig. 1.13 are labeled relative to the alphanumeric markers denoting the peaks in Fig. 1.12, in which each row corresponds to

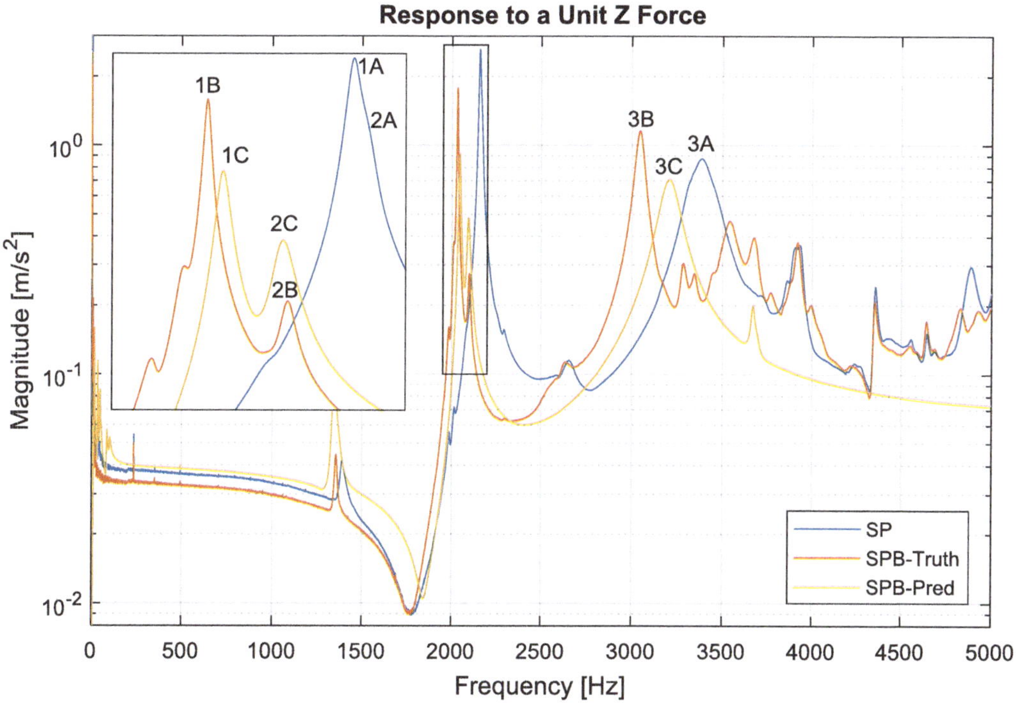

Fig. 1.12 SPB response as a function of frequency for a uniform unit Z force

the shaker & plate, shaker, plate & block, and the TSM model respectively. While it is of no surprise that all the shapes are some form of drum mode, given that a uniform Z force would highly excite these, the most interesting result is that the TSM correctly predicts the mode shape at peak 2C relative to the shape extracted from the truth data at peak 2B. This is especially notable since the mode shape of the original shaker & plate, from peak 2A, is completely different. If a test were planned based on a model of the shaker & plate (i.e. without considering how the test part would change the dynamics of the system), one might choose to locate a control accelerometer in the center of the plate, and that sensor would see very little motion as the shaker plate actually moved in a two-lobe motion with a zero near the centerline, potentially over testing any components mounted off of this centerline.

1.7 Conclusion

This paper presented a method for generating an experimentally based model of a shaker using the transmission simulator method of modal substructuring. By implementing this technique, the effects of the internal shaker components and the linear stiffness and damping from the joints where the adapter plate attaches to the shaker may be captured experimentally while allowing for arbitrary FEM components to be coupled to the shaker. Functionally, this will enable the dynamic behavior of the shaker to be predicted during the pretest phase so one can avoid over- or under-testing components when mounting them to the shaker at various locations on the adapter plate. To verify this method, a shaker table with a simple adapter plate and block to mimic a test article were used to predict the change in response of the shaker. After determining the modal properties of each subsystem and performing the substructuring procedure, the predicted results were compared to experimental truth data. The results were promising, in that the model closely anticipates the variation in the response of the shaker and plate due to the presence of the block from 0–3000 Hz, which is the manufacturer's recommended operable frequency range. In the results shown so far, the FEM of the plate and block was not calibrated to match the test on the plate and block, and it was found to be in error above 3 kHz. Additional work should be done to calibrate the FEM, and to determine whether this improves the range over which the substructuring predictions are accurate. Also, further work with more complex test structures and adapter plates or adapter cubes is required to determine if the method can be applied to an arbitrary test article.

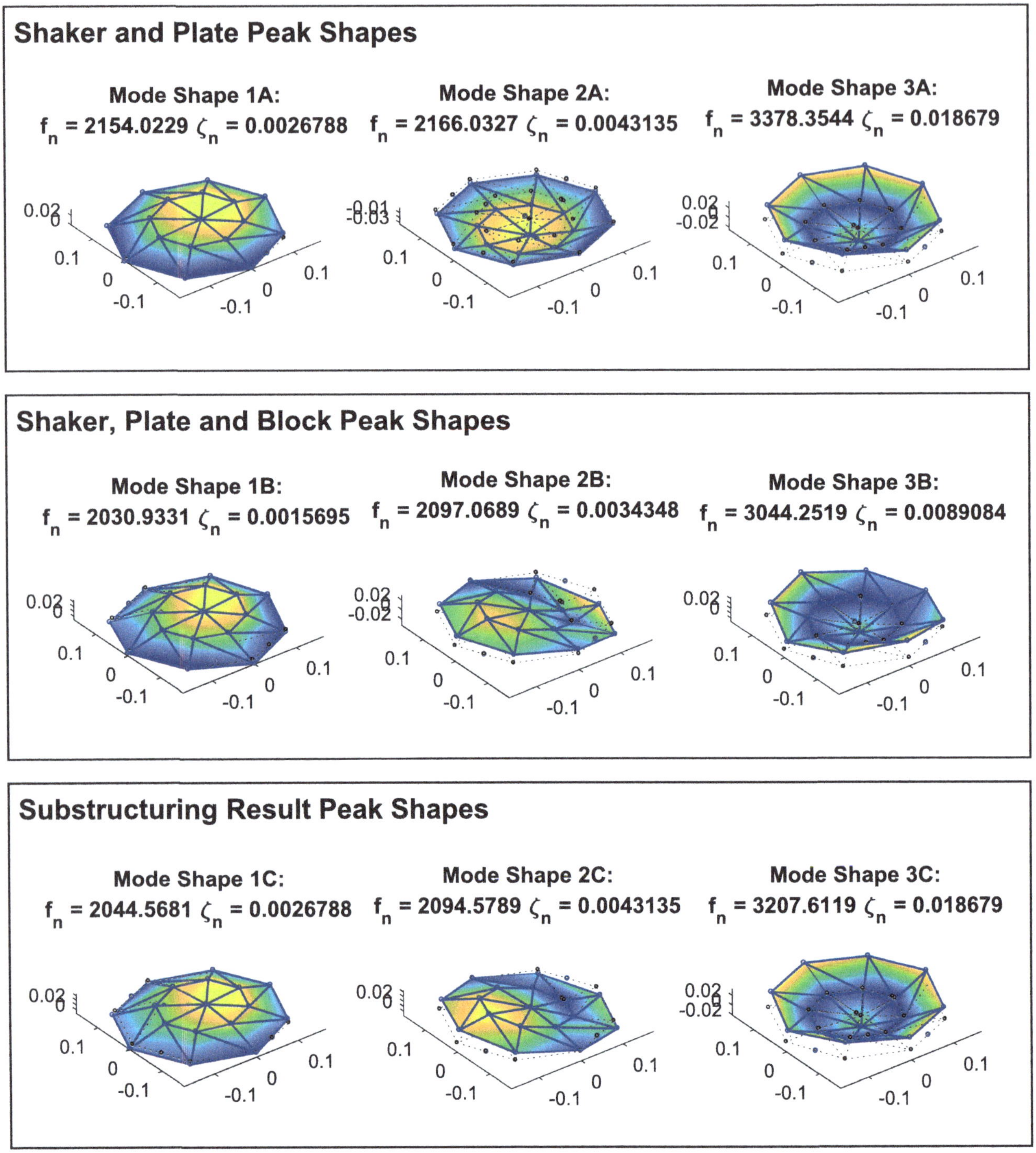

Fig. 1.13 Peak deformation shapes to the unit Z force

Appendix

Appendix A: Plate Mode Shapes

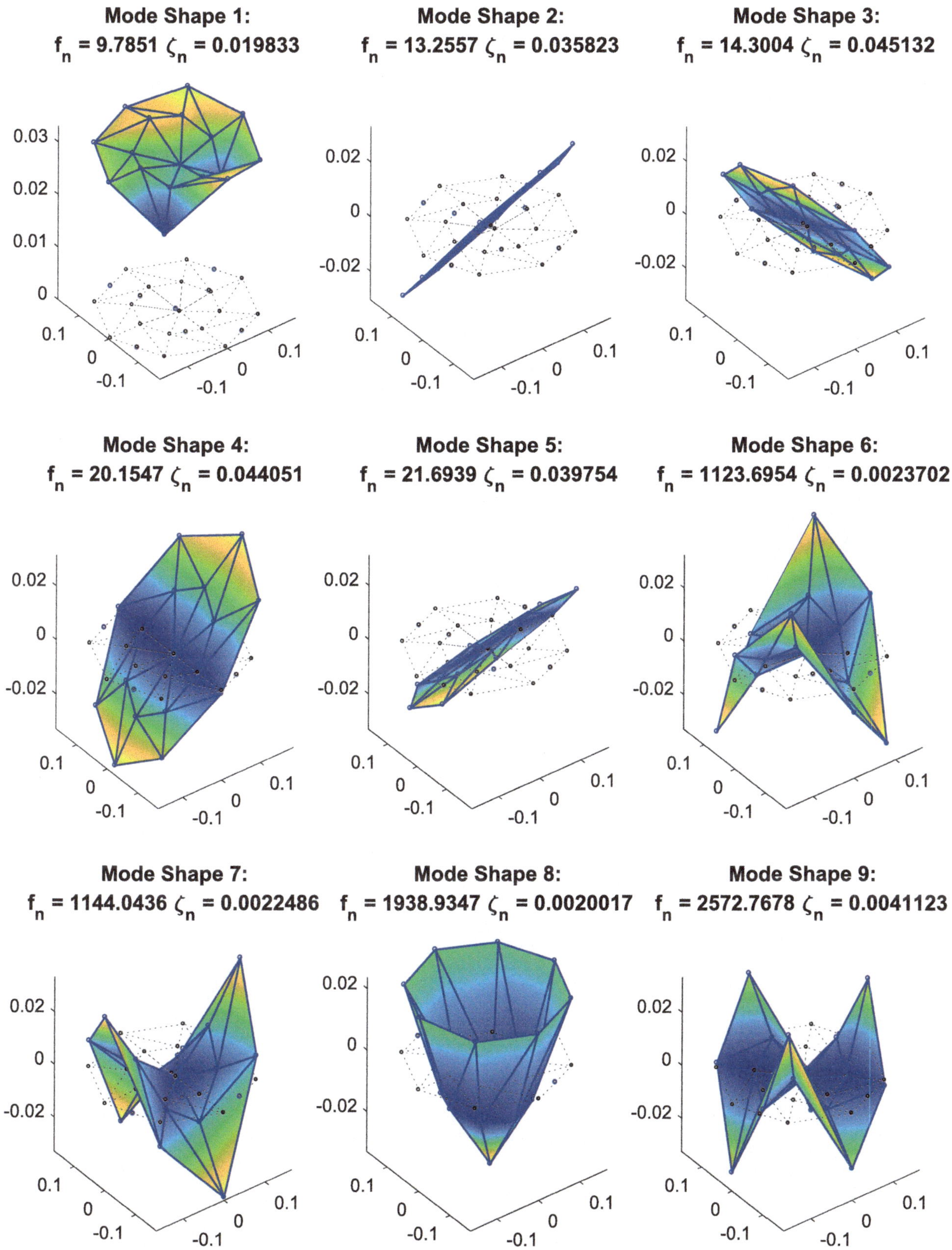

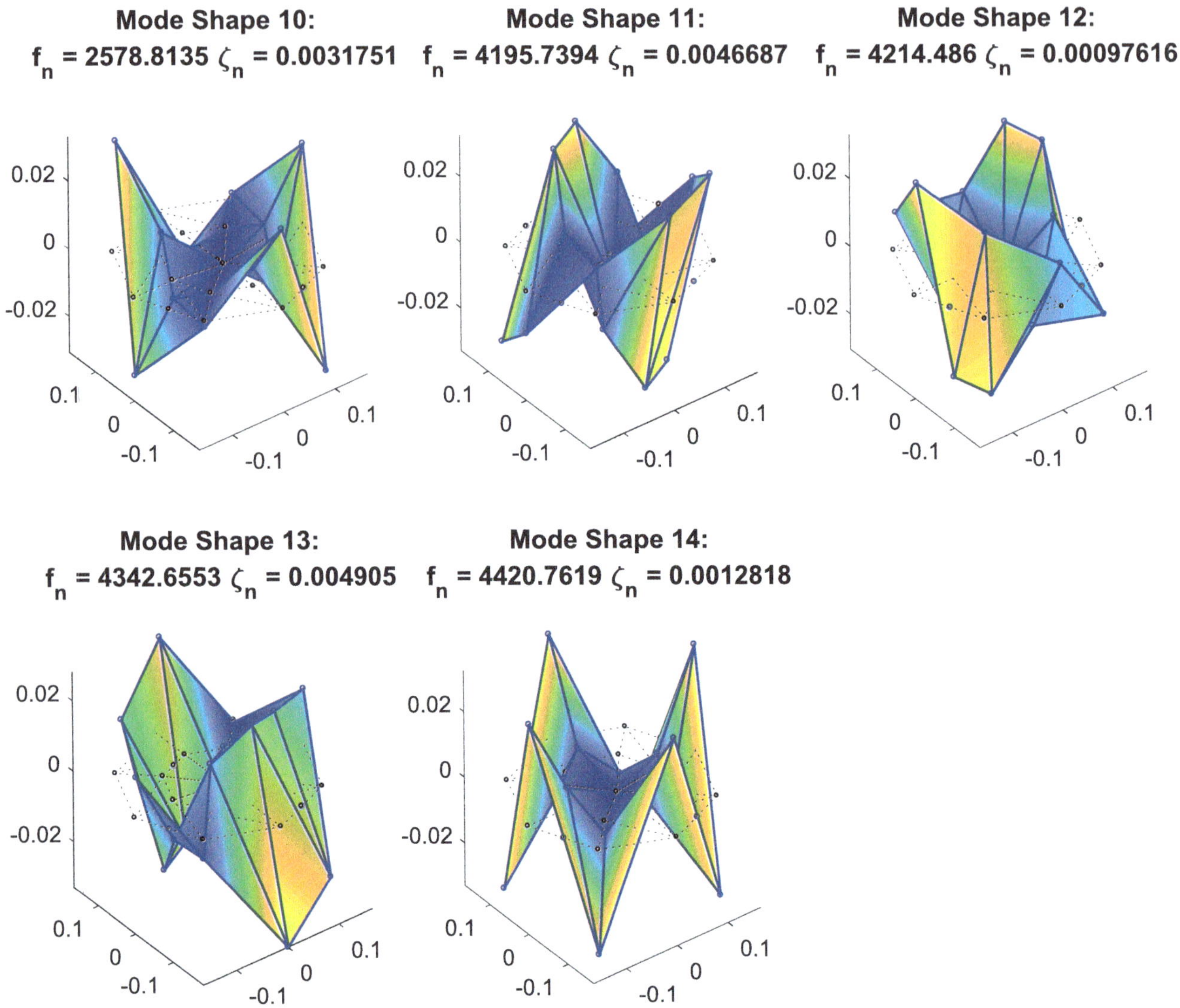
Mode Shape 10:
f_n = 2578.8135 ζ_n = 0.0031751
Mode Shape 11:
f_n = 4195.7394 ζ_n = 0.0046687
Mode Shape 12:
f_n = 4214.486 ζ_n = 0.00097616
Mode Shape 13:
f_n = 4342.6553 ζ_n = 0.004905
Mode Shape 14:
f_n = 4420.7619 ζ_n = 0.0012818

Appendix B: Plate & Block Mode Shapes

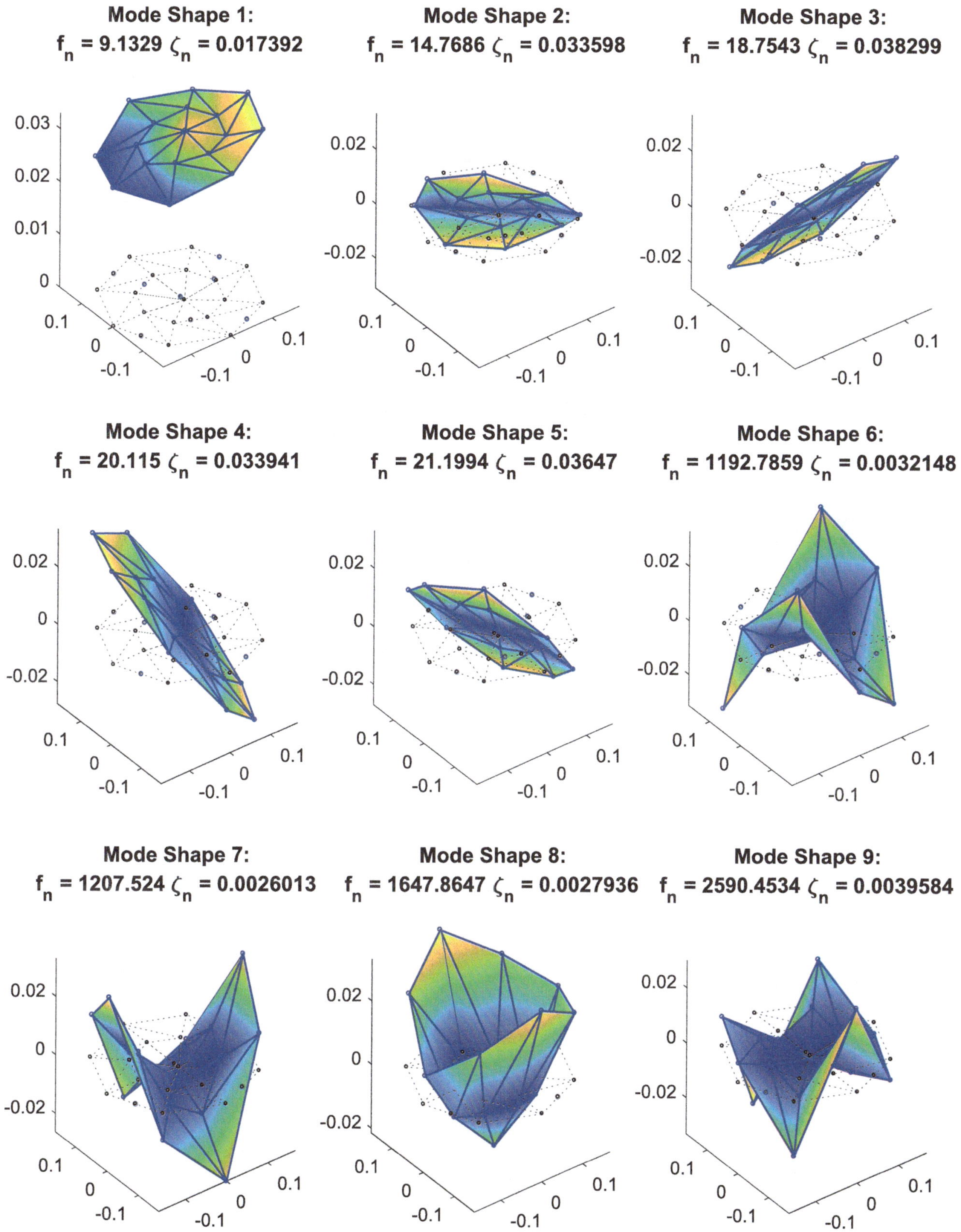

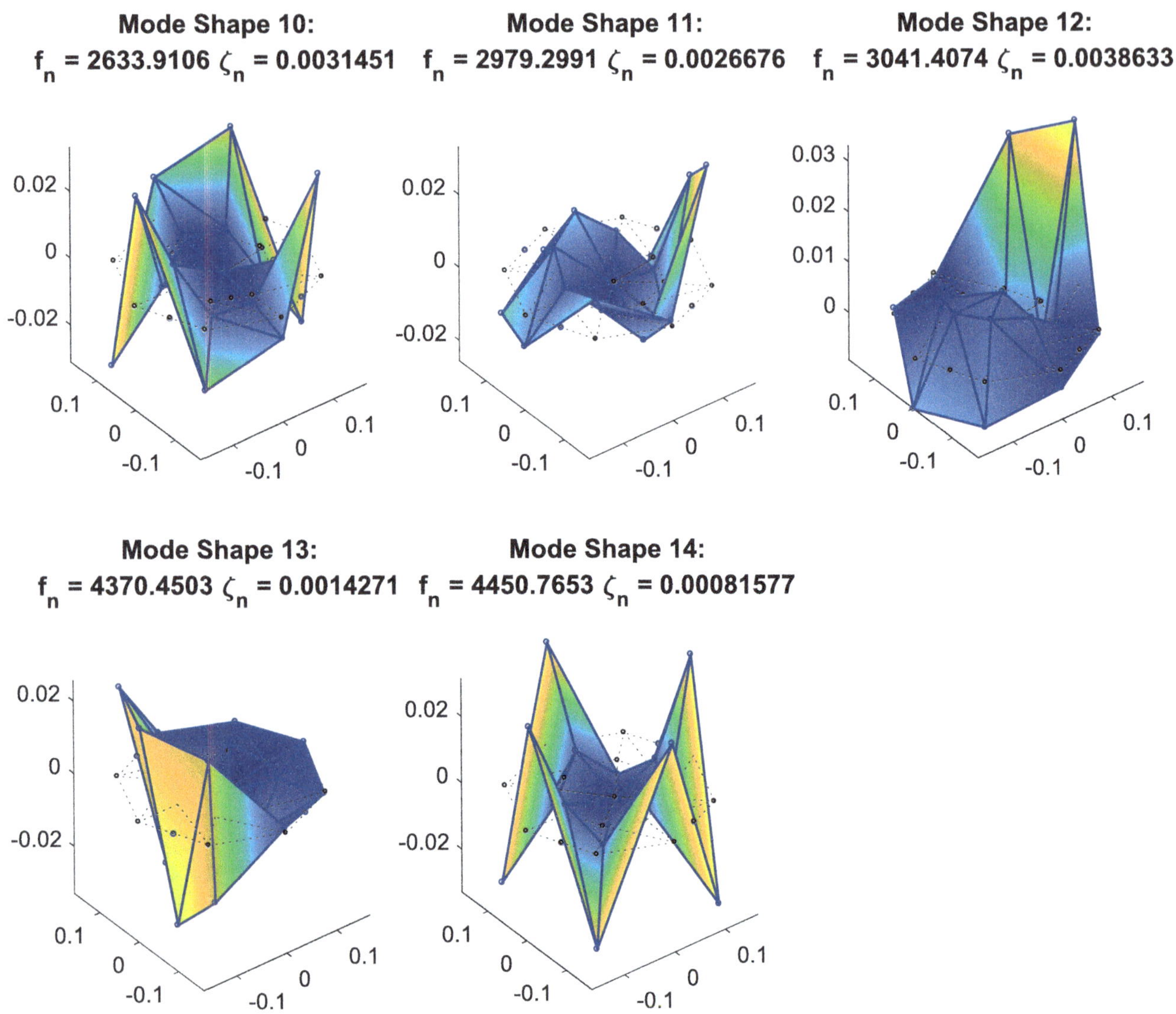
Mode Shape 10:
f_n = 2633.9106 ζ_n = 0.0031451
Mode Shape 11:
f_n = 2979.2991 ζ_n = 0.0026676
Mode Shape 12:
f_n = 3041.4074 ζ_n = 0.0038633
Mode Shape 13:
f_n = 4370.4503 ζ_n = 0.0014271
Mode Shape 14:
f_n = 4450.7653 ζ_n = 0.00081577
0.03
0.02
0.01
0
-0.02
0.1
-0.1

Appendix C: Shaker and Plate Mode Shapes

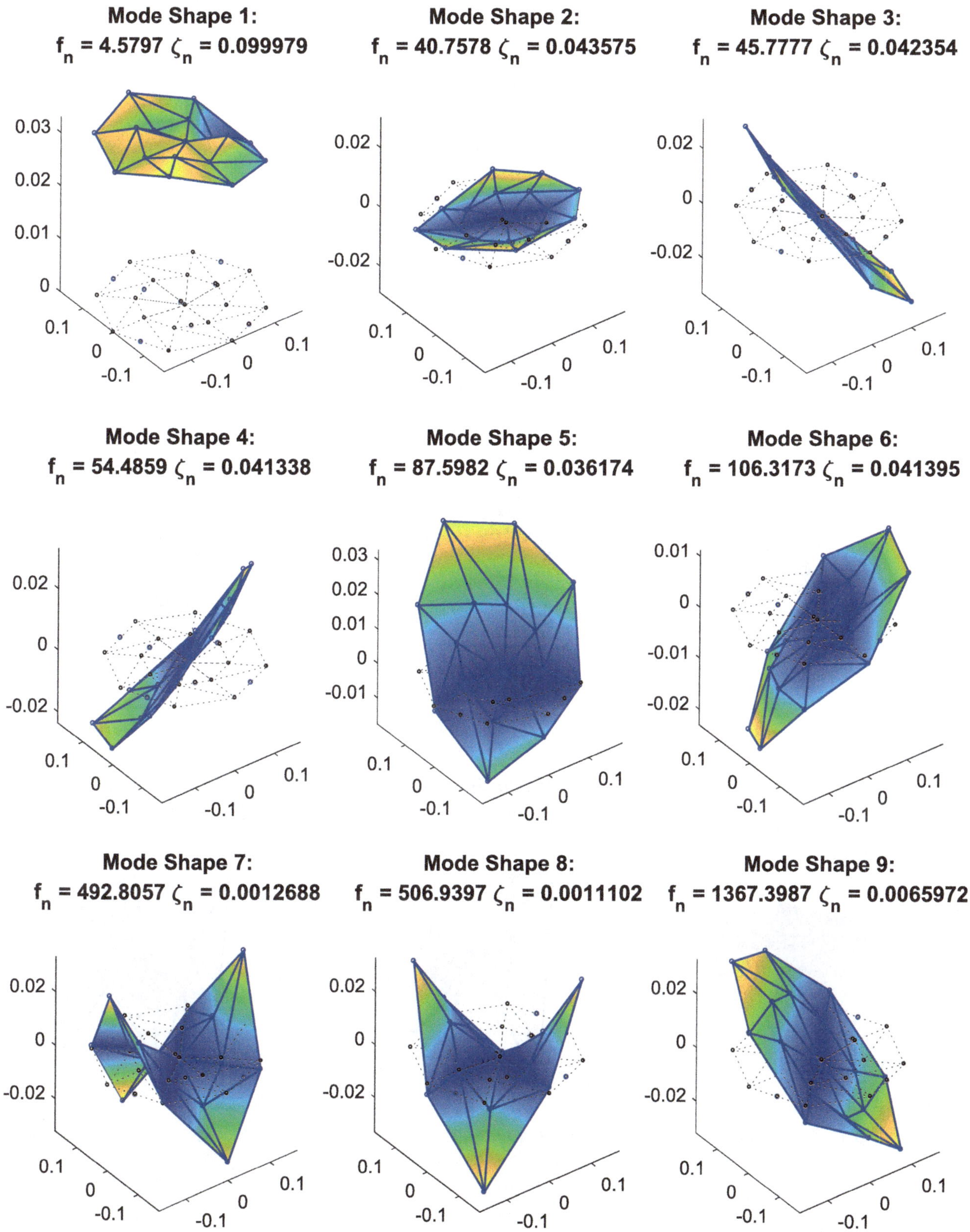

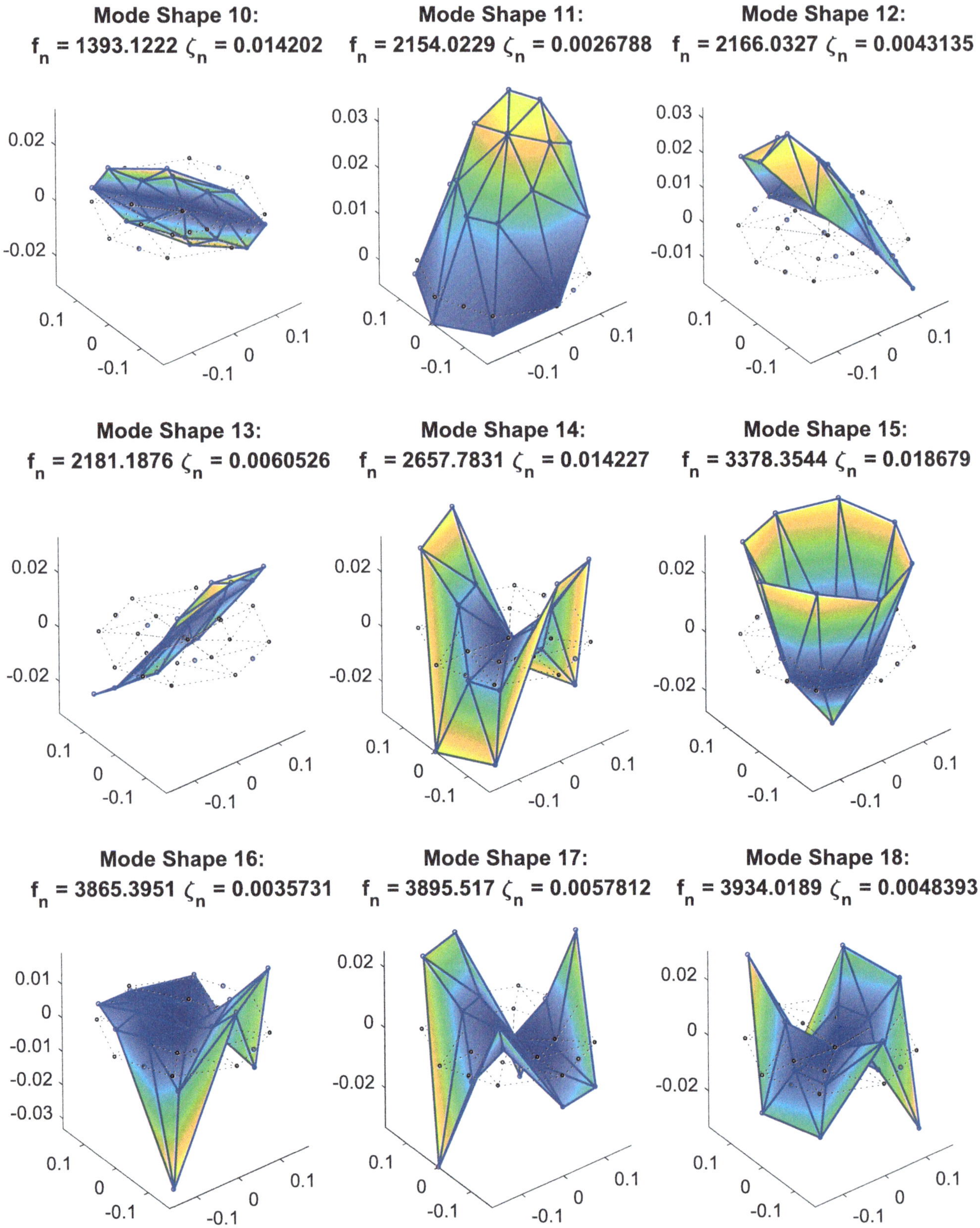
Mode Shape 10:
f$_n$ = 1393.1222 ζ$_n$ = 0.014202
Mode Shape 11:
f$_n$ = 2154.0229 ζ$_n$ = 0.0026788
Mode Shape 12:
f$_n$ = 2166.0327 ζ$_n$ = 0.0043135
Mode Shape 13:
f$_n$ = 2181.1876 ζ$_n$ = 0.0060526
Mode Shape 14:
f$_n$ = 2657.7831 ζ$_n$ = 0.014227
Mode Shape 15:
f$_n$ = 3378.3544 ζ$_n$ = 0.018679
Mode Shape 16:
f$_n$ = 3865.3951 ζ$_n$ = 0.0035731
Mode Shape 17:
f$_n$ = 3895.517 ζ$_n$ = 0.0057812
Mode Shape 18:
f$_n$ = 3934.0189 ζ$_n$ = 0.0048393

Appendix D: Shaker, Plate and Block Mode Shapes

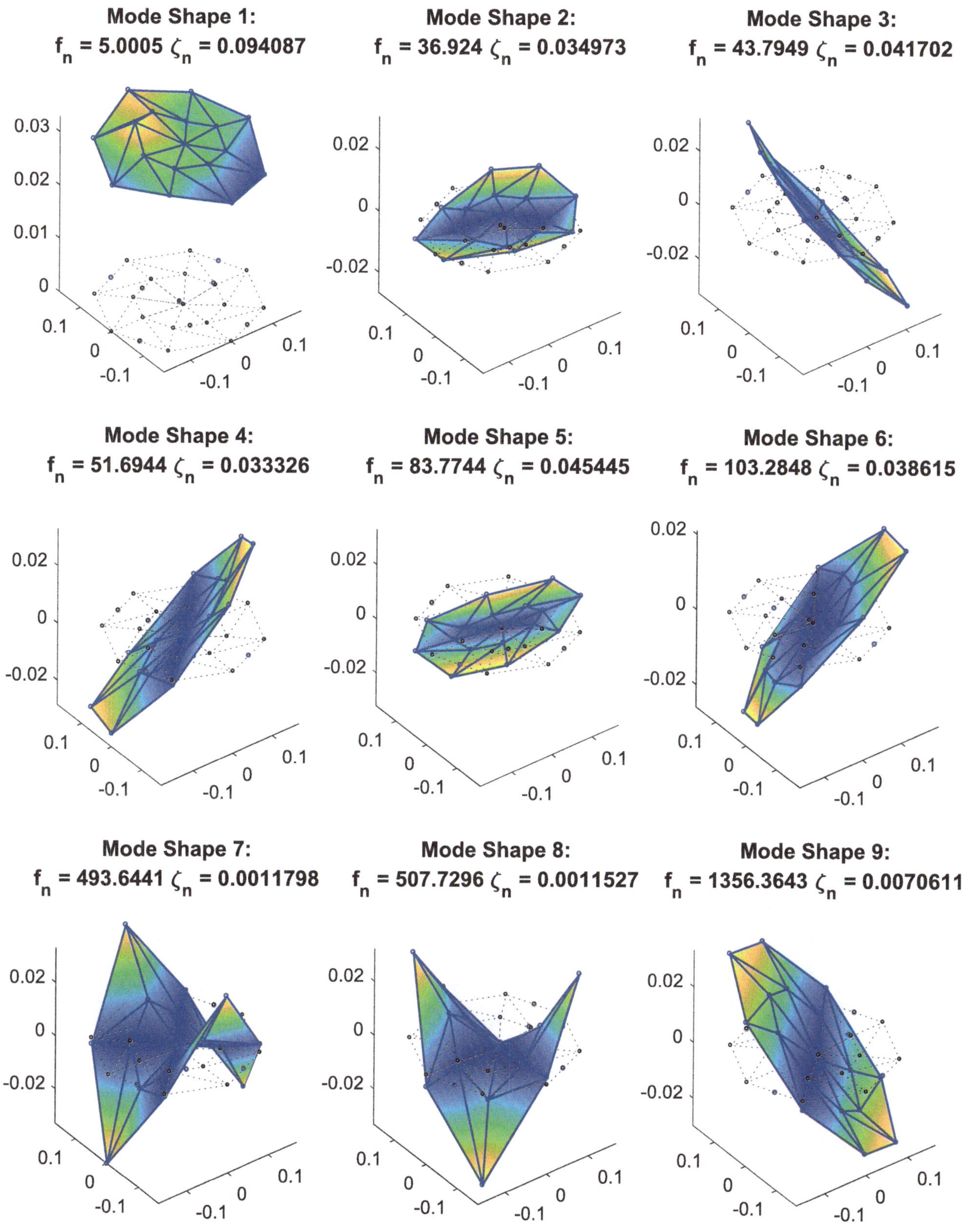

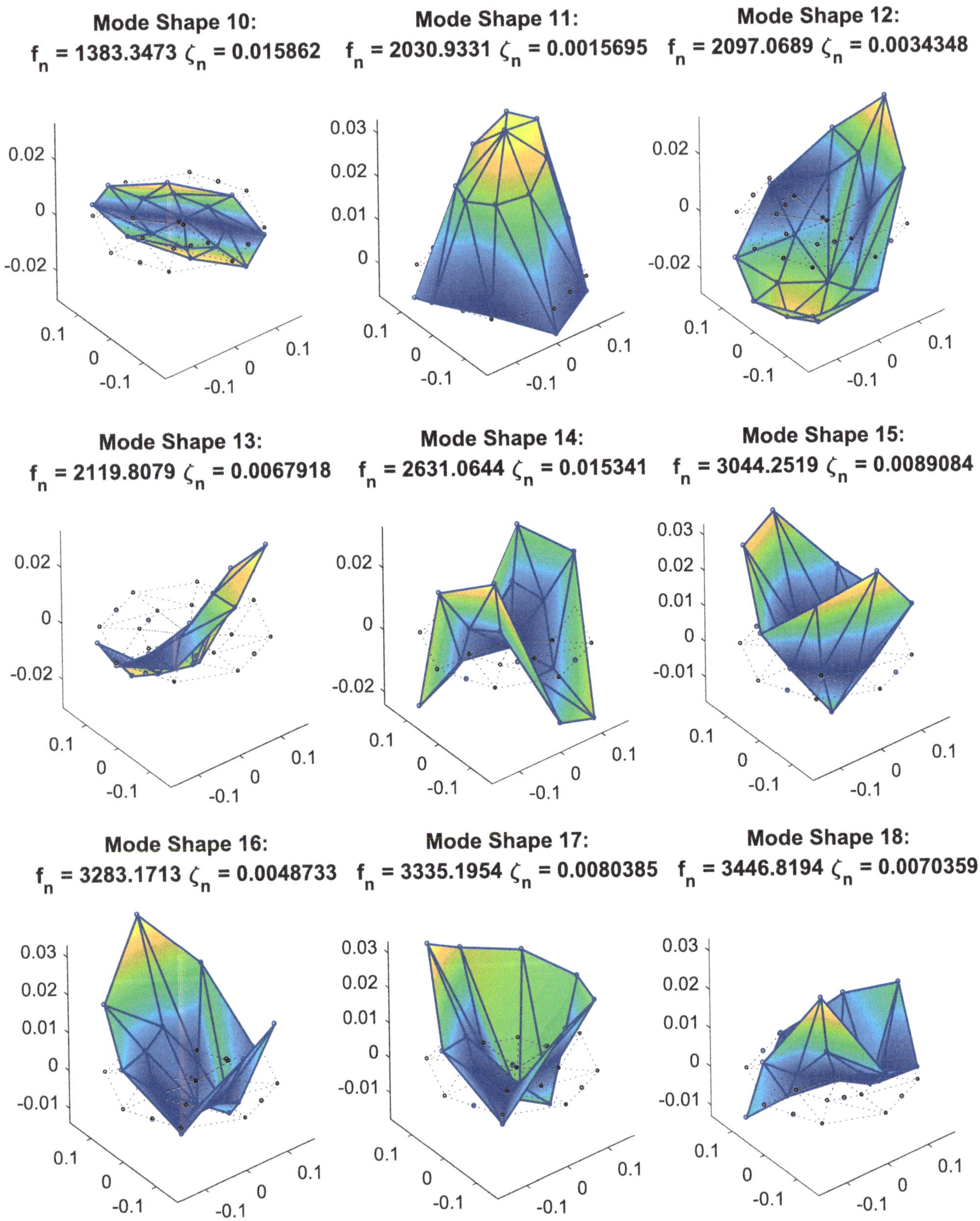
Mode Shape 10:
f_n = 1383.3473 ζ_n = 0.015862
Mode Shape 11:
f_n = 2030.9331 ζ_n = 0.0015695
Mode Shape 12:
f_n = 2097.0689 ζ_n = 0.0034348
Mode Shape 13:
f_n = 2119.8079 ζ_n = 0.0067918
Mode Shape 14:
f_n = 2631.0644 ζ_n = 0.015341
Mode Shape 15:
f_n = 3044.2519 ζ_n = 0.0089084
Mode Shape 16:
f_n = 3283.1713 ζ_n = 0.0048733
Mode Shape 17:
f_n = 3335.1954 ζ_n = 0.0080385
Mode Shape 18:
f_n = 3446.8194 ζ_n = 0.0070359

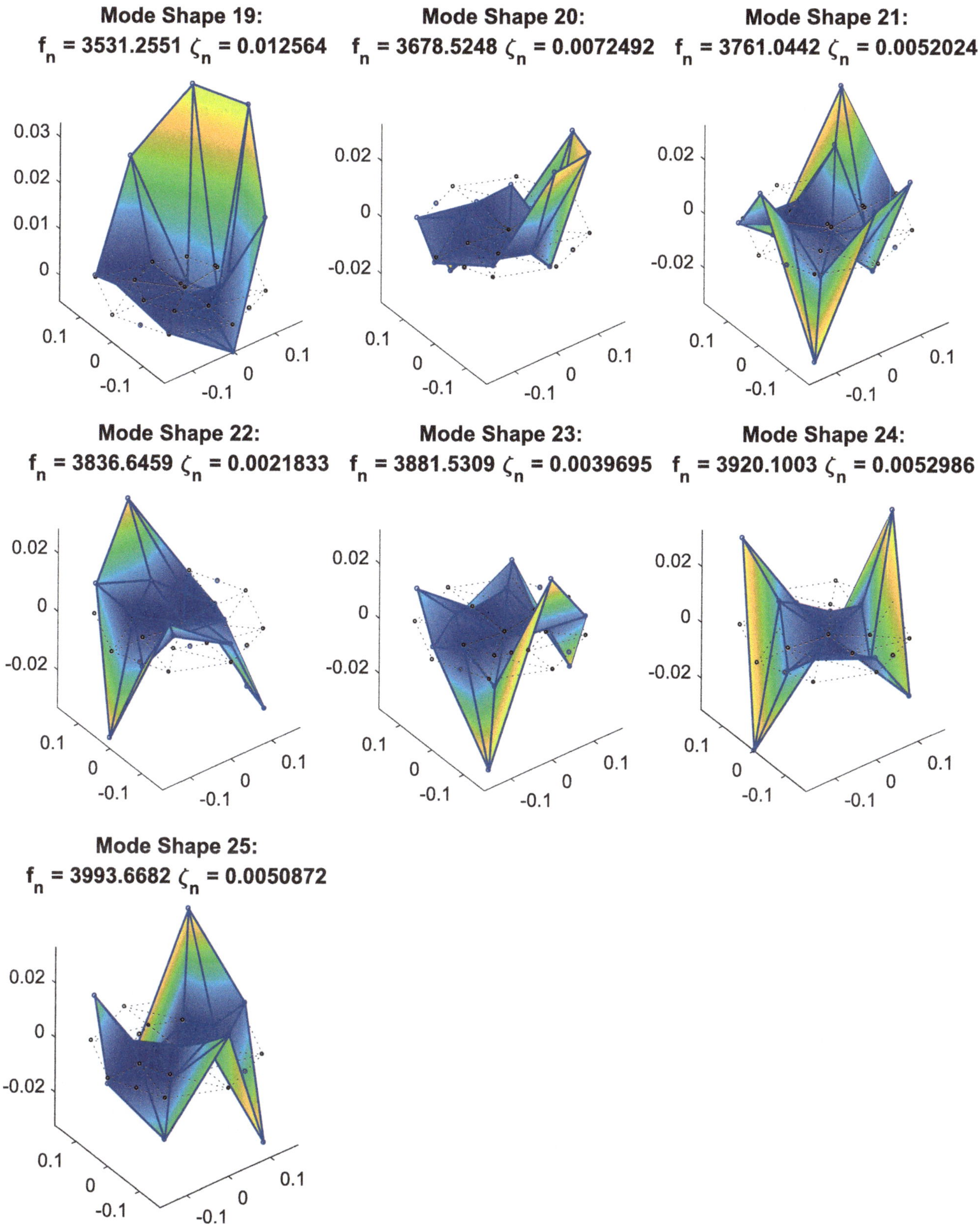
Mode Shape 19:
f n = 3531.2551 ζ n = 0.012564
Mode Shape 20:
f n = 3678.5248 ζ n = 0.0072492
Mode Shape 21:
f n = 3761.0442 ζ n = 0.0052024
Mode Shape 22:
f n = 3836.6459 ζ n = 0.0021833
Mode Shape 23:
f n = 3881.5309 ζ n = 0.0039695
Mode Shape 24:
f n = 3920.1003 ζ n = 0.0052986
Mode Shape 25:
f n = 3993.6682 ζ n = 0.0050872

References

1. Delima, W., Jones, R., Dodgen, E., Ambrose, M.: A numerical approach to system model identification of random vibration test. In: Proceeding of 35th IMAC, a Conference and Exposition on Structural Dynamics, Garden Grove, California (2017)
2. Mayes, R.L., Arviso, M.: Design studies for the transmission simulator method of experimental dynamic substructuring. In: ISMA2010, Lueven, Belgium (2010)
3. Allen, M.S., Mayes, R.L., Bergman, E.J.: Experimental modal substructuring to couple and uncouple substructures with flexible fixtures and multi-point connections. J. Sound Vib. **329**, 4891–4906 (2010)
4. Roettgen, D., Allen, M., Mayes, R.: Ampair 600 wind turbine three-bladed assembly substructuring using the transmission simulator method. J. Sound Vib. **383**, 14–17 (2016)
5. Allen, M., Ginsberg, J.H.: A global, single-input-multi-output (SIMO) implementation of the algorithm of mode isolation and applications to analytical and experimental data. Mech. Syst. Signal Process. **20**, 1090–1111 (2006)
6. Allen, M., Kammer, D., Mayes, R.: Metrics for diagnosing negative mass and stiffness when uncoupling experimental and analytical substructures. J. Sound Vib. **331**, 5435–5448 (2012)

Chapter 2
Predicting Assembly Effective Mass from Two Component Effective Mass Models

Benjamin R. Pacini, Randall L. Mayes, and Patrick S. Hunter

Abstract Effective mass models are powerful tools that allow for a convenient means to calculate the energy associated with vibration response of a structure to a base input acceleration in a particular direction. This is useful for hardware qualification activities and margin assessment. Traditionally, these models are generated from purely analytical means such as a finite element model. However, experimental methods have recently been introduced as an intriguing alternative, particularly for applications where no finite element model is available. In this work, an effective mass modal model of a cable-connector assembly is desired, and neither component has a finite element model. Moreover, there can be multiple cable-connector combinations making analytical modeling as well as explicit testing of each combination impractical. This work develops the capability to combine an experimentally derived connector effective mass model with a simplified and easily extensible analytical cable model. The experimental connector effective mass model is generated through specialized modal testing. The simplified cable model is a Timoshenko beam finite element model whose properties are empirically derived from pinned-pinned cable modal data. The modeled length of the cable is appropriately adjusted for each configuration. Finally, the cable and connector component models can be combined to form the final assembly modal effective mass model for a given translational direction. This method lends itself to developing catalogues of connector and cable data, which can then be easily combined to form any number of assembly configurations without having to explicitly test/model them.

Keywords Effective mass · Cable Dynamics · Structural Dynamics · Margin quantification · Modal model

Nomenclature

m_i	Effective mass of mode i
$\{\phi\}_R$	Rigid body row vector
$\mathbf{M}$	Mass matrix
$\{\phi\}$	Mode shape vector
k_i	Effective stiffness of mode i
f_n	Natural frequency
$f_{transition}$	Transition frequency
EI	Bending stiffness
KAG	Shear stiffness
L	Length of beam element

2.1 Introduction and Motivation

An effective mass model is a modal model that simulates the response of a structure due to a base acceleration input in one direction, see Fig. 2.1. It can be used to calculate the actual energy in the structure during the base acceleration environment. This is a useful metric for comparing energy at failure relative to that in a qualification test (i.e. margin quantification).

Sandia National Laboratories is a multimission laboratory managed and operated by National Technology and Engineering Solutions of Sandia, LLC., a wholly owned subsidiary of Honeywell International, Inc., for the U.S. Department of Energy's National Nuclear Security Administration under contract DE-NA-0003525.

B. R. Pacini · R. L. Mayes (✉) · P. S. Hunter
Structural Dynamics Department, Sandia National Laboratories, Albuquerque, NM, USA
e-mail: brpacin@sandia.gov; randy.mayes@awe.co.uk; pshunte@sandia.gov

A. Linderholt et al. (eds.), *Dynamics of Coupled Structures, Volume 4*, Conference Proceedings of the Society for Experimental Mechanics Series, https://doi.org/10.1007/978-3-319-74654-8_2

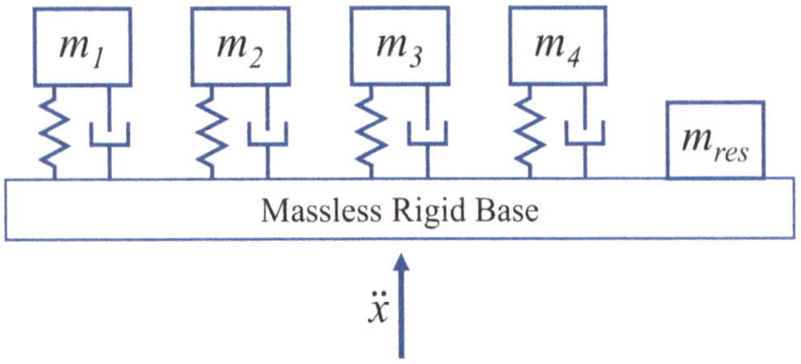

Fig. 2.1 Effective mass modal model

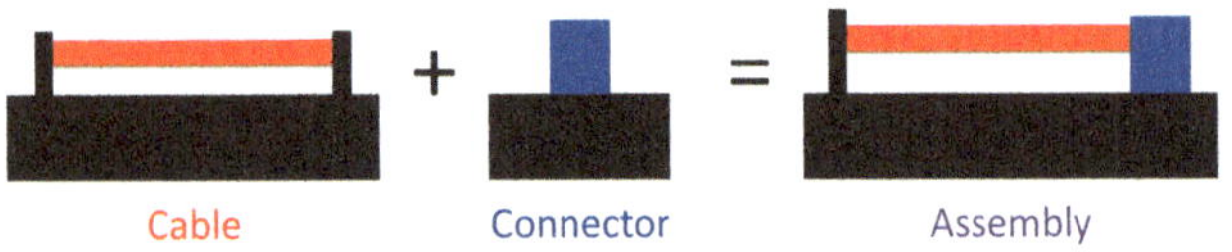

Fig. 2.2 Combining connector and cable components to create cable-connector assembly effective mass model

Standard methods for calculating an effective mass modal models require a finite element model [1]. Recently, however, methodologies have been developed to extract these models experimentally of a component on a fixture [2]. Typically, there are three effective mass modal models for a component, one for each translational direction (X, Y, and Z). Rotation directions are usually ignored since standard laboratory tests focus on a single translational direction.

Developing effective mass models for cable-connector assemblies poses several challenges. The first is that finite element models (FEMs) are typically not developed for the assemblies nor their components. Additionally, there can be multiple cable-connector combinations that make analytical modeling as well as explicit testing of each combination impractical. The goal of this work is to develop a method that can calculate the effective mass modal model of a cable-connector assembly using limited dynamic information of the individual connector and cable components, see Fig. 2.2. This is accomplished by combining an experimentally derived connector effective mass modal model with a calibrated numerical model of the cable. The practical implementation would require modal testing each cable family at minimum and maximum lengths to create a catalog of empirically derived cable stiffness properties. Additionally, each connector type would have to be modal tested in order to develop a catalog of connector effective mass modal models. Thus as cable-connector designs change (either switching out the connector or changing the cable length), the important dynamic information (i.e. effective mass and corresponding natural frequencies) can be readily calculated using the process developed herein without the need to recreate a complex FEM nor re-test the combined physical hardware. This allows for the capability to quantify product margin during qualification activities with minimal effort.

The remainder of this report is outlined as follows. Section 2.2 describes the proposed method for calculating the effective mass modal model of a cable-connector assembly by combining a beam model representing the cable to an effective mass modal model of the connector. The methodology is verified with an analytical model. The proposed method is then employed on physical hardware in Sect. 2.3. Included in the section is a discussion of the connector effective mass modal model extracted from a modal test (2.3.1). Additionally, the cable stiffness properties are empirically derived from cable modal test data in Sect. 2.3.2. Section 2.3.3 uses the connector and cable information from 2.3.1 and 2.3.2 to calculate the cable-connector effective mass modal model and associated frequencies. These are then compared with the results of a truth test conducted on a physical cable-connector assembly. Included in these results are estimations of the uncertainties associated with the effective mass and natural frequencies for both the truth test data and the results of the proposed method. Conclusions are presented in Sect. 2.4.

2.2 Analytical Verification for Combining Effective Mass Models

The goal of this work is to have a method to calculate the natural frequencies and associated effective masses of the modes with high effective mass in a specific translational direction for a cable-connector assembly without the use of a complex finite element model and only knowing the cable type, cable length, and connector type. From experiments, it has been shown that the cable mode shapes can be accurately modeled as pinned-pinned beam mode shapes (see Sect. 2.3.2.3). Experiments can also be used to extract the connector effective mass. The analytical verification was to determine if we could couple the two separate results and achieve the analytical truth model response.

The remainder of this section discusses the analytical verification. Section 2.2.1 describes the truth model that was used to develop and validate the proposed method which is described in Sect. 2.2.2. The results of the verification are presented in 2.2.3.

2.2.1 Analytical Truth Model

An analytical truth model (see Fig. 2.3) was created to investigate the efficacy of the method used to compute cable-connector assembly effective masses and natural frequencies using cable and connector component dynamic information. It consisted of a cantilevered beam for the connector and a pinned beam for the cable. The two beams are coupled via a pinned connection.

Natural frequencies and mode shapes were computed by solving the eigen value problem. The effective mass for each translational direction was calculated using Eq. (2.1) [1].

$$m_i = P_i^2 \tag{2.1}$$

where m_i is the effective mass of mode i and P_i is the modal participation factor for mode i. Equation (2.2) was used to compute P_i for each mode [1]:

$$P_i = \{\phi\}_R \mathbf{M} \{\phi\}_i \tag{2.2}$$

where $\{\phi\}_R$ is a rigid body row vector with a "1" at every entry corresponding to DOF oriented in the translation direction of interest and zero otherwise, $\mathbf{M}$ is the mass matrix, and $\{\phi\}_i$ is the shape of the mode for which the effective mas is calculated. Note that Eqs. (2.1 and 2.2) are assuming unit modal mass. Typically, effective mass is presented as a percentage of the total mass of the object.

The proposed method described in the following subsection was designed to match the natural frequencies and effective masses of the dominant modes of this truth model for transverse beam axes motion. Dominant modes are those that have large effective mass for the given direction (>5% of the total mass). The reason for this is that modes with the largest effective mass have the greatest potential to cause damage during qualification/operation environments.

As previously discussed, an effective mass model is produced for each translational direction. Effective mass and frequencies in the axial coordinate of the cable (X direction) were determined by another method. Thus the remainder of this paper focuses on the remaining transverse cable directions (Y and Z).

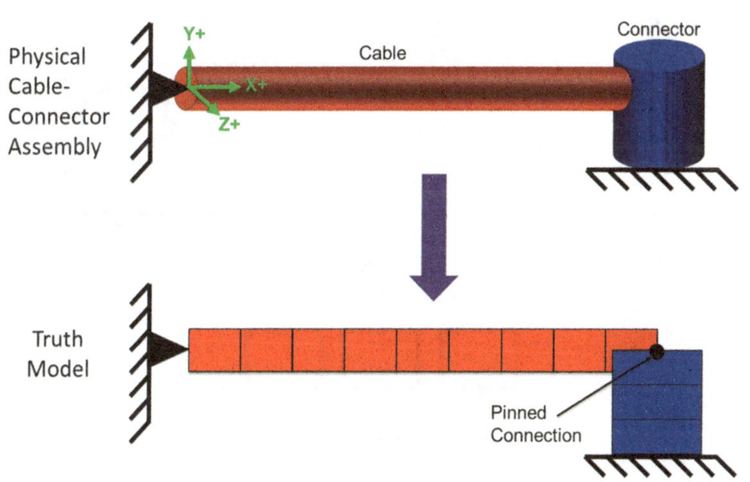

Fig. 2.3 Truth model for effective mass connection models

2.2.2 *Creating Assembly Effective Mass Models from Component Dynamic Information*

The proposed method assumes that an effective mass modal model of the connector and stiffness properties of the cable will be available. As discussed in the introduction, in practice these data will be contained in catalogs created from previously-conducted connector and cable testing. The basic concept of the proposed method is to create a simple numerical model of the cable beam from the desired length and stiffness properties then use the connector effective mass model to appropriately load the tip of the cable as it does in the cable-connector assembly. This is accomplished by sequentially connecting individual effective masses (m_{coi}) and stiffnesses (k_{coi}) of the connector to the tip element of the cable model, see Fig. 2.4. Mathematically, this is approximated by adding m_{coi} to the element in the mass matrix that corresponds to the cable tip DOF for the given translational direction and similarly adding k_{coi} in the stiffness matrix. The eigenvalue problem is then solved and the effective mass model is derived from the results and Eq. (2.1) to create the assembly submodel.

There could be multiple connector modes in the bandwidth of interest that are coupled to the cable. In this scenario, the first connector effective mass and stiffness are coupled to the cable and its effective mass model is calculated to create an assembly submodel. The second connector effective mass and stiffness are then used without the first (i.e. the connector effective mass added to the cable tip is just m_{co2} and not $m_{co1} + m_{co2}$) to create a second assembly submodel. This continues for every connector mode with large effective mass (see discussion below) in the bandwidth of interest. The final effective mass model of the assembly (called the final assembly model) is stitched together from each assembly submodel. The results from the first assembly submodel are included in the final assembly model up until a transition frequency. The results from the second assembly submodel are then used until a second transition frequency. For this work, each transition frequency was selected to be half-way between the subsequent connector natural frequencies. This process is depicted in Fig. 2.5 when two connector modes are used to compute the final assembly model.

It is important to note that only those modes of the connecter with large effective masses are used in this process. The reason for this is, if in a particular direction, a connector mode has a very small effective mass, it will correspondingly have a very low effective stiffness. When coupled to the cable as shown in Fig. 2.4, the tip of the cable would be virtually

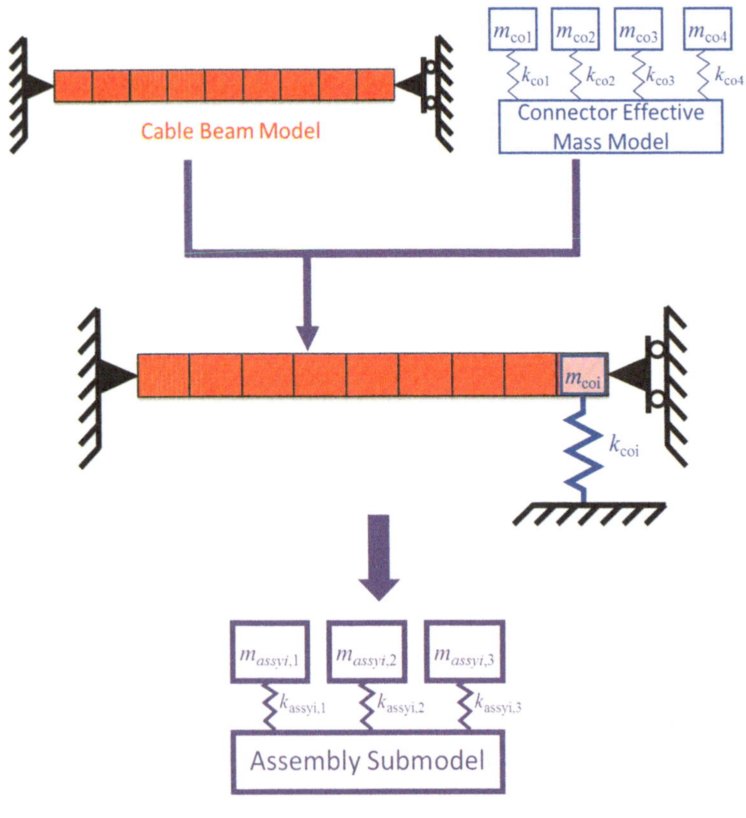

Fig. 2.4 Proposed method for calculating cable-connector assembly effective mass and natural frequencies from dynamic information from individual components

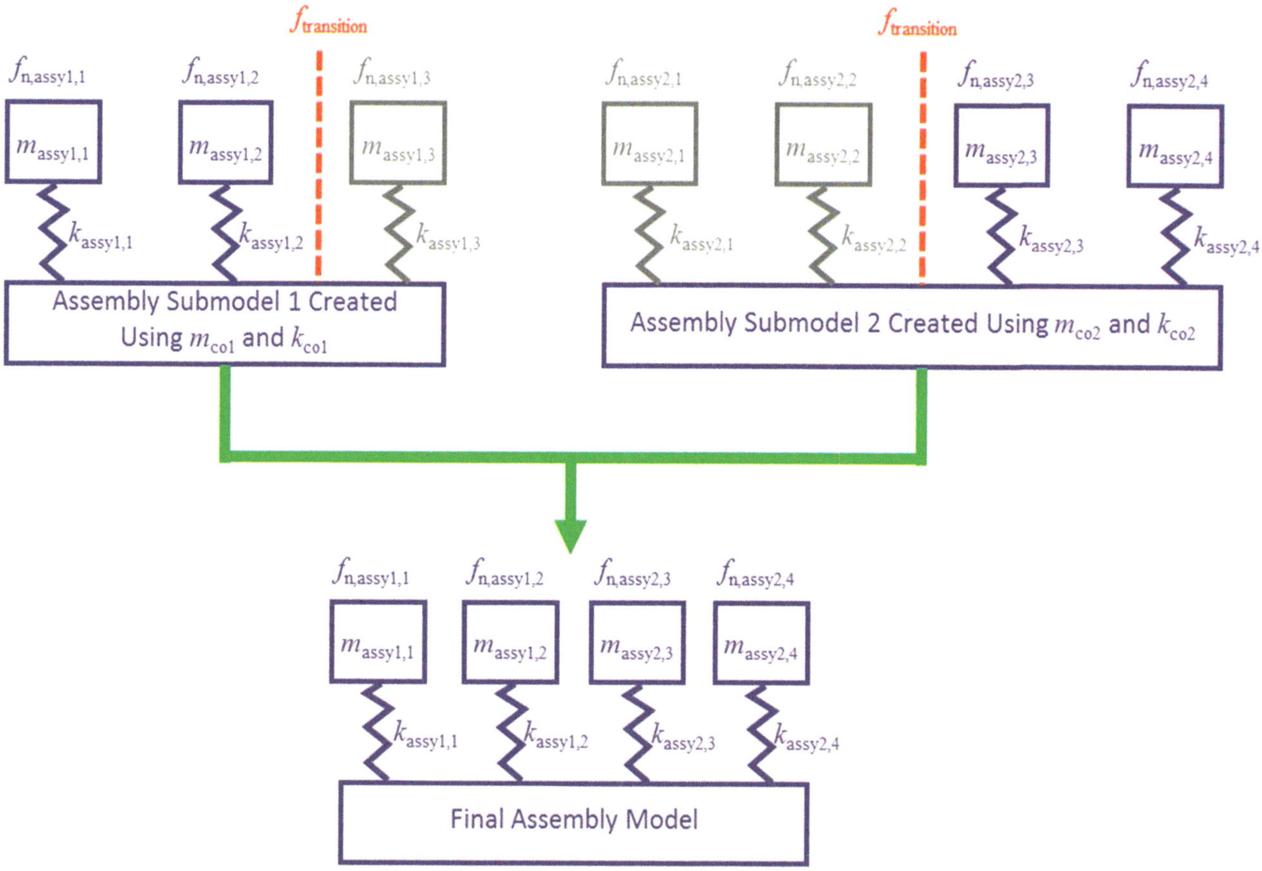

Fig. 2.5 Example of creating the final assembly model using different assembly submodels

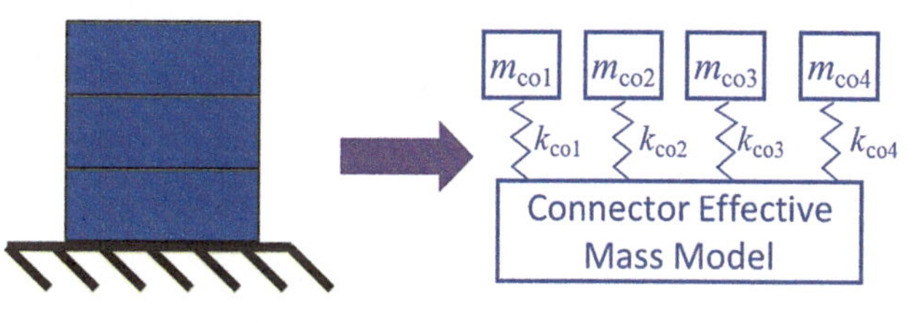

Fig. 2.6 Connector beam used to create the effective mass modal model for analytical verification

unconstrained and behave like a pinned-free beam which does not accurately represent the conditions of the true assembly. Therefore, only connector modes with large effective masses are included in this process.

In practice, an experimental effective mass modal model will be available for the connector. For the analytical verification work, however, the effective mass modal model of the connector was extracted from a cantilevered beam representing the connector in the truth model, see Fig. 2.6.

2.2.3 *Analytical Verification Results*

To demonstrate the capabilities and limitations of this process, stiffness and geometrical parameters were selected for the individual connector and cable beams shown in Fig. 2.3 and for the truth model, were coupled accordingly. The effective mass modal model was created for the connector using Eq. (2.1) in substitution for an experimental effective mass modal model which would be utilized in practice. The cable was then coupled to this model as described in Sect. 2.2.2 to create the final assembly model for the Y-direction and then for the Z-direction. For both directions, only two connector modes with significant effective mass were within the desired bandwidth of 7000 Hz. A comparison of the effective masses for each

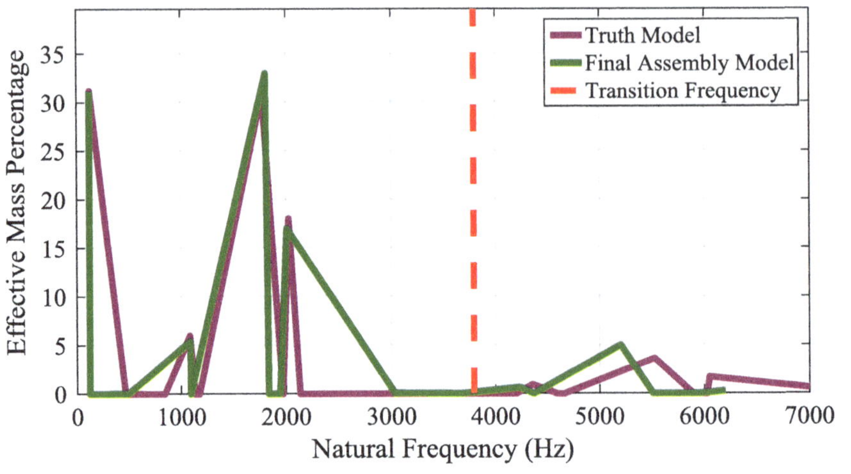

Fig. 2.7 Analytical verification results, truth model vs final assembly model comparison for Y-direction

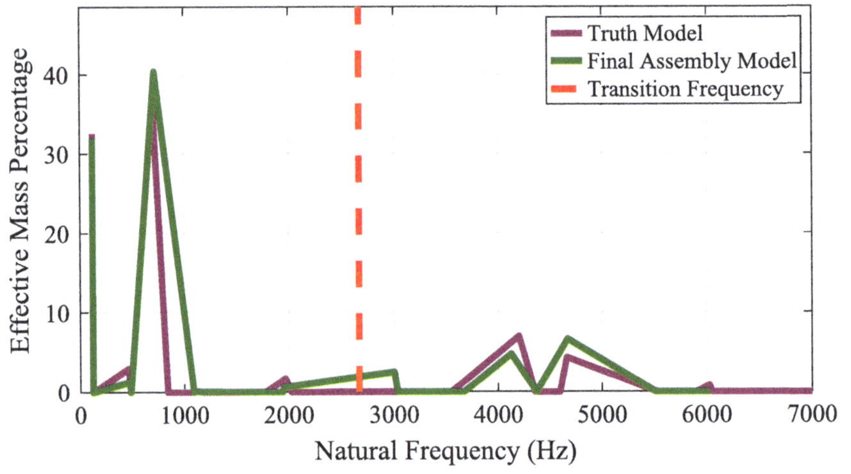

Fig. 2.8 Analytical verification results, truth model vs final assembly model comparison for Z-direction

mode and corresponding natural frequency for the truth model and the final assembly models are shown in Figs. 2.7 and 2.8 for the Y- and Z-directions, respectively. The key quantity to notice is how close the vertices of each plot align, as they show both the frequency and effective mass of one of the modes. Additionally, the final assembly model is a reduced order model and as such has fewer DOFs than the truth model, resulting in fewer modes. Lastly, since only two connector modes were used to create the final assembly model, there is only one transition frequency for each direction.

In general, the final assembly model matches the effective mass and frequency of the truth model well for modes with large effective mass for both directions. It is also worth noting that the cable deformation in the mode shapes match well. For the modes in the vertical direction with greater than 4% effective mass, the final assembly model had a maximum frequency error of less than 6% (the predicted frequency was 5200 Hz when the truth was 5520 Hz) and the maximum effective mass error was less than 2%.

Likewise, the results for the Z-direction agree well for modes with effective mass larger than 4%. There appears to be larger effective mass errors for the modes near 4200 Hz and 4660 Hz. In the truth model, these modes correspond to the connector 2nd bending in phase and out of phase with the cable 6th bending. This coupling distributes the effective mass in the truth model between these two modes in a fashion that cannot replicated in the final assembly model since the connector is absorbed into the mass of the cable tip. However, it is worth noting that the sum of the effective masses of these two modes is nearly identical (~0.2% difference) for the truth and final assembly models. This indicates that the final assembly model is able to capture the general dynamic effect of this mode pair. These results were sufficiently accurate that this process was implemented on physical hardware.

Table 2.1 Experimentally extracted connector effective mass model results

Mode	Frequency (Hz)	Description	Normalized effective mass (%)		
			X	Y	Z
1	768	Connector bending in Z	0	2.5	52.3[a]
2	785	Connector bending in X	44.4[a]	1.4	0
3	1170	Cable stub bending in Y	10.1	7	0
4	1264	Cable stub bending in Z	0	0.1	0.4
5	1926	Connector axial	0	80.2	0.1

[a]These values were reconfigured so that effective mass of connector bending is either purely in the X or Z direction

2.3 Proof of Concept on Physical Hardware

This section discusses the calculation of the effective mass modal model and corresponding natural frequencies for a physical cable-connector using the process developed in the previous section. The experimental effective mass modal model for the connector is first extracted from a modal test (Sect. 2.3.1). The stiffness parameters for the cable model are then empirically fit from modal tests of two different cable lengths in Sect. 2.3.2. The process developed in Sect. 2.2.2 is then employed to combine the connector effective mass modal model and the simplified cable FEM created from the cable stiffness and geometrical parameters. The results are compared to a truth test in Sect. 2.3.3.

2.3.1 Connector Effective Mass Experiment

This section discusses the extraction of the connector effective mass modal model from experimental data. The connector was installed into a stiff fixture and modal tested. Natural frequencies, damping values, and mode shapes were extracted. The method described in [2] was then used to extract the effective mass modal model of the connector, see Table 2.1 for the results. For brevity, the details of the technique are not given here, but [2] provides a detailed procedure on its implementation.

Note that the mode shapes showed that the bending modes 1 and 2 were not directly aligned with the coordinate axes, e.g. the connector did not bend purely in X, but also had some Z direction motion. This resulted in the effective mass for each of the first two modes being distributed between these two directions. The analytical verification from Section 2.2 had the connector modes aligned with the coordinate axes. Additionally, modes 1 and 2 have similar frequencies, so they could be oriented in any two orthogonal directions. Therefore, the bending axes were realigned to match the modeling axes, but with the same amount of effective mass is still retained in the appropriate direction.

2.3.2 Cable Model as a Timoshenko Beam

2.3.2.1 Creation of the Cable Model as a Timoshenko Beam

Previous work by Goodding, et al. shows that cables can be represented by the bending (EI) and shear (KAG) stiffness terms of the Timoshenko beam without the rotary inertia term [3]. The EI and KAG values are used to create the stiffness matrix for the cable model. This cable is of the same length as that of the physical cable-connector assembly tested in the truth experiment in Sect. 2.3.3. The first step is to calculate the stiffness matrix of the Bernoulli-Euler beam, $\mathbf{K}_{\mathbf{cable,B}}$, using EI and the length [4]. This is performed in one transverse direction using the translation and rotation at the first node and then the translation and rotation at the second node of the beam element. The shear beam stiffness matrix, $\mathbf{K}_{\mathbf{cable,S}}$, is constructed from the elemental form given in (2.3) which was calculated from the wave equation [5].

$$\begin{bmatrix} \frac{\mathrm{KAG}}{\mathrm{L}} & 0 & -\frac{\mathrm{KAG}}{\mathrm{L}} & 0 \\ 0 & 0 & 0 & 0 \\ -\frac{\mathrm{KAG}}{\mathrm{L}} & 0 & \frac{\mathrm{KAG}}{\mathrm{L}} & 0 \\ 0 & 0 & 0 & 0 \end{bmatrix} \tag{2.3}$$

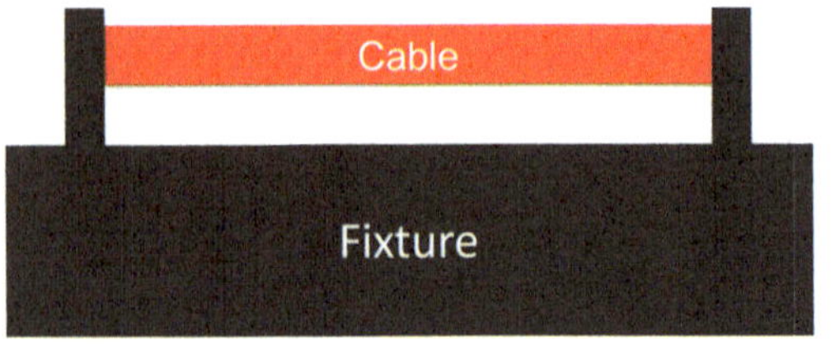

Fig. 2.9 Configuration for experiments used to empirically fit cable stiffness parameters

Table 2.2 Results of cable stiffness fit, experiment vs model

Length (in)	Mode	% Difference in f_n
8	1	9
	2	2
	3	−2
	4	5
6	1	−10
	2	0
	3	−11

where L is the elemental length of the cable beam. Note that this elemental matrix will produce a $\mathbf{K}_{\mathbf{cable,S}}$ where all of the diagonal values will be 2KAG/L except the first and last. Additionally, the zero terms in (2.3) correspond to the rotational DOFs of the cable beam. The total stiffness matrix of the cable beam, $\mathrm{K}_{\mathbf{cable}}$, is then the sum of $\mathbf{K}_{\mathbf{cable,B}}$ and $\mathbf{K}_{\mathbf{cable,S}}$. Once $\mathbf{K}_{\mathbf{cable}}$ has been calculated (along with the cable mass matrix, $\mathbf{M}_{\mathbf{cable}}$), the cable is constrained with pinned boundary conditions for the stiffness parameter calibration to experimental data, see Sect. 2.3.2.2. The $\mathbf{M}_{\mathbf{cable}}$ and $\mathbf{K}_{\mathbf{cable}}$ comprise the cable model that is used in Sect. 2.3.3 along with the connector effective mass modal model from Sect. 2.3.1 to calculate the effective mass model and corresponding natural frequencies of a physical cable-connector assembly using the method outlined in Sect. 2.2.2.

There is the problem of shear locking over-stiffening the result using the formulation in Eq. (2.3) [6]. However, with this class of problem, the connector is sufficiently stiff that it minimizes the effect due to shear locking.

2.3.2.2 Modal Test of the Cable

The proposed method discussed in Sect. 2.2.2 requires the EI and KAG stiffness properties of the cable by itself in order to compute the simple FEM ($\mathbf{M}_{\mathbf{cable}}$ and $\mathbf{K}_{\mathbf{cable}}$). Since a cable is a complex, non-homogenous combination of wire strands and insulation material, the stiffness properties were empirically fit from modal testing of the cable clamped at either end, see Fig. 2.9. In order to achieve robust stiffness properties, two cable lengths were tested, and natural frequencies, damping values, and mode shapes were extracted for each.

2.3.2.3 Fitting the Stiffness Parameters for the Timoshenko Beam

For the Timoshenko beam, two stiffness parameters need to be fit: EI (bending stiffness) and KAG (shear stiffness). This was accomplished by performing a sensitivity analysis according to the following:

$$\{f_{n,\mathrm{exp}}\} = \{f_{n,model}\} + \begin{bmatrix} \dfrac{\partial \{f_{n,model}\}}{\partial \mathrm{EI}} & \dfrac{\partial \{f_{n,model}\}}{\partial \mathrm{KAG}} \end{bmatrix} \begin{Bmatrix} \Delta \mathrm{EI} \\ \Delta \mathrm{KAG} \end{Bmatrix} \tag{2.4}$$

where $\{f_{n,exp}\}$ is a vector containing the experimental natural frequencies of the cable, and $\{f_{n,model}\}$ is a vector of the model natural frequencies calculated using an initial estimate of EI and KAG. Each $\{f_{n,model}\}$ were computed from an eigenvalue analysis of a Timoshenko beam defined by $\mathbf{M}_{\mathbf{cable}}$ and $\mathbf{K}_{\mathbf{cable}}$. Note that the models were created using pinned-pinned boundary conditions.

Table 2.2 displays the difference between the experimental natural frequencies of the cable and those of the Timoshenko beam generated from the converged EI and KAG values. It is important to note that several cable modes were affected by motion of clamps on the fixture which secure the two ends of the cable during the experiment. The influence on the frequencies could not be exactly quantified but are estimated to potentially shift frequencies up to 10%.

The boundary conditions for the cable experiment were initially assumed to be clamped-clamped. However, no reasonable fit to the experimental frequencies could be achieved. The use of pinned boundary conditions was supported by the fact that the deflection of the ends of the cable in the experimental mode shapes visually appeared to replicate the motion of a pinned-pinned beam. Note that the KAG and EI values were one to two orders of magnitude below the values that one would obtain with the copper wire with the measured diameter. This reinforced that the experimental process of fitting these two parameters was required.

2.3.3 Comparison with Truth Test Results

The results from Sect. 2.3.2 were used to create a cable model of the physical hardware and this was coupled to the connector experimental effective mass modal model presented in Sect. 2.3.1 using the method proposed in Sect. 2.2.2 to create a final assembly model of the physical cable-connector assembly. During the course of this work, it was discovered that the cable lengths used in Sect. 2.3.2 to empirically fit the stiffness parameters did not bound the cable length used in the truth test. Thus the results presented below represent an extrapolation on the length effect, a poor engineering practice that lends itself to larger-than-expected errors. Schedule did not permit performing an additional cable test on the shorter length. Figures 2.10 and 2.11 show the truth test results and the final assembly model calculations for the effective mass and natural frequencies for the cable-connector assembly. Note that for either direction only modes with the two largest effective masses are presented.

These results include estimated uncertainty bounds on both the effective mass and natural frequency for the truth test and final assembly model. They assume cable lengths interpolated within the tested bounds (an assumption that was not appropriately observed here). Uncertainties are valuable when computing the failure margin of a product. Due to cost and schedule limitations, an exhaustive uncertainty quantification effort could not be performed, so the error bounds in Figs. 2.10 and 2.11 were determined using engineering judgement and multiple experiments from previous work. For the truth test results, the frequency and effective mass uncertainties were chosen to be, respectively, ±10% of the extracted natural frequency and ±7% of the total mass of the cable-connector assembly. These numbers have been shown to be typical for effective mass experiments. The frequency and effective mass uncertainties for the final assembly model were selected to be a root-sum-square of the individual uncertainties from each source as listed in Table 2.3.

Ideally, the final assembly model results in Figs. 2.10 and 2.11 would encompass the test results. The test frequencies are nearly within the model frequency bands, but this is not the case with the effective mass, particularly in the Z direction. The cable stiffness extrapolation, especially for shorter cables as explained above is impacting the model results, but at this stage we do not know if it is the only error. One phenomenon that was observed during the analytical verification work was that

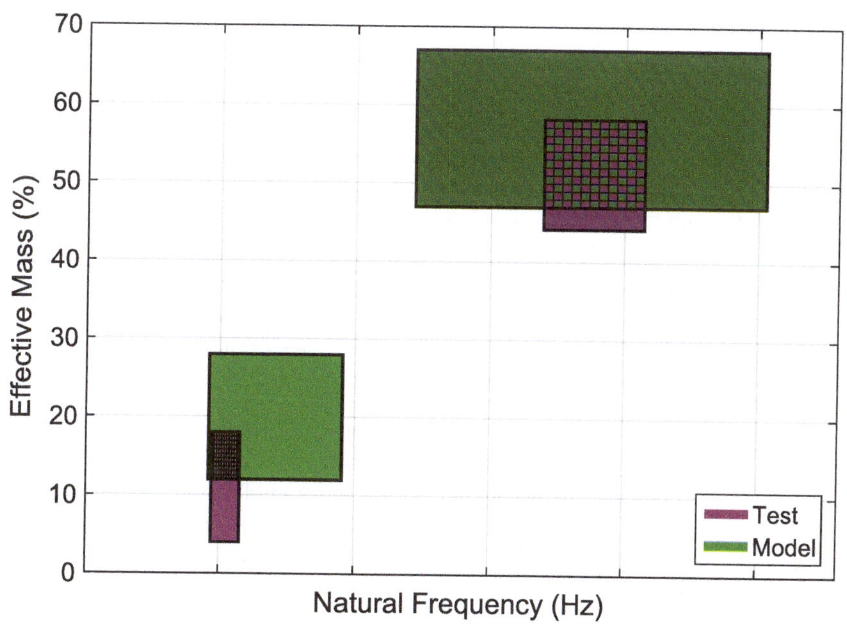

Fig. 2.10 Truth test vs final assembly model results, Y-direction

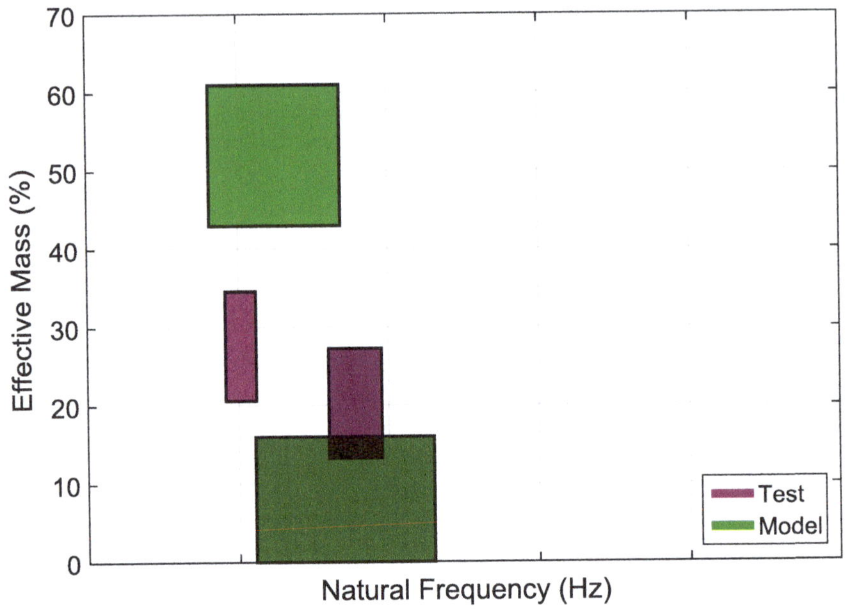

Fig. 2.11 Truth test vs final assembly model results, Z-direction

Table 2.3 Sources of uncertainty in final assembly model

	Source		
Parameter	Connecter experiment	Max error from analytical verification results	Unit-to-unit variability
Frequency	±10% of natural frequency	±6% of natural frequency	±30% of natural frequency
Effective mass	±7% of total mass	±3% of total mass	±10% of calculated effective mass

when modes of the individual connector and cable were close in frequency, they coupled in the cable-connector assembly which then distributed the effective mass between the two modes in a fashion that the final assembly model was not able to reproduce. However, the sum of the effective masses of these two modes was captured by the model. This is similar to the situation for the two modes in the Z direction.

2.4 Conclusion

A method for calculating the effective mass modal model of a cable-connector assembly using dynamic information of the components was developed. An analytical verification model showed that the method appeared to be sufficiently accurate to implement on physical hardware. Using a connector effective mass modal model extracted from experimental data and a simple cable FEM created from empirically fit stiffness parameters, the effective mass modal model of a simplified assembly was computed. The results were compared to a truth experiment with uncertainties applied to both the effective mass and natural frequencies for the model and experiment. While the model did not completely encompass the experimental results, its performance was encouraging especially when considering the influence of unanticipated extrapolation. Closely coupled cable and connector modes may also induce more error to the process.

Notice This manuscript has been authored by National Technology and Engineering Solutions of Sandia, LLC. under Contract No. DE-NA0003525 with the U.S. Department of Energy/National Nuclear Security Administration. The United States Government retains and the publisher, by accepting the article for publication, acknowledges that the United States Government retains a non-exclusive, paid-up, irrevocable, world-wide license to publish or reproduce the published form of this manuscript, or allow others to do so, for United States Government purposes.

References

1. Mayes, R.L., Linehan, D.W.: Measuring effective mass of a circuit board. In: International Modal Analysis Conference 32, Orlando, Orlando FL (2014)
2. Mayes, R.L., Hunter, P.S.: A simpler formulation for effective mass calculated from experimental free mode shapes of a test article on a fixture. In: International Modal Analysis Conference 35, Garden Grove, CA (2017)
3. Goodding, J.C., Ardelean, E.V., Babuska, V., Robertson, L.M., Lane, S.A.: Experimental techniques and structural parameter estimation studies of spacecraft cables. J. Spacecr. Rocket. **48**, 942–957 (2011)
4. Cook, R.D., Malkus, D.S., Plesha, M.E.: Concepts and Applications of Finite Element Analysis, 3rd edn, pp. 113–114. Wiley, New York (1989)
5. Blevins, R.D.: Formulas for Dynamics, Acoustics and Vibration, the Atrium, Southern Gate, Chichester, West Sussex, PO19 8SQ, p. 167. Wiley, United Kingdom (2016)
6. Felippa, C.: Nonlinear finite element methods course notes; chapter 11: The TL Plane Beam Element: Formulation, University of Colorado at Boulder

Chapter 3
Comparison of Craig-Bampton Approaches for Systems with Arbitrary Viscous Damping in Dynamic Substructuring

Fabian M. Gruber and Daniel Rixen

Abstract The classical Craig-Bampton method does not take any damping effects into account for the model order reduction of damped systems. There is generally no justification to neglect damping effects. If damping significantly influences the dynamic behavior of the system, the approximation accuracy can be very poor. One procedure to handle arbitrary damped systems is to transform the second-order differential equations into twice the number of first-order differential equations resulting in state-space representation of the system. Solving the corresponding eigenvalue problem allows the damped equations for the internal degrees of freedom of the substructures to be decoupled, but complex eigenmodes and eigenvalues occur.

Hasselman and Kaplan presented a coupling procedure for damped systems that employs complex component modes. Beliveau and Soucy proposed another version that modifies the classical Craig-Bampton method to include damping by replacing the real fixed interface normal modes of the second-order system with the corresponding complex modes of the first-order system. Additionally, they suggest an adaption of the method of Hasselman and Kaplan. A report of de Kraker gives another description of the Craig-Bampton method using complex normal modes and modified static modes. The derivation of all the different Craig-Bampton substructuring methods for viscously damped systems is presented in a comprehensible consistent manner. A comparison between the different formulations will be given. The presented theory and the comparison between the methods are illustrated by an example.

Keywords Dynamic substructuring · Component mode synthesis · Craig-Bampton method · Damped systems · State-space formulation · Complex modes

3.1 Introduction

Dynamic substructuring techniques reduce the size of large models very efficiently. The large model is thereby divided into N substructures; each substructure is analyzed and reduced separately and then assembled into a low-order reduced model. This low-order reduced model approximates the original large model's behavior. During this process, each substructure's degrees of freedom (DOFs) are divided into internal DOFs (those not shared with any adjacent substructure) and boundary or interface DOFs (those shared with adjacent substructures and therefore forming the model's interface DOFs) [1]. Many substructuring methods that work with second-order equations of motion,

$$\boldsymbol{M}^{(s)}\ddot{\boldsymbol{u}}^{(s)} + \boldsymbol{C}^{(s)}\dot{\boldsymbol{u}}^{(s)} + \boldsymbol{K}^{(s)}\boldsymbol{u}^{(s)} = \boldsymbol{f}^{(s)} + \boldsymbol{g}^{(s)}, \qquad s = 1, \ldots, N, \tag{3.1}$$

have been proposed in the past [2–9]. The equations of motion (3.1) of substructure s have $n^{(s)}$ DOFs. The superscript (s) is the label of the particular substructure s. $\boldsymbol{M}^{(s)}$, $\boldsymbol{C}^{(s)}$, $\boldsymbol{K}^{(s)}$, and $\boldsymbol{u}^{(s)}$ are, respectively, the substructure's mass matrix, damping matrix, stiffness matrix, and displacement vector. $\boldsymbol{f}^{(s)}$ is the external force vector and $\boldsymbol{g}^{(s)}$ is the vector of internal forces connecting adjacent substructures at their boundary DOFs. Only the substructures' mass and stiffness properties are commonly taken into account for the reduction by all those methods. The undamped equations of motion,

$$\boldsymbol{M}^{(s)}\ddot{\boldsymbol{u}}^{(s)} + \boldsymbol{K}^{(s)}\boldsymbol{u}^{(s)} = \boldsymbol{f}^{(s)} + \boldsymbol{g}^{(s)}, \qquad s = 1, \ldots, N, \tag{3.2}$$

F. M. Gruber (✉) · D. Rixen
Faculty of Mechanical Engineering, Technical University of Munich, Garching, Germany
e-mail: fabian.gruber@tum.de; rixen@tum.de

A. Linderholt et al. (eds.), *Dynamics of Coupled Structures, Volume 4*, Conference Proceedings of the Society for Experimental Mechanics Series, https://doi.org/10.1007/978-3-319-74654-8_3

are assumed to correctly describe the substructure dynamics. It is assumed that there is no damping or that damping effects are completely negligible when building the reduction basis, i.e., $\boldsymbol{C}^{(s)} = \boldsymbol{0}$. Most substructuring methods work with the undamped equations of motion (3.2) and afford great approximation accuracy if the underlying system is damped only slightly or not at all [10]. The most popular approach is a fixed interface method, the Craig-Bampton method [2], which is based on fixed interface vibration modes and interface constraint modes. The substructures are assembled using interface displacements, which is referred to as primal assembly.

The classical Craig-Bampton method [2] does not consider any damping effects. If damping significantly influences the dynamic behavior of the system under consideration, then the approximation accuracy of the Craig-Bampton method can be very poor. There is generally no justification to neglect damping effects. The damped equations of motion (3.1) have to be taken into account to incorporate damping effects. One procedure to handle and decouple damped systems is to transform the second-order differential equations (3.1) into twice the number of first-order differential equations, resulting in a state-space representation of the system. Solving the corresponding eigenvalue problem allows the damped equations of motion to be decoupled, but complex eigenmodes and eigenvalues occur. Hasselman and Kaplan [11] presented a coupling procedure for damped substructured systems, which is an extension of the Craig-Bampton method by employing complex component modes. Beliveau and Soucy [12] proposed another version that modifies the classical Craig-Bampton method to include damping by replacing the real fixed interface normal modes of the second-order system with the corresponding complex modes of the first-order system. A report of de Kraker [13] gives another description of the Craig-Bampton method for damped systems using complex normal modes and modified static modes.

In this contribution, we want to present the derivation of all these different Craig-Bampton substructuring approaches for viscously damped systems in a comprehensible consistent manner. The methods will be briefly derived and a comparison between the different formulations will be given. The presented theory and the comparison between the methods will be illustrated on a simple example taken from [12].

Section 3.2 recalls the primal assembly procedure, and the terminology and notation used throughout this paper is set up. In Sect. 3.3, the different Craig-Bampton approaches are derived. Starting in Sect. 3.3.1, the classical Craig-Bampton approach is presented and used for damped systems as well. Following this, the approaches that have taken damping effects explicitly into account are derived: Hasselman and Kaplan's method in Sect. 3.3.2, Beliveau and Soucy's method in Sect. 3.3.3, de Kraker's method in Sect. 3.3.4 and an adaption of Hasselman and Kaplan's method in Sect. 3.3.5. The properties of the presented methods are subsequently illustrated in detail in Sect. 3.4 using an example. Finally, a brief summary of findings and conclusions is given in Sect. 3.5.

3.2 Primal Assembly

Consider the second-order equations of motion (3.1). The local displacements $\boldsymbol{u}^{(s)}$ of each substructure can be divided in $n_i^{(s)}$ local internal DOFs $\boldsymbol{u}_i^{(s)}$ and $n_b^{(s)}$ boundary DOFs $\boldsymbol{u}_b^{(s)}$:

$$\boldsymbol{u}^{(s)} = \begin{bmatrix} \boldsymbol{u}_b^{(s)} \\ \boldsymbol{u}_i^{(s)} \end{bmatrix} = \begin{bmatrix} \boldsymbol{L}_b^{(s)} & \boldsymbol{0} \\ \boldsymbol{0} & \boldsymbol{I} \end{bmatrix} \begin{bmatrix} \boldsymbol{u}_b \\ \boldsymbol{u}_i^{(s)} \end{bmatrix} \tag{3.3}$$

The boundary DOFs $\boldsymbol{u}_b^{(s)} = \boldsymbol{L}_b^{(s)} \boldsymbol{u}_b$ of substructure s are a subset of the n_b assembled boundary DOFs $\boldsymbol{u}_b$ of the global domain. $\boldsymbol{L}_b^{(s)}$ is a Boolean localization matrix (connectivity matrix) relating the assembled boundary DOFs $\boldsymbol{u}_b$ of the global domain to the substructure boundary DOFs $\boldsymbol{u}_b^{(s)}$ [1, 5]. Consequently, the equations of motion (3.2) of one substructure s partitioned in the same manner is written as

$$\begin{bmatrix} \boldsymbol{M}_{bb}^{(s)} & \boldsymbol{M}_{bi}^{(s)} \\ \boldsymbol{M}_{ib}^{(s)} & \boldsymbol{M}_{ii}^{(s)} \end{bmatrix} \begin{bmatrix} \ddot{\boldsymbol{u}}_b^{(s)} \\ \ddot{\boldsymbol{u}}_i^{(s)} \end{bmatrix} + \begin{bmatrix} \boldsymbol{C}_{bb}^{(s)} & \boldsymbol{C}_{bi}^{(s)} \\ \boldsymbol{C}_{ib}^{(s)} & \boldsymbol{C}_{ii}^{(s)} \end{bmatrix} \begin{bmatrix} \dot{\boldsymbol{u}}_b^{(s)} \\ \dot{\boldsymbol{u}}_i^{(s)} \end{bmatrix} + \begin{bmatrix} \boldsymbol{K}_{bb}^{(s)} & \boldsymbol{K}_{bi}^{(s)} \\ \boldsymbol{K}_{ib}^{(s)} & \boldsymbol{K}_{ii}^{(s)} \end{bmatrix} \begin{bmatrix} \boldsymbol{u}_b^{(s)} \\ \boldsymbol{u}_i^{(s)} \end{bmatrix} = \begin{bmatrix} \boldsymbol{f}_b^{(s)} \\ \boldsymbol{f}_i^{(s)} \end{bmatrix} + \begin{bmatrix} \boldsymbol{g}_b^{(s)} \\ \boldsymbol{0} \end{bmatrix}. \tag{3.4}$$

Eqs. (3.2) and (3.3) of all substructures N can be assembled in a primal way as

$$\boldsymbol{M}\ddot{\boldsymbol{u}} + \boldsymbol{C}\dot{\boldsymbol{u}} + \boldsymbol{K}\boldsymbol{u} = \boldsymbol{f}, \tag{3.5}$$

where

$$M = \begin{bmatrix} \hat{M}_{bb} & \hat{M}_{bi}^{(1)} & \cdots & \hat{M}_{bi}^{(N)} \\ \hat{M}_{ib}^{(1)} & M_{ii}^{(1)} & & 0 \\ \vdots & & \ddots & \\ \hat{M}_{ib}^{(N)} & 0 & & M_{ii}^{(N)} \end{bmatrix}, \quad C = \begin{bmatrix} \hat{C}_{bb} & \hat{C}_{bi}^{(1)} & \cdots & \hat{C}_{bi}^{(N)} \\ \hat{C}_{ib}^{(1)} & C_{ii}^{(1)} & & 0 \\ \vdots & & \ddots & \\ \hat{C}_{ib}^{(N)} & 0 & & C_{ii}^{(N)} \end{bmatrix}, \tag{3.6}$$

$$K = \begin{bmatrix} \hat{K}_{bb} & \hat{K}_{bi}^{(1)} & \cdots & \hat{K}_{bi}^{(N)} \\ \hat{K}_{ib}^{(1)} & K_{ii}^{(1)} & & 0 \\ \vdots & & \ddots & \\ \hat{K}_{ib}^{(N)} & 0 & & K_{ii}^{(N)} \end{bmatrix}, \quad u = \begin{bmatrix} u_b \\ u_i^{(1)} \\ \vdots \\ u_i^{(N)} \end{bmatrix}, \quad f = \begin{bmatrix} \hat{f}_b \\ f_i^{(1)} \\ \vdots \\ f_i^{(N)} \end{bmatrix} \tag{3.7}$$

with

$$\hat{M}_{bb} = \sum_{s=1}^{N} L_b^{(s)^T} M_{bb}^{(s)} L_b^{(s)}, \qquad \hat{M}_{ib}^{(s)} = M_{ib}^{(s)} L_b^{(s)}, \qquad \hat{M}_{bi}^{(s)} = L_b^{(s)^T} M_{bi}^{(s)},$$

$$\hat{C}_{bb} = \sum_{s=1}^{N} L_b^{(s)^T} C_{bb}^{(s)} L_b^{(s)}, \qquad \hat{C}_{ib}^{(s)} = C_{ib}^{(s)} L_b^{(s)}, \qquad \hat{C}_{bi}^{(s)} = L_b^{(s)^T} C_{bi}^{(s)},$$

$$\hat{K}_{bb} = \sum_{s=1}^{N} L_b^{(s)^T} K_{bb}^{(s)} L_b^{(s)}, \qquad \hat{K}_{ib}^{(s)} = K_{ib}^{(s)} L_b^{(s)}, \qquad \hat{K}_{bi}^{(s)} = L_b^{(s)^T} K_{bi}^{(s)}, \qquad \hat{f}_b = \sum_{s=1}^{N} L_b^{(s)^T} f_b^{(s)}.$$

The reaction forces $g^{(s)}$ on the interfaces of the substructures cancel out during assembly. This assembly procedure is called primal assembly since the compatibility between the substructures is enforced using the same boundary displacements for adjacent substructures [1, 5].

3.3 Craig-Bampton Reduction Approaches for Damped Systems

This section gives a derivation of all the different Craig-Bampton substructuring methods for viscously damped systems and is organized as follows: Sect. 3.3.1 recalls briefly the classical Craig-Bampton approach. Subsequently, substructuring methods explicitly accounting for damping effects are presented: Hasselman and Kaplan's method in Sect. 3.3.2, Beliveau and Soucy's method in Sect. 3.3.3, de Kraker's method in Sect. 3.3.4 and an adaption of Hasselman and Kaplan's method in Sect. 3.3.5.

3.3.1 Craig-Bampton Method (CB)

3.3.1.1 Method for Undamped systems

The classical Craig-Bampton method [2] works with the undamped equations of motion (3.2). Considering the partitioned equations of motion (3.4) without damping, the internal DOFs $u_i^{(s)}$ of every substructure s can be seen as being excited by its boundary DOFs $u_b^{(s)}$, namely

$$M_{ii}^{(s)} \ddot{u}_i^{(s)} + K_{ii}^{(s)} u_i^{(s)} = f_i^{(s)} - K_{ib}^{(s)} u_b^{(s)} - M_{ib}^{(s)} \ddot{u}_b^{(s)}. \tag{3.8}$$

This indicates that the internal displacements $u_i^{(s)}$ of each substructure can be approximated by a superposition of a static response and of eigenmodes associated with $M_{ii}^{(s)}$ and $K_{ii}^{(s)}$. The static response is given by

$$u_{i,\text{stat}}^{(s)} = -K_{ii}^{(s)^{-1}} K_{ib}^{(s)} u_b^{(s)} = \Psi_{ib}^{(s)} u_b^{(s)}, \tag{3.9}$$

where the columns of matrix $\boldsymbol{\Psi}_{ib}^{(s)} = -\boldsymbol{K}_{ii}^{(s)^{-1}}\boldsymbol{K}_{ib}^{(s)}$ are the static response modes also called constraint modes [2]. The fixed interface normal modes $\boldsymbol{\phi}_k^{(s)}$ are obtained as eigensolutions of the generalized eigenproblem

$$\boldsymbol{K}_{ii}^{(s)}\boldsymbol{\phi}_k^{(s)} = \omega_k^{(s)^2}\boldsymbol{M}_{ii}^{(s)}\boldsymbol{\phi}_k^{(s)}. \tag{3.10}$$

The columns of the $n_i^{(s)} \times n_\phi^{(s)}$ matrix $\boldsymbol{\Phi}_{i\phi}^{(s)}$ contain the first $n_\phi^{(s)}$ fixed interface normal modes $\boldsymbol{\phi}_k^{(s)}$, which can also be considered as the free vibration modes of the substructure s clamped on its boundary DOFs $\boldsymbol{u}_b^{(s)}$. The fixed interface normal modes $\boldsymbol{\phi}_k^{(s)}$ are orthonormalized with respect to the mass matrix $\boldsymbol{M}_{ii}^{(s)}$. The approximation of $\boldsymbol{u}_i^{(s)}$ therefore writes

$$\boldsymbol{u}_i^{(s)} \approx \boldsymbol{u}_{i,\text{stat}}^{(s)} + \boldsymbol{\Phi}_{i\phi}^{(s)}\boldsymbol{\eta}^{(s)} = \boldsymbol{\Psi}_{ib}^{(s)}\boldsymbol{u}_b^{(s)} + \boldsymbol{\Phi}_{i\phi}^{(s)}\boldsymbol{\eta}^{(s)} \tag{3.11}$$

and the displacements of the substructure are approximated by

$$\boldsymbol{u}^{(s)} = \begin{bmatrix}\boldsymbol{u}_b^{(s)}\\ \boldsymbol{u}_i^{(s)}\end{bmatrix} \approx \underbrace{\begin{bmatrix}\boldsymbol{I} & \boldsymbol{0}\\ \boldsymbol{\Psi}_{ib}^{(s)} & \boldsymbol{\Phi}_{i\phi}^{(s)}\end{bmatrix}}_{\boldsymbol{T}_{\text{CB}}^{(s)}}\begin{bmatrix}\boldsymbol{u}_b^{(s)}\\ \boldsymbol{\eta}^{(s)}\end{bmatrix}. \tag{3.12}$$

The local Craig-Bampton reduction matrix $\boldsymbol{T}_{\text{CB}}^{(s)}$ is used to compute each substructure's reduced mass and stiffness matrix [1, 2]:

$$\boldsymbol{M}_{\text{red,CB}}^{(s)} = \boldsymbol{T}_{\text{CB}}^{(s)^T}\boldsymbol{M}^{(s)}\boldsymbol{T}_{\text{CB}}^{(s)} = \begin{bmatrix}\tilde{\boldsymbol{M}}_{bb}^{(s)} & \tilde{\boldsymbol{M}}_{b\phi}^{(s)}\\ \tilde{\boldsymbol{M}}_{\phi b}^{(s)} & \boldsymbol{I}_{\phi\phi}\end{bmatrix}, \qquad \boldsymbol{K}_{\text{red,CB}}^{(s)} = \boldsymbol{T}_{\text{CB}}^{(s)^T}\boldsymbol{K}^{(s)}\boldsymbol{T}_{\text{CB}}^{(s)} = \begin{bmatrix}\tilde{\boldsymbol{K}}_{bb}^{(s)} & \boldsymbol{0}\\ \boldsymbol{0} & \boldsymbol{\Omega}_{\phi\phi}^{(s)^2}\end{bmatrix} \tag{3.13}$$

Matrix $\boldsymbol{\Omega}_{\phi\phi}^{(s)}$ contains the kept $n_\phi^{(s)}$ eigenvalues $\omega_k^{(s)}$ on the diagonal. The lower-right partitions of the reduced mass $\boldsymbol{M}_{\text{red,CB}}^{(s)}$ and stiffness matrix $\boldsymbol{K}_{\text{red,CB}}^{(s)}$ in Eq. (3.13) are diagonal due to the orthogonality conditions of the kept fixed interface normal modes $\boldsymbol{\Phi}_{i\phi}^{(s)}$.

Primal assembly of all N substructures' reduced matrices gives the assembled reduced matrices $\boldsymbol{M}_{\text{red,CB}}$ and $\boldsymbol{K}_{\text{red,CB}}$ in the configuration space. The global Craig-Bampton reduction matrix $\boldsymbol{T}_{\text{CB}}$ for the approximation of the displacements $\boldsymbol{u}$ writes

$$\boldsymbol{u} \approx \underbrace{\begin{bmatrix}\boldsymbol{I} & \boldsymbol{0} & \cdots & \boldsymbol{0}\\ \boldsymbol{\Psi}_{ib}^{(1)}\boldsymbol{L}_b^{(1)} & \boldsymbol{\Phi}_{i\phi}^{(1)} & & \boldsymbol{0}\\ \vdots & & \ddots & \\ \boldsymbol{\Psi}_{ib}^{(N)}\boldsymbol{L}_b^{(N)} & \boldsymbol{0} & & \boldsymbol{\Phi}_{i\phi}^{(N)}\end{bmatrix}}_{\boldsymbol{T}_{\text{CB}}}\begin{bmatrix}\boldsymbol{u}_b\\ \boldsymbol{\eta}^{(1)}\\ \vdots\\ \boldsymbol{\eta}^{(N)}\end{bmatrix} \tag{3.14}$$

and the assembled Craig-Bampton reduced mass and stiffness matrix are

$$\boldsymbol{M}_{\text{red,CB}} = \boldsymbol{T}_{\text{CB}}^T\boldsymbol{M}\boldsymbol{T}_{\text{CB}}, \qquad \boldsymbol{K}_{\text{red,CB}} = \boldsymbol{T}_{\text{CB}}^T\boldsymbol{K}\boldsymbol{T}_{\text{CB}}. \tag{3.15}$$

3.3.1.2 Application to Systems with Arbitrary Viscous Damping

If the damped equations of motion (3.1) are considered, the Craig-Bampton reduction matrix $\boldsymbol{T}_{\text{CB}}^{(s)}$ of Eq. (3.12) can also be used for the reduction of the substructure's damping matrix $\boldsymbol{C}^{(s)}$. The substructure's reduced mass and stiffness matrix are given in Eq. (3.13) and the reduced damping matrix writes

$$\boldsymbol{C}_{\text{red,CB}}^{(s)} = \boldsymbol{T}_{\text{CB}}^{(s)^T}\boldsymbol{C}^{(s)}\boldsymbol{T}_{\text{CB}}^{(s)} = \begin{bmatrix}\tilde{\boldsymbol{C}}_{bb}^{(s)} & \tilde{\boldsymbol{C}}_{b\phi}^{(s)}\\ \tilde{\boldsymbol{C}}_{\phi b}^{(s)} & \tilde{\boldsymbol{C}}_{\phi\phi}^{(s)}\end{bmatrix}. \tag{3.16}$$

In general, $\tilde{C}^{(s)}_{\phi\phi}$ is a fully populated matrix and is not diagonalized by the kept normal modes. The corresponding parts of the Craig-Bampton reduced mass matrix $M^{(s)}_{\text{red,CB}}$ and stiffness matrix $K^{(s)}_{\text{red,CB}}$ in Eq. (3.13) are diagonal. For the reduction matrix $T^{(s)}_{\text{CB}}$ only properties of the mass and stiffness matrix are taken into account, but the properties of damping matrix $C^{(s)}$ are not used. This is the main reason for very poor approximation accuracy if damping significantly influences the dynamic behavior of the system, and the classical Craig-Bampton method is used as reduction method. The assembled damping matrix $C_{\text{red,CB}}$ of the global reduced system can be built according to Eq. (3.15).

3.3.2 Hasselman and Kaplan's Method (HK)

Hasselman and Kaplan [11] proposed a method for damped systems that employs complex component modes. This is in contrast to the substructure's real normal modes $\Phi^{(s)}_{i\phi}$ used by the classical Craig-Bampton approach in Sect. 3.3.1. The physical substructure quantities $M^{(s)}$, $C^{(s)}$, $K^{(s)}$ and $f^{(s)}$ are partitioned in internal DOFs $u^{(s)}_i$ and boundary DOFs $u^{(s)}_b$ according to Eq. (3.4). Primal assembly is used to get the global assembled (unreduced) matrices M, C and K as described in Sect. 3.2. The corresponding second-order equations (3.5) are transformed into twice the number of first-order equations in state-space format [1, 14–16]. The state-space vector is

$$z(t) = \begin{bmatrix} u(t) \\ v(t) \end{bmatrix} \tag{3.17}$$

with $v(t) = \dot{u}(t)$. Adding the redundant equations $Mv(t) = M\dot{u}(t)$ to the equations of motion (3.5), the generalized state-space symmetric form

$$\mathcal{A}\dot{z} + \mathcal{B}z = \mathcal{F} \tag{3.18}$$

is obtained with

$$\mathcal{A} = \begin{bmatrix} C & M \\ M & 0 \end{bmatrix}, \quad \mathcal{B} = \begin{bmatrix} K & 0 \\ 0 & -M \end{bmatrix}, \quad \mathcal{F} = \begin{bmatrix} f \\ 0 \end{bmatrix}. \tag{3.19}$$

Analogously, the substructure state-space vector $z^{(s)}$ is

$$z^{(s)} = \begin{bmatrix} u^{(s)} \\ v^{(s)} \end{bmatrix} \tag{3.20}$$

and z is the primarily assembled state-space vector of all substructure state-space vectors $z^{(s)}$, as described in Sect. 3.2 for the displacements u.

Hasselman and Kaplan transform the homogeneous equations for each constrained substructure ($u^{(s)}_b = 0, f^{(s)}_i = 0$)

$$M^{(s)}_{ii}\ddot{u}^{(s)}_i + C^{(s)}_{ii}\dot{u}^{(s)}_i + K^{(s)}_{ii}u^{(s)}_i = 0, \tag{3.21}$$

into first-order form

$$\mathcal{A}^{(s)}_{ii}\dot{z}^{(s)}_i + \mathcal{B}^{(s)}_{ii}z^{(s)}_i = 0 \tag{3.22}$$

with the state-space matrices $\mathcal{A}^{(s)}_{ii}$ and $\mathcal{B}^{(s)}_{ii}$:

$$\mathcal{A}^{(s)}_{ii} = \begin{bmatrix} C^{(s)}_{ii} & M^{(s)}_{ii} \\ M^{(s)}_{ii} & 0 \end{bmatrix}, \quad \mathcal{B}^{(s)}_{ii} = \begin{bmatrix} K^{(s)}_{ii} & 0 \\ 0 & -M^{(s)}_{ii} \end{bmatrix} \tag{3.23}$$

Solution of the corresponding first-order eigenproblem

$$\left(\mathbf{\Lambda}^{(s)}\boldsymbol{\mathcal{A}}_{ii}^{(s)} + \boldsymbol{\mathcal{B}}_{ii}^{(s)}\right)\mathbf{\Phi}_{\mathrm{ss}}^{(s)} = \mathbf{0} \tag{3.24}$$

for the dynamic modes $\mathbf{\Phi}_{\mathrm{ss}}^{(s)}$, which are also called complex fixed interface normal modes, leads to representation of the internal DOFs $z_i^{(s)}$ of the state-space vector

$$z_i^{(s)} = \begin{bmatrix} \boldsymbol{u}_i^{(s)} \\ \dot{\boldsymbol{u}}_i^{(s)} \end{bmatrix} = \mathbf{\Phi}_{\mathrm{ss}}^{(s)}\eta = \begin{bmatrix} \mathbf{\Phi}_{\mathrm{ss},u}^{(s)} \\ \mathbf{\Phi}_{\mathrm{ss},v}^{(s)} \end{bmatrix}\eta = \begin{bmatrix} \mathbf{\Phi}_{\mathrm{ss},u}^{(s)} \\ \mathbf{\Phi}_{\mathrm{ss},u}^{(s)}\mathbf{\Lambda}^{(s)} \end{bmatrix}\eta \tag{3.25}$$

with $\mathbf{\Lambda}^{(s)}$ being the diagonal eigenvalue matrix. Thereby, in Eq. (3.25), $\mathbf{\Phi}_{\mathrm{ss}}^{(s)}$ is split into displacement part $\mathbf{\Phi}_{\mathrm{ss},u}^{(s)}$ and velocity part $\mathbf{\Phi}_{\mathrm{ss},v}^{(s)}$ using the relation $\mathbf{\Phi}_{\mathrm{ss},v}^{(s)} = \mathbf{\Phi}_{\mathrm{ss},u}^{(s)}\mathbf{\Lambda}^{(s)}$ for complex eigenmodes [1]. The subscript $(\bullet)_{\mathrm{ss}}$ indicates that state-space quantities are considered. One substructure is represented as

$$z^{(s)} = \begin{bmatrix} \boldsymbol{u}^{(s)} \\ \dot{\boldsymbol{u}}^{(s)} \end{bmatrix} = \begin{bmatrix} \boldsymbol{u}_b^{(s)} \\ \boldsymbol{u}_i^{(s)} \\ \dot{\boldsymbol{u}}_b^{(s)} \\ \dot{\boldsymbol{u}}_i^{(s)} \end{bmatrix} = \begin{bmatrix} \boldsymbol{I} & \mathbf{0} & \mathbf{0} \\ \mathbf{0} & \mathbf{0} & \mathbf{\Phi}_{\mathrm{ss},u}^{(s)} \\ \mathbf{0} & \boldsymbol{I} & \mathbf{0} \\ \mathbf{0} & \mathbf{0} & \mathbf{\Phi}_{\mathrm{ss},u}^{(s)}\mathbf{\Lambda}^{(s)} \end{bmatrix} \begin{bmatrix} \boldsymbol{u}_b^{(s)} \\ \dot{\boldsymbol{u}}_b^{(s)} \\ \eta^{(s)} \end{bmatrix} \tag{3.26}$$

and the assembled state-space vector z becomes

$$z = \begin{bmatrix} \boldsymbol{u} \\ \dot{\boldsymbol{u}} \end{bmatrix} = \begin{bmatrix} \boldsymbol{u}_b \\ \boldsymbol{u}_i \\ \dot{\boldsymbol{u}}_b \\ \dot{\boldsymbol{u}}_i \end{bmatrix} = \begin{bmatrix} \boldsymbol{I} & \mathbf{0} & \mathbf{0} \\ \mathbf{0} & \mathbf{0} & \mathbf{\Phi}_{\mathrm{ss},u} \\ \mathbf{0} & \boldsymbol{I} & \mathbf{0} \\ \mathbf{0} & \mathbf{0} & \mathbf{\Phi}_{\mathrm{ss},u}\mathbf{\Lambda} \end{bmatrix} \begin{bmatrix} \boldsymbol{u}_b \\ \dot{\boldsymbol{u}}_b \\ \eta \end{bmatrix} \tag{3.27}$$

with

$$\mathbf{\Lambda} = \begin{bmatrix} \mathbf{\Lambda}^{(1)} & & \mathbf{0} \\ & \ddots & \\ \mathbf{0} & & \mathbf{\Lambda}^{(N)} \end{bmatrix}, \quad \mathbf{\Phi}_{\mathrm{ss},u} = \begin{bmatrix} \mathbf{\Phi}_{\mathrm{ss},u}^{(1)} & & \mathbf{0} \\ & \ddots & \\ \mathbf{0} & & \mathbf{\Phi}_{\mathrm{ss},u}^{(N)} \end{bmatrix}. \tag{3.28}$$

Hasselman and Kaplan express the internal DOFs $z_i^{(s)} = [\boldsymbol{u}_i^{(s)} \;\; \dot{\boldsymbol{u}}_i^{(s)}]^T$ of the state-space vector by state-space normal modes $\mathbf{\Phi}_{\mathrm{ss}}^{(s)}$ and corresponding modal parameters η, but leave the boundary DOFs $z_b^{(s)} = [\boldsymbol{u}_b^{(s)} \;\; \dot{\boldsymbol{u}}_b^{(s)}]^T$ unchanged [11]. Note that there is no coupling between boundary DOFs and internal DOFs in Eq. (3.27).

An approximation of $z_i^{(s)}$ in Eq. (3.25) is obtained by keeping only the first $n_{\mathrm{ss},\phi}^{(s)}$ dynamic modes $\mathbf{\Phi}_{\mathrm{ss}}^{(s)}$. Note that Matrix $\mathbf{\Phi}_{\mathrm{ss},\phi}^{(s)}$ contains those $n_{\mathrm{ss},\phi}^{(s)}$ dynamic modes, and internal DOFs $z_i^{(s)}$ of the substructure's state-space vector are approximated by

$$z_i^{(s)} \approx \mathbf{\Phi}_{\mathrm{ss},\phi}^{(s)}\eta^{(s)} = \begin{bmatrix} \mathbf{\Phi}_{\mathrm{ss},\phi,u}^{(s)} \\ \mathbf{\Phi}_{\mathrm{ss},\phi,v}^{(s)} \end{bmatrix}\eta^{(s)} = \begin{bmatrix} \mathbf{\Phi}_{\mathrm{ss},\phi,u}^{(s)} \\ \mathbf{\Phi}_{\mathrm{ss},\phi,u}^{(s)}\mathbf{\Lambda}^{(s)} \end{bmatrix}\eta^{(s)}. \tag{3.29}$$

Consequently, the assembled state-space vector is approximated by

$$z = \begin{bmatrix} \boldsymbol{u} \\ \dot{\boldsymbol{u}} \end{bmatrix} = \begin{bmatrix} \boldsymbol{u}_b \\ \boldsymbol{u}_i \\ \dot{\boldsymbol{u}}_b \\ \dot{\boldsymbol{u}}_i \end{bmatrix} \approx \underbrace{\begin{bmatrix} \boldsymbol{I} & \mathbf{0} & \mathbf{0} \\ \mathbf{0} & \mathbf{0} & \mathbf{\Phi}_{\mathrm{ss},\phi,u} \\ \mathbf{0} & \boldsymbol{I} & \mathbf{0} \\ \mathbf{0} & \mathbf{0} & \mathbf{\Phi}_{\mathrm{ss},\phi,u}\mathbf{\Lambda} \end{bmatrix}}_{\boldsymbol{T}_{\mathrm{HK}}} \begin{bmatrix} \boldsymbol{u}_b \\ \dot{\boldsymbol{u}}_b \\ \eta \end{bmatrix} \tag{3.30}$$

and the reduced state-space matrices as proposed by Hasselman and Kaplan are

$$\boldsymbol{\mathcal{A}}_{\mathrm{red,HK}} = \boldsymbol{T}_{\mathrm{HK}}^T\,\boldsymbol{\mathcal{A}}\,\boldsymbol{T}_{\mathrm{HK}}, \qquad \boldsymbol{\mathcal{B}}_{\mathrm{red,HK}} = \boldsymbol{T}_{\mathrm{HK}}^T\,\boldsymbol{\mathcal{B}}\,\boldsymbol{T}_{\mathrm{HK}}. \tag{3.31}$$

$\mathcal{A}_{\text{red,HK}}$ and $\mathcal{B}_{\text{red,HK}}$ are complex matrices. The substructures' reduced matrices obtained by the reduction in Eq. (3.29) for the internal DOFs are diagonal. Therefore, the submatrices of $\mathcal{A}_{\text{red,HK}}$ and $\mathcal{B}_{\text{red,HK}}$ corresponding to the modal parameters $\boldsymbol{\eta}$ are also diagonal.

Note that Hasselman and Kaplan do not use any static modes. The boundary DOFs $\boldsymbol{u}_b$ and $\dot{\boldsymbol{u}}_b$ are retained unchanged in the reduced system Eq. (3.31), but there is no coupling between boundary DOFs and internal DOFs as written in Eq. (3.30). The internal DOFs $\boldsymbol{u}_i$ and $\dot{\boldsymbol{u}}_i$ are exclusively approximated by modal parameters $\boldsymbol{\eta}$ with corresponding dynamic modes $\boldsymbol{\Phi}_{\text{ss},\phi,u}$ for the displacement part and $\boldsymbol{\Phi}_{\text{ss},\phi,v} = \boldsymbol{\Phi}_{\text{ss},\phi,u}\boldsymbol{\Lambda}$ for the velocity part.

3.3.3 Beliveau and Soucy's Method (BS)

Beliveau and Soucy [12] start from the point that, in general, the Craig-Bampton reduction matrix $\boldsymbol{T}_{\text{CB}}^{(s)}$ in Eq. (3.12) with the real matrix $\boldsymbol{\Phi}_{i\phi}^{(s)}$ does not diagonalize the damping matrix $\boldsymbol{C}^{(s)}$ for arbitrary viscous damping. The Craig-Bampton reduction matrix in Eq. (3.12) is modified to the following:

$$\boldsymbol{T}_{\text{BS}}^{(s)} = \begin{bmatrix} \boldsymbol{I} & \boldsymbol{0} \\ \boldsymbol{\Psi}_{ib}^{(s)} & \boldsymbol{\Phi}_{\text{damped}}^{(s)} \end{bmatrix} \tag{3.32}$$

The only difference between Beliveau and Soucy's method and the classical Craig-Bampton method is the use of damped, i.e., complex, mode shapes $\boldsymbol{\Phi}_{\text{damped}}^{(s)}$ rather than real normal modes $\boldsymbol{\Phi}_{i\phi}^{(s)}$. To determine the damped mode shapes $\boldsymbol{\Phi}_{\text{damped}}^{(s)}$ for Beliveau and Soucy's method, the second-order substructure quantities $\boldsymbol{M}^{(s)}$, $\boldsymbol{C}^{(s)}$, $\boldsymbol{K}^{(s)}$ and $\boldsymbol{f}^{(s)}$ are transformed into state-space form as they are for Hasselman and Kaplan's method, see Eqs. (3.20), (3.21), (3.22), and (3.23). The corresponding first-order eigenproblem (3.24) is solved. For underdamped systems, the eigenvalues $\boldsymbol{\Lambda}^{(s)}$ and corresponding state-space eigenvectors $\boldsymbol{\Phi}_{\text{ss}}^{(s)}$ occur in complex conjugate pairs and the matrices of eigenvalues and eigenvectors are partitioned as follows [12]:

$$\boldsymbol{\Lambda}^{(s)} = \begin{bmatrix} \boldsymbol{\Lambda}^{(s)+} & \boldsymbol{0} \\ \boldsymbol{0} & \boldsymbol{\Lambda}^{(s)-} \end{bmatrix}, \qquad \boldsymbol{\Phi}_{\text{ss}}^{(s)} = \begin{bmatrix} \boldsymbol{\Phi}_{\text{ss}}^{(s)+} & \boldsymbol{\Phi}_{\text{ss}}^{(s)-} \end{bmatrix} \tag{3.33}$$

Thereby, $\boldsymbol{\Lambda}^{(s)-}$ is the complex conjugate of $\boldsymbol{\Lambda}^{(s)+}$ and $\boldsymbol{\Phi}^{(s)-}$ is the complex conjugate of $\boldsymbol{\Phi}^{(s)+}$, i.e.,

$$\boldsymbol{\Lambda}^{(s)+} = \overline{\boldsymbol{\Lambda}^{(s)-}}, \qquad \boldsymbol{\Phi}^{(s)+} = \overline{\boldsymbol{\Phi}^{(s)+}} \tag{3.34}$$

with the overbar $\overline{(\bullet)}$ representing the complex conjugate. The superscript $(\bullet)^+$ indicates the eigenvalue of a complex conjugate pair with a positive imaginary part and the superscript $(\bullet)^-$ indicates the eigenvalue of a complex conjugate pair with a negative imaginary part.

Furthermore, the complex state-space eigenvector $\boldsymbol{\Phi}_{\text{ss}}^{(s)}$ of Eq. (3.33) is split into displacement and velocity parts similarly to Eq. (3.25):

$$\boldsymbol{\Phi}_{\text{ss}}^{(s)} = \begin{bmatrix} \boldsymbol{\Phi}_{\text{ss},u}^{(s)+} & \boldsymbol{\Phi}_{\text{ss},u}^{(s)-} \\ \boldsymbol{\Phi}_{\text{ss},v}^{(s)+} & \boldsymbol{\Phi}_{\text{ss},v}^{(s)-} \end{bmatrix} = \begin{bmatrix} \boldsymbol{\Phi}_{\text{ss},u}^{(s)+} & \boldsymbol{\Phi}_{\text{ss},u}^{(s)-} \\ \boldsymbol{\Phi}_{\text{ss},u}^{(s)+}\boldsymbol{\Lambda}^{(s)+} & \boldsymbol{\Phi}_{\text{ss},u}^{(s)-}\boldsymbol{\Lambda}^{(s)-} \end{bmatrix} \tag{3.35}$$

Beliveau and Soucy use $\boldsymbol{\Phi}_{\text{ss},u}^{(s)+}$ as damped mode shapes $\boldsymbol{\Phi}_{\text{damped}}^{(s)}$ in Eq. (3.32). If $\boldsymbol{\Phi}_{\text{ss},u,\phi}^{(s)+}$ denotes the matrix containing only the first $n_\phi^{(s)}$ modes of matrix $\boldsymbol{\Phi}_{\text{ss},u}^{(s)+}$ as columns, Beliveau and Soucy's reduction matrix $\boldsymbol{T}_{\text{BS}}^{(s)}$ is

$$\begin{bmatrix} \boldsymbol{u}_b^{(s)} \\ \boldsymbol{u}_i^{(s)} \end{bmatrix} \approx \underbrace{\begin{bmatrix} \boldsymbol{I} & \boldsymbol{0} \\ \boldsymbol{\Psi}_{ib}^{(s)} & \boldsymbol{\Phi}_{\text{ss},u,\phi}^{(s)+} \end{bmatrix}}_{\boldsymbol{T}_{\text{BS}}^{(s)}} \begin{bmatrix} \boldsymbol{u}_b^{(s)} \\ \boldsymbol{\eta}^{(s)} \end{bmatrix}. \tag{3.36}$$

Beliveau and Soucy replace the real normal modes $\boldsymbol{\Phi}_{i\phi}^{(s)}$ in the classical Craig-Bampton reduction matrix $\boldsymbol{T}_{\text{CB}}^{(s)}$ in Eq. (3.12) by the complex matrix $\boldsymbol{\Phi}_{\text{ss},u,\phi}^{(s)+}$ that corresponds to the displacement part of the one half of the complex conjugate eigenmodes of the eigenproblem (3.24). Applying the reduction matrix $\boldsymbol{T}^{(s)}_{\mathrm{BS}}$ to the configuration space quantities $\boldsymbol{M}^{(s)}$, $\boldsymbol{C}^{(s)}$ and $\boldsymbol{K}^{(s)}$ provides the reduced substructure matrices:

$$\boldsymbol{M}^{(s)}_{\mathrm{red,BS}} = \boldsymbol{T}^{(s)\,H}_{\mathrm{BS}}\,\boldsymbol{M}^{(s)}\,\boldsymbol{T}^{(s)}_{\mathrm{BS}}, \qquad \boldsymbol{C}^{(s)}_{\mathrm{red,BS}} = \boldsymbol{T}^{(s)\,H}_{\mathrm{BS}}\,\boldsymbol{C}^{(s)}\,\boldsymbol{T}^{(s)}_{\mathrm{BS}}, \qquad \boldsymbol{K}^{(s)}_{\mathrm{red,BS}} = \boldsymbol{T}^{(s)\,H}_{\mathrm{BS}}\,\boldsymbol{K}^{(s)}\,\boldsymbol{T}^{(s)}_{\mathrm{BS}} \tag{3.37}$$

Thereby, $(\bullet)^H$ represents the Hermitian transpose or conjugate transpose, but not simply the transpose $(\bullet)^T$. Beliveau and Soucy state that "[...] a Hermitian system is obtained by using the Hermitian [...], rather than the transpose" [12]. The substructures' reduced matrices $\boldsymbol{M}^{(s)}_{\mathrm{red,BS}}$, $\boldsymbol{C}^{(s)}_{\mathrm{red,BS}}$ and $\boldsymbol{K}^{(s)}_{\mathrm{red,BS}}$ are complex since the reduction matrix $\boldsymbol{T}^{(s)}_{\mathrm{BS}}$ is a complex matrix.

Primal assembly of all substructures' reduced matrices $\boldsymbol{M}^{(s)}_{\mathrm{red,BS}}$, $\boldsymbol{C}^{(s)}_{\mathrm{red,BS}}$ and $\boldsymbol{K}^{(s)}_{\mathrm{red,BS}}$ as described in Sect. 3.2 provides the assembled reduced matrices $\boldsymbol{M}_{\mathrm{red,BS}}$, $\boldsymbol{C}_{\mathrm{red,BS}}$ and $\boldsymbol{K}_{\mathrm{red,BS}}$ in configuration space. These matrices are transformed to state-space form resulting in the reduced state-space matrices $\mathcal{A}_{\mathrm{red,BS}}$ and $\mathcal{B}_{\mathrm{red,BS}}$ as suggested by Beliveau and Soucy:

$$\mathcal{A}_{\mathrm{red,BS}} = \begin{bmatrix} \boldsymbol{C}_{\mathrm{red,BS}} & \boldsymbol{M}_{\mathrm{red,BS}} \\ \boldsymbol{M}_{\mathrm{red,BS}} & \boldsymbol{0} \end{bmatrix}, \qquad \mathcal{B}_{\mathrm{red,BS}} = \begin{bmatrix} \boldsymbol{K}_{\mathrm{red,BS}} & \boldsymbol{0} \\ \boldsymbol{0} & -\boldsymbol{M}_{\mathrm{red,BS}} \end{bmatrix} \tag{3.38}$$

$\mathcal{A}_{\mathrm{red,BS}}$ and $\mathcal{B}_{\mathrm{red,BS}}$ are Hermitian matrices, i.e., their Hermitian transpose (or complex conjugate transpose) is equal to the matrices [12]. Note that Beliveau and Soucy's method is only suited for underdamped systems since the method requires that eigenvalues and corresponding eigenvectors occur as complex conjugate pairs. If that is not the case, i.e., overdamped eigenvalues and corresponding eigenvalues occur, the derivation of Eqs. (3.33), (3.34), (3.35), and (3.36) does not work anymore. It is not clear from Beliveau and Soucy's derivation in [12] how overdamped systems have to be handled.

3.3.4 *De Kraker's Method* (**DK**)

As for the other methods, de Kraker [13] transforms the second-order substructure quantities $\boldsymbol{M}^{(s)}$, $\boldsymbol{C}^{(s)}$, $\boldsymbol{K}^{(s)}$ and $\boldsymbol{f}^{(s)}$ into state-space format

$$\mathcal{A}^{(s)}\dot{z}^{(s)} + \mathcal{B}^{(s)}z^{(s)} = \mathcal{F}^{(s)} \tag{3.39}$$

with

$$\mathcal{A}^{(s)} = \begin{bmatrix} \boldsymbol{C}^{(s)} & \boldsymbol{M}^{(s)} \\ \boldsymbol{M}^{(s)} & \boldsymbol{0} \end{bmatrix}, \qquad \mathcal{B}^{(s)} = \begin{bmatrix} \boldsymbol{K}^{(s)} & \boldsymbol{0} \\ \boldsymbol{0} & -\boldsymbol{M}^{(s)} \end{bmatrix}, \qquad z^{(s)} = \begin{bmatrix} \boldsymbol{u}^{(s)} \\ \boldsymbol{v}^{(s)} \end{bmatrix}, \qquad \mathcal{F}^{(s)} = \begin{bmatrix} \boldsymbol{f}^{(s)} \\ \boldsymbol{0} \end{bmatrix}. \tag{3.40}$$

Equation (3.39) is again partitioned into internal DOFs $z^{(s)}_i$ and boundary DOFs $z^{(s)}_b$:

$$\begin{bmatrix} \mathcal{A}^{(s)}_{bb} & \mathcal{A}^{(s)}_{bi} \\ \mathcal{A}^{(s)}_{ib} & \mathcal{A}^{(s)}_{ii} \end{bmatrix} \begin{bmatrix} \dot{z}^{(s)}_b \\ \dot{z}^{(s)}_i \end{bmatrix} + \begin{bmatrix} \mathcal{B}^{(s)}_{bb} & \mathcal{B}^{(s)}_{bi} \\ \mathcal{B}^{(s)}_{ib} & \mathcal{B}^{(s)}_{ii} \end{bmatrix} \begin{bmatrix} z^{(s)}_b \\ z^{(s)}_i \end{bmatrix} = \begin{bmatrix} \mathcal{F}^{(s)}_b \\ \mathcal{F}^{(s)}_i \end{bmatrix} \tag{3.41}$$

Once more, the corresponding eigenproblem for the internal DOFs

$$\left(\boldsymbol{\Lambda}^{(s)}\mathcal{A}^{(s)}_{ii} + \mathcal{B}^{(s)}_{ii}\right)\boldsymbol{\Phi}^{(s)}_{\mathrm{ss}} = \boldsymbol{0} \tag{3.42}$$

is solved to get the complex fixed interface normal modes $\boldsymbol{\Phi}^{(s)}_{\mathrm{ss}}$. Considering the partitioned equations of motion (3.41), the second row is used to get the static modes [13], which is different from the other methods:

$$\mathcal{B}^{(s)}_{ib}z^{(s)}_b + \mathcal{B}^{(s)}_{ii}z^{(s)}_i = \boldsymbol{0} \tag{3.43}$$

The static solution $z^{(s)}_{i,\mathrm{stat}}$ for the internal DOFs $z^{(s)}_i$ in Eq. (3.43) writes

$$z^{(s)}_{i,\mathrm{stat}} = -\mathcal{B}^{(s)\,-1}_{ii}\mathcal{B}^{(s)}_{ib}z^{(s)}_b = \boldsymbol{\Psi}^{(s)}_{\mathrm{ss}}z^{(s)}_b \tag{3.44}$$

with the state-space constraint modes $\boldsymbol{\Psi}_{\mathrm{ss}}^{(s)}$:

$$\boldsymbol{\Psi}_{\mathrm{ss}}^{(s)} = -\boldsymbol{\mathcal{B}}_{ii}^{(s)^{-1}} \boldsymbol{\mathcal{B}}_{ib}^{(s)} \tag{3.45}$$

By keeping only the first $n_{\mathrm{ss},\phi}^{(s)}$ dynamic modes $\boldsymbol{\Phi}_{\mathrm{ss}}^{(s)}$ in matrix $\boldsymbol{\Phi}_{\mathrm{ss},\phi}^{(s)}$ and using the state-space constraint modes $\boldsymbol{\Psi}_{\mathrm{ss}}^{(s)}$ of Eq. (3.45), the substructure approximation writes

$$\begin{bmatrix} z_b^{(s)} \\ z_i^{(s)} \end{bmatrix} \approx \underbrace{\begin{bmatrix} \boldsymbol{I} & \boldsymbol{0} \\ \boldsymbol{\Psi}_{\mathrm{ss}}^{(s)} & \boldsymbol{\Phi}_{\mathrm{ss},\phi}^{(s)} \end{bmatrix}}_{\boldsymbol{T}_{\mathrm{DK}}^{(s)}} \begin{bmatrix} z_b^{(s)} \\ \eta^{(s)} \end{bmatrix} \tag{3.46}$$

and the reduced substructure matrices of de Kraker's method are:

$$\boldsymbol{\mathcal{A}}_{\mathrm{red,DK}}^{(s)} = \boldsymbol{T}_{\mathrm{DK}}^{(s)^T} \boldsymbol{\mathcal{A}}^{(s)} \boldsymbol{T}_{\mathrm{DK}}^{(s)}, \qquad \boldsymbol{\mathcal{B}}_{\mathrm{red,DK}}^{(s)} = \boldsymbol{T}_{\mathrm{DK}}^{(s)^T} \boldsymbol{\mathcal{B}}^{(s)} \boldsymbol{T}_{\mathrm{DK}}^{(s)} \tag{3.47}$$

The coordinates of the boundary DOFs are still present in Eq. (3.47) and primal assembly is straightforward, as described in Sect. 3.2. $\boldsymbol{\mathcal{A}}_{\mathrm{red,DK}}^{(s)}$ and $\boldsymbol{\mathcal{B}}_{\mathrm{red,DK}}^{(s)}$ are complex matrices. The submatrices of $\boldsymbol{\mathcal{A}}_{\mathrm{red,DK}}^{(s)}$ and $\boldsymbol{\mathcal{B}}_{\mathrm{red,DK}}^{(s)}$ corresponding to the modal parameters $\eta^{(s)}$ are diagonal. The number $n_{\mathrm{ss},\phi}^{(s)}$ of kept dynamic modes $\boldsymbol{\Phi}_{\mathrm{ss}}^{(s)}$ should always be chosen such that both parts of complex conjugate pairs are kept (if there are any). Keeping only one half of a complex conjugate pairs results in a poor approximation accuracy.

In the following, the difference of de Kraker's method to the other methods due to the computation of the state-space constraint modes is illustrated in detail. Therefore, consider the static solution $z_{\mathrm{stat}}^{(s)}$ for the substructure state-space vector:

$$z_{\mathrm{stat}}^{(s)} = \begin{bmatrix} z_b^{(s)} \\ z_{i,\mathrm{stat}}^{(s)} \end{bmatrix} = \begin{bmatrix} \boldsymbol{I} \\ \boldsymbol{\Psi}_{\mathrm{ss}}^{(s)} \end{bmatrix} z_b^{(s)} = \begin{bmatrix} \boldsymbol{I} \\ -\boldsymbol{\mathcal{B}}_{ii}^{(s)^{-1}} \boldsymbol{\mathcal{B}}_{ib}^{(s)} \end{bmatrix} z_b^{(s)} \tag{3.48}$$

Consider also the matrix $\boldsymbol{\mathcal{B}}^{(s)}$ in Eq. (3.41) together with relation of Eq. (3.40):

$$\boldsymbol{\mathcal{B}}^{(s)} = \begin{bmatrix} \boldsymbol{\mathcal{B}}_{bb}^{(s)} & \boldsymbol{\mathcal{B}}_{bi}^{(s)} \\ \boldsymbol{\mathcal{B}}_{ib}^{(s)} & \boldsymbol{\mathcal{B}}_{ii}^{(s)} \end{bmatrix} = \begin{bmatrix} \boldsymbol{K}_{bb}^{(s)} & \boldsymbol{0} & \boldsymbol{K}_{bi}^{(s)} & \boldsymbol{0} \\ \boldsymbol{0} & -\boldsymbol{M}_{bb}^{(s)} & \boldsymbol{0} & -\boldsymbol{M}_{bi}^{(s)} \\ \boldsymbol{K}_{ib}^{(s)} & \boldsymbol{0} & \boldsymbol{K}_{ii}^{(s)} & \boldsymbol{0} \\ \boldsymbol{0} & -\boldsymbol{M}_{ib}^{(s)} & \boldsymbol{0} & -\boldsymbol{M}_{ii}^{(s)} \end{bmatrix} \tag{3.49}$$

According to Eq. (3.49), matrix $\boldsymbol{\mathcal{B}}_{ii}^{(s)^{-1}}$ in Eq. (3.48) writes

$$\boldsymbol{\mathcal{B}}_{ii}^{(s)^{-1}} = \begin{bmatrix} \boldsymbol{K}_{ii}^{(s)^{-1}} & \boldsymbol{0} \\ \boldsymbol{0} & -\boldsymbol{M}_{ii}^{(s)^{-1}} \end{bmatrix}. \tag{3.50}$$

Therefore, matrix $\boldsymbol{\Psi}_{\mathrm{ss}}^{(s)}$ is expressed as

$$\boldsymbol{\Psi}_{\mathrm{ss}}^{(s)} = -\boldsymbol{\mathcal{B}}_{ii}^{(s)^{-1}} \boldsymbol{\mathcal{B}}_{ib}^{(s)} = -\begin{bmatrix} \boldsymbol{K}_{ii}^{(s)^{-1}} \boldsymbol{K}_{ib}^{(s)} & \boldsymbol{0} \\ \boldsymbol{0} & \boldsymbol{M}_{ii}^{(s)^{-1}} \boldsymbol{M}_{ib}^{(s)} \end{bmatrix}, \tag{3.51}$$

which provides the static solution $z_{i,\mathrm{stat}}^{(s)}$ for the internal DOFs $z_i^{(s)}$

$$z_{i,\mathrm{stat}}^{(s)} = \begin{bmatrix} \boldsymbol{u}_{i,\mathrm{stat}}^{(s)} \\ \dot{\boldsymbol{u}}_{i,\mathrm{stat}}^{(s)} \end{bmatrix} \tag{3.52}$$

the two relations

$$\boldsymbol{u}_{i,\text{stat}}^{(s)} = -\boldsymbol{K}_{ii}^{(s)^{-1}} \boldsymbol{K}_{ib}^{(s)} \boldsymbol{u}_b^{(s)}, \tag{3.53}$$

$$\dot{\boldsymbol{u}}_{i,\text{stat}}^{(s)} = -\boldsymbol{M}_{ii}^{(s)^{-1}} \boldsymbol{M}_{ib}^{(s)} \dot{\boldsymbol{u}}_b^{(s)}. \tag{3.54}$$

It is interesting to notice that de Kraker uses another relation between boundary and internal DOFs for the velocities than for the displacements. For the displacements in Eq. (3.53), the equilibrium of static forces $\boldsymbol{K}_{ii}^{(s)} \boldsymbol{u}_{i,\text{stat}}^{(s)} + \boldsymbol{K}_{ib}^{(s)} \boldsymbol{u}_b^{(s)} = \boldsymbol{0}$ is used for the computation of static internal displacements $\boldsymbol{u}_{i,\text{stat}}^{(s)}$ due to boundary displacements $\boldsymbol{u}_b^{(s)}$. On the other side, for the velocities in Eq. (3.54), the equilibrium of impulses $\boldsymbol{M}_{ii}^{(s)} \dot{\boldsymbol{u}}_{i,\text{stat}}^{(s)} + \boldsymbol{M}_{ib}^{(s)} \dot{\boldsymbol{u}}_b^{(s)} = \boldsymbol{0}$ is used. All the other methods including static modes use the relation of Eq. (3.53) instead of Eq. (3.54) for the velocity parts.

3.3.5 *Adaption of Hasselman and Kaplan's Method* (**HKadapt**)

Beliveau and Soucy suggest an adaption of Hasselman and Kaplan's method, but do not show all steps of their adaption. On the one hand, they add static modes $\boldsymbol{\Psi}_{ib}^{(s)}$ compared to Hasselman and Kaplan's method. On the other hand, they replace the dynamic modes used by Hasselman and Kaplan with the complex dynamic modes $\boldsymbol{\Phi}_{\text{ss},u,\phi}^{(s)+}$, which they use for their method [12]:

$$z^{(s)} = \begin{bmatrix} \boldsymbol{u}^{(s)} \\ \dot{\boldsymbol{u}}^{(s)} \end{bmatrix} = \begin{bmatrix} \boldsymbol{u}_b^{(s)} \\ \boldsymbol{u}_i^{(s)} \\ \dot{\boldsymbol{u}}_b^{(s)} \\ \dot{\boldsymbol{u}}_i^{(s)} \end{bmatrix} \approx \begin{bmatrix} \boldsymbol{I} & \boldsymbol{0} & \boldsymbol{0} \\ \boldsymbol{\Psi}_{ib}^{(s)} & \boldsymbol{0} & \boldsymbol{\Phi}_{\text{ss},u,\phi}^{(s)+} \\ \boldsymbol{0} & \boldsymbol{I} & \boldsymbol{0} \\ \boldsymbol{0} & \boldsymbol{\Psi}_{ib}^{(s)} & \boldsymbol{\Phi}_{\text{ss},u,\phi}^{(s)+} \boldsymbol{\Lambda}^{(s)} \end{bmatrix} \begin{bmatrix} \boldsymbol{u}_b^{(s)} \\ \dot{\boldsymbol{u}}_b^{(s)} \\ \boldsymbol{\eta}^{(s)} \end{bmatrix} \tag{3.55}$$

Using the approximation of Eq. (3.55) as suggested by Beliveau and Soucy, we get large errors in our computations independent of using the transpose or the Hermitian transpose when building the reduced matrices:

- Beliveau and Soucy state again that "The complex conjugate transpose of the [reduction matrix] is used, however, rather than the transpose [...]." Using the Hermitian transpose, the reduced system does not possess complex conjugate pairs anymore.
- Using the transpose instead of the Hermitian transpose, we get better results, but, nevertheless, the approximation accuracy is very bad compared to the other methods.

In the following, the adaption of Hasselman and Kaplan's method as suggested by Beliveau and Soucy is therefore not further considered.

Another adaption of Hasselman and Kaplan's method is obtained as follows: Add static modes $\boldsymbol{\Psi}_{ib}^{(s)}$ in Eq. (3.30) of Hasselman and Kaplan's method as done in Eq. (3.55) by Beliveau and Soucy and use the same dynamic modes (complex fixed interface normal modes) for the approximation of the internal DOFs

$$z_i^{(s)} \approx \boldsymbol{\Phi}_{\text{ss},\phi}^{(s)} \boldsymbol{\eta}^{(s)} = \begin{bmatrix} \boldsymbol{\Phi}_{\text{ss},\phi,u}^{(s)} \\ \boldsymbol{\Phi}_{\text{ss},\phi,v}^{(s)} \end{bmatrix} \boldsymbol{\eta}^{(s)} = \begin{bmatrix} \boldsymbol{\Phi}_{\text{ss},\phi,u}^{(s)} \\ \boldsymbol{\Phi}_{\text{ss},\phi,u}^{(s)} \boldsymbol{\Lambda}^{(s)} \end{bmatrix} \boldsymbol{\eta}^{(s)} \tag{3.56}$$

as for Hasselman and Kaplan's (HK) method in Eq. (3.27) and for de Kraker's (DK) method in Eq. (3.46), but not the dynamic modes $\boldsymbol{\Phi}_{\text{ss},u,\phi}^{(s)+}$ as used by Beliveau and Soucy in Eq. (3.36). Taking these two points into account, the substructure reduction matrix is

$$z^{(s)} = \begin{bmatrix} \boldsymbol{u}^{(s)} \\ \dot{\boldsymbol{u}}^{(s)} \end{bmatrix} = \begin{bmatrix} \boldsymbol{u}_b^{(s)} \\ \boldsymbol{u}_i^{(s)} \\ \dot{\boldsymbol{u}}_b^{(s)} \\ \dot{\boldsymbol{u}}_i^{(s)} \end{bmatrix} \approx \underbrace{\begin{bmatrix} \boldsymbol{I} & \boldsymbol{0} & \boldsymbol{0} \\ \boldsymbol{\Psi}_{ib}^{(s)} & \boldsymbol{0} & \boldsymbol{\Phi}_{\text{ss},\phi,u}^{(s)} \\ \boldsymbol{0} & \boldsymbol{I} & \boldsymbol{0} \\ \boldsymbol{0} & \boldsymbol{\Psi}_{ib}^{(s)} & \boldsymbol{\Phi}_{\text{ss},\phi,u}^{(s)} \boldsymbol{\Lambda}^{(s)} \end{bmatrix}}_{\boldsymbol{T}_{\text{HKadapt}}^{(s)}} \begin{bmatrix} \boldsymbol{u}_b^{(s)} \\ \dot{\boldsymbol{u}}_b^{(s)} \\ \boldsymbol{\eta}^{(s)} \end{bmatrix}. \tag{3.57}$$

The reduced substructure matrices are

$$\mathcal{A}^{(s)}_{\text{red,HKadapt}} = \boldsymbol{T}^{(s)^T}_{\text{HKadapt}} \mathcal{A}^{(s)} \boldsymbol{T}^{(s)}_{\text{HKadapt}}, \qquad \mathcal{B}^{(s)}_{\text{red,HKadapt}} = \boldsymbol{T}^{(s)^T}_{\text{HKadapt}} \mathcal{B}^{(s)} \boldsymbol{T}^{(s)}_{\text{HKadapt}}. \tag{3.58}$$

The coordinates of the boundary DOFs are still present in Eq. (3.58) and primal assembly is again straightforward. Equation (3.57) uses the same relation between boundary and internal DOFs for the displacements and velocities

$$\boldsymbol{u}^{(s)}_i = \boldsymbol{\Psi}^{(s)}_{ib} \boldsymbol{u}^{(s)}_b = -\boldsymbol{K}^{(s)^{-1}}_{ii} \boldsymbol{K}^{(s)}_{ib} \boldsymbol{u}^{(s)}_b, \tag{3.59}$$

$$\dot{\boldsymbol{u}}^{(s)}_i = \boldsymbol{\Psi}^{(s)}_{ib} \dot{\boldsymbol{u}}^{(s)}_b = -\boldsymbol{K}^{(s)^{-1}}_{ii} \boldsymbol{K}^{(s)}_{ib} \dot{\boldsymbol{u}}^{(s)}_b, \tag{3.60}$$

which is in contrast to de Kraker's relations in Eqs. (3.53) and (3.54).

3.4 Example

A simply coupled structure with 20 displacement DOFs is used to compare the methods described in Sect. 3.3. This example has already been used in [12]. As illustrated in Fig. 3.1, the structure consists of 20 masses m_i connected by springs k_i and dampers c_i for $i = 1, \ldots, 20$. The stiffnesses of the springs and the masses are arbitrarily set equal to 1 and the damping values are given in Table 3.1. The state-space model of this system has twice the number of DOFs, i.e., 40 DOFs. The 40 complex eigenvalues of the full (unreduced) system are computed for reference, which are in fact 20 complex conjugate pairs since the system is underdamped.

The system in Fig. 3.1 is decomposed in 4 substructures: DOF u_5 is the boundary DOF of substructures 1 and 2, DOF u_{10} is the boundary DOF of substructures 2 and 3 and DOF u_{15} is the boundary DOF of substructures 3 and 4. Therefore, the system has $n_b = 3$ displacement boundary DOFs $\boldsymbol{u}_b$. See [12] for details concerning the model.

Figure 3.2 depicts all 40 eigenvalues in the complex plane for the coupled unreduced system, which are labeled as 'Exact'. Eigenvalues of the coupled unreduced (full) system will be called λ_{full}. It can be seen that the system has only complex conjugate eigenvalue pairs. There are no real eigenvalues, i.e., the system does not have any overdamped modes.

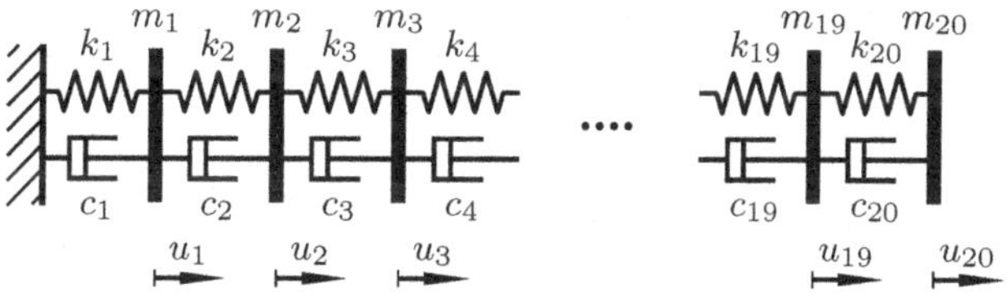

Fig. 3.1 Structure with 20 displacements [12]. The structure is decomposed in 4 substructures, each having 5 displacement DOFs. The stiffnesses of the springs and the masses are arbitrarily set equal to 1 and the damping values are given in Table 3.1

Table 3.1 Viscous damping coefficient of system in Fig. 3.1 [12]

Damper number	Viscous damping	Damper number	Viscous damping
c_1	0.10	c_{11}	0.09
c_2	0.50	c_{12}	0.50
c_3	0.08	c_{13}	0.07
c_4	0.60	c_{14}	0.11
c_5	0.09	c_{15}	0.50
c_6	0.11	c_{16}	0.09
c_7	0.50	c_{17}	0.07
c_8	0.12	c_{18}	0.40
c_9	0.09	c_{19}	0.11
c_{10}	0.07	c_{20}	0.50

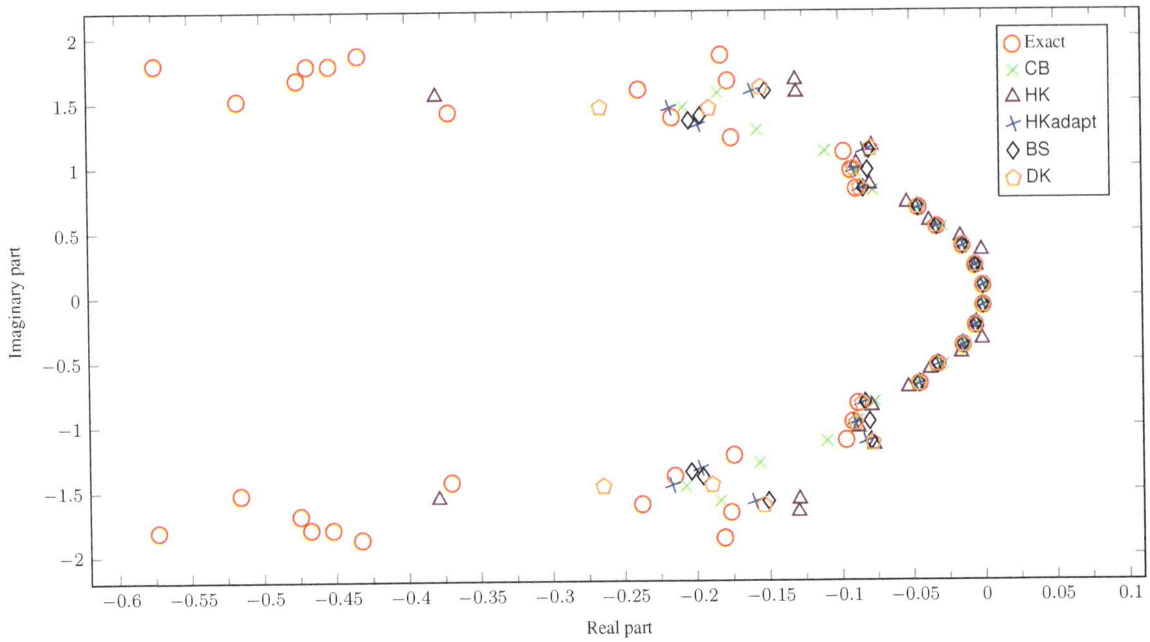

Fig. 3.2 Exact eigenvalues λ_{full} of the coupled unreduced system and eigenvalues λ_{red} of the reduced systems in the complex plane. The eigenvalues are approximated using the Craig-Bampton method (CB), Hasselman and Kaplan's method (HK), the adaption of Hasselman and Kaplan's method (HKadapt), Beliveau and Soucy's method (BS) and de Kraker's method (DK)

3.4.1 Approximation of Eigenvalues

The structure's lowest eigenvalues are approximated using the five methods for the damped systems of Sect. 3.3:

- Classical Craig-Bampton method (CB, Sect. 3.3.1)
- Hasselman and Kaplan's method (HK, Sect. 3.3.2)
- Adaption of Hasselman and Kaplan's method (HKadapt, Sect. 3.3.5)
- Beliveau and Soucy's method (BS, Sect. 3.3.3)
- De Kraker's method (DK, Sect. 3.3.4)

We always keep as many dynamic modes as necessary to obtain a reduced system with the same number of $n_{\text{ss,red}} = 22$ DOFs in state-space format. The reduced matrices $\boldsymbol{\mathcal{A}}_{\text{red}}$ and $\boldsymbol{\mathcal{B}}_{\text{red}}$ will have the dimension 22×22 for all methods.

For example, $n_\phi^{(1)} = n_\phi^{(2)} = n_\phi^{(3)} = n_\phi^{(4)} = 2$ real fixed interface normal modes are kept for each substructure applying the classical Craig-Bampton method (CB). This provides $n_\phi^{(1)} + n_\phi^{(2)} + n_\phi^{(3)} + n_\phi^{(4)} + n_b = 11$ DOFs for the reduced system in configuration space resulting in a reduced system with $n_{\text{ss,red}} = 22$ DOFs in state-space format.

On the other hand, de Kraker's method keeps $n_{\text{ss},\phi}^{(s)} = 4$ complex dynamic modes $\boldsymbol{\Phi}_{\text{ss}}^{(s)}$ for each of the four substructures. Keeping $n_{\text{ss},\phi}^{(s)} = 4$ means that the complex modes corresponding to the $n_{\text{ss},\phi}^{(s)} = 4$ eigenvalues with lowest absolute magnitude are kept for substructure s. Since the eigenvalues occur in complex conjugate pairs, for an even number $n_{\text{ss},\phi}^{(s)}$ always conjugated complex eigenvector pairs are kept. It became apparent during the simulations that no good approximations are obtained by keeping an odd number of $n_{\text{ss},\phi}^{(s)}$ dynamic modes, i.e., keeping only one half of a conjugate complex eigenvector pair.

Figure 3.2 shows the 22 approximated eigenvalues λ_{red} in the complex plane for the different reduction methods. The approximation accuracy of eigenvalues with small absolute magnitude seems to be good for all methods. All reduction methods generate complex conjugate eigenvalue pairs. To gain better insight, Fig. 3.3 provides a zoom plot of Fig. 3.2 for imaginary parts $|\Im(\lambda_{\text{red}})| \leq 1.2$, which confirms the relatively good approximation accuracy of the methods under consideration for eigenvalues with small absolute magnitude: Beliveau and Soucy's method (BS), the adaption of Hasselman and Kaplan's method (HKadapt) and de Kraker's method (DK) provide very good approximations. The three aforementioned methods use all complex fixed interface normal modes for the approximation. The classical Craig-Bampton approach (CB), using real fixed interface normal modes, approximates the three complex conjugate eigenvalue pairs with the smallest absolute magnitude accurately, but already shows significant deviations from the exact eigenvalues λ_{full} for the other

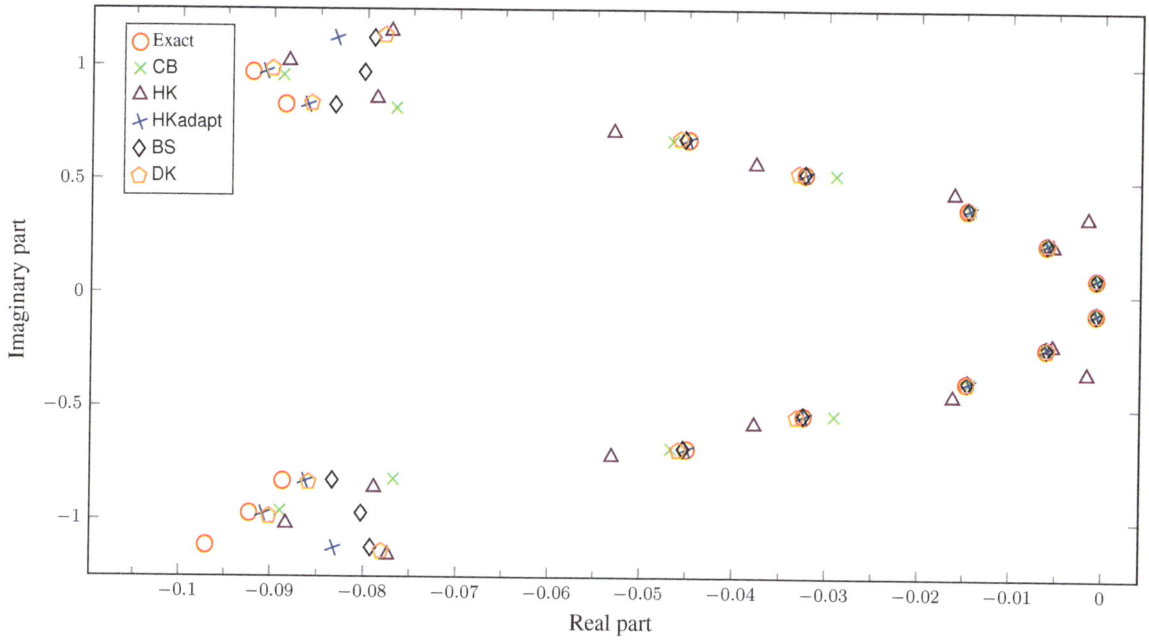

Fig. 3.3 Zoom plot of Fig. 3.2

eigenvalues. Especially, the deviation of the real parts from the exact eigenvalues are obviously larger for the classical Craig-Bampton approach compared to the three aforementioned methods using complex fixed interface normal modes. This demonstrates that the correct approximation of damping properties using the classical Craig-Bampton approach is inadequate since the Craig-Bampton method does not take damping properties into account when building the reduction basis (see Sect. 3.3.1). The approximation accuracy for eigenvalues with small absolute magnitude is poor using Hasselman and Kaplan's method (HK), despite the use of complex fixed interface normal modes. This is due to the fact that Hasselman and Kaplan do not use static modes for their approximation.

To illustrate and assess the approximation accuracy of the different methods better, the relative errors of the approximated eigenvalues are considered in Sect. 3.4.2.

3.4.2 *Relative Errors of Approximated Eigenvalues*

To further compare the approximation accuracy of the different methods, the relative errors $\varepsilon_{\text{rel},\Re}$ of the eigenvalues' real parts and relative errors $\varepsilon_{\text{rel},\Im}$ of the eigenvalues' imaginary parts are now considered. The relative errors $\varepsilon_{\text{rel},\Re,j}$ of the real parts and the relative errors $\varepsilon_{\text{rel},\Im,j}$ of the imaginary parts of the j-th eigenvalue are computed according to

$$\varepsilon_{\text{rel},\Re,j} = \frac{|\Re(\lambda_{\text{red},j}) - \Re(\lambda_{\text{full},j})|}{\Re(\lambda_{\text{full},j})} \quad \text{and} \quad \varepsilon_{\text{rel},\Im,j} = \frac{|\Im(\lambda_{\text{red},j}) - \Im(\lambda_{\text{full},j})|}{\Im(\lambda_{\text{full},j})}. \tag{3.61}$$

In Eq. (3.61), $\lambda_{\text{full},j}$ is the j-th eigenvalue of the full (unreduced and coupled) system and $\lambda_{\text{red},j}$ is the j-th eigenvalue of the reduced system. Figure 3.4 shows the relative errors $\varepsilon_{\text{rel},\Re}$ of the eigenvalues' real parts (Fig. 3.4a) and the relative errors $\varepsilon_{\text{rel},\Im}$ of the eigenvalues' imaginary parts (Fig. 3.4b) for the 16 eigenvalues with smallest absolute magnitude of the considered system. The number of DOFs $n_{\text{ss,red}} = 22$ of the reduced systems is the same for all methods. The relative errors for both real and imaginary parts for eigenvalue 17, 18, 19, ... are approximated similarly poor by all methods with errors larger than 10% and therefore not shown in Fig. 3.4. Figure 3.4 displays the relative errors of all eigenvalues, i.e., eigenvalues 1 and 2 are one complex conjugate eigenvalue pair, eigenvalues 3 and 4 are another complex conjugate eigenvalue pair and so forth.

Hasselman and Kaplan's method (HK) causes very large errors, especially for the eigenvalue pair with small absolute magnitude. Eigenvalues with larger absolute magnitudes are approximated better, but the relative error is still around 10%. The method suffers from the fact that no static modes are used, which explains the large errors for the lowest eigenvalues as well. This demonstrates the importance of using static modes for the approximation of damped systems.

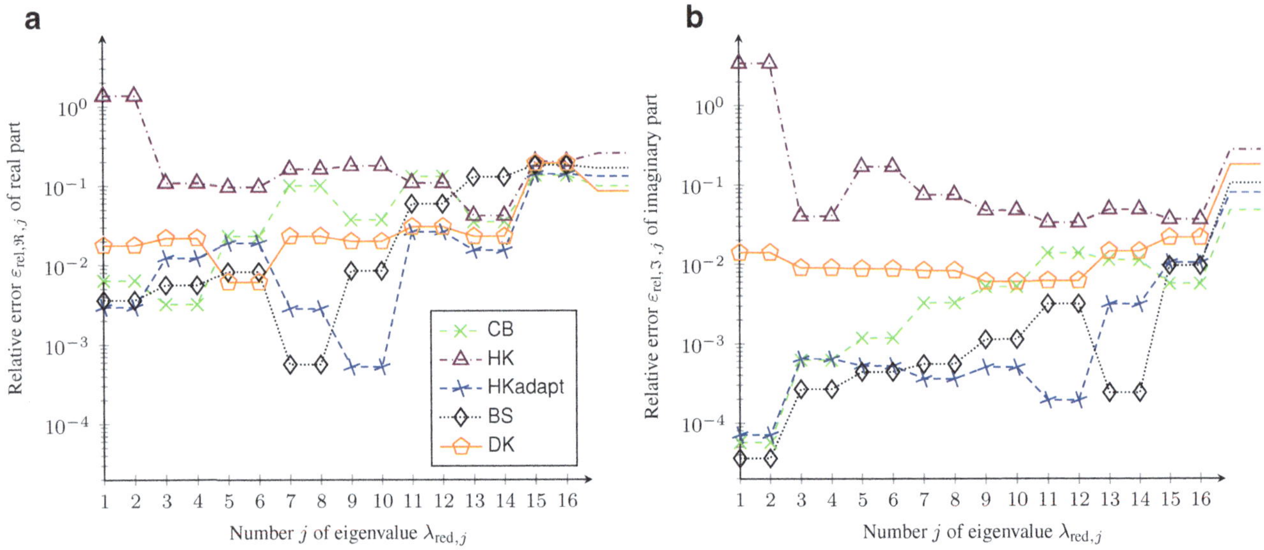

Fig. 3.4 Relative error of real and imaginary parts of approximated eigenvalues λ_{red}. The relative errors of the lowest 16 eigenvalues are illustrated using the Craig-Bampton method (CB), Hasselman and Kaplan's method (HK), the adaption of Hasselman and Kaplan's method (HKadapt), Beliveau and Soucy's method (BS) and de Kraker's method (DK). The number of DOFs of the reduced systems, $n_{\mathrm{ss,red}} = 22$, is the same for all methods. (**a**) Relative error $\varepsilon_{\mathrm{rel},\Re,j}$ of the real part of eigenvalue $\lambda_{\mathrm{red},j}$. (**b**) Relative error $\varepsilon_{\mathrm{rel},\Im,j}$ of the imaginary part of eigenvalue $\lambda_{\mathrm{red},j}$

The classical Craig-Bampton method (CB), which does not take any damping properties into account to build the reduction basis, provides relatively small errors for the 4 lowest eigenvalues. For eigenvalues with larger magnitudes (eigenvalues 5 to 12) the relative errors increase to a level of around 10%, which is comparable to the poor approximation accuracy of Hasselman and Kaplan's method (HK). Especially, the relative errors of the real parts of the eigenvalues in Fig. 3.4a demonstrate that the classical Craig-Bampton method represents the damping effects very inaccurately: The relative errors of the real parts of eigenvalues 1 to 4 are small. This is due to the fact that these eigenvalues have very small real parts indicating that those eigenvalues are only slightly damped, as can be seen in Fig. 3.3. When the real parts become larger, i.e., damping is higher, the relative errors of the real parts increase dramatically using the classical Craig-Bampton method. On the other hand, the relative errors of the imaginary parts are pretty small for low damped eigenvalues using the classical Craig-Bampton approach (see Fig. 3.4b). As expected, the classical Craig-Bampton method is not well-suited to represent the damping effects accurately.

De Kraker's method (DK) results in errors (around 1%) for both real and imaginary parts for the first 14 eigenvalues. It is interesting that the relative errors stay at a nearly constant level for these eigenvalues. For eigenvalues with small damping (eigenvalues 1 to 4), de Kraker's method produces larger relative errors than the classical Craig-Bampton method (CB). For eigenvalues with higher damping values (starting with eigenvalue 5), de Kraker's method approximates the real part better than the classical Craig-Bampton method. This indicates that higher damping properties can be represented accurately by using complex normal modes, but not by using real normal modes.

Beliveau and Soucy's method (BS) and the adaption of Hasselman and Kaplan's method (HKadapt) result in the best approximation accuracy of the methods under consideration. Both methods use complex normal modes as in de Kraker's method (DK). The relative errors of the real parts are comparably small for both methods and stay below 1% for the first 14 eigenvalues. The imaginary parts are approximated even better starting with relative errors around 0.01% for the first two eigenvalues and always remain below 0.1% for the first 10 eigenvalues.

Comparing all methods, Hasselman and Kaplan's method (HK) and the classical Craig-Bampton method (CB) do not provide good approximations. De Kraker's method (DK) provides acceptable approximation accuracy while Beliveau and Soucy's method (BS) and the adaption of Hasselman and Kaplan's method (HKadapt) perform best.

3.5 Conclusions

In this paper, a derivation and comparison of five different Craig-Bampton approaches for systems with arbitrary viscous damping was provided. The investigated methods are the classical Craig-Bampton method, Hasselman and Kaplan's method, an adaption of Hasselman and Kaplan's method, Beliveau and Soucy's method and de Kraker's method. The approximation accuracy of Hasselman and Kaplan's method suffers from the fact that no static modes are incorporated. The classical Craig-Bampton method suffers as a result of not accounting for any damping effect when approximating a damped system. Using different static modes for displacement and velocity degrees of freedom makes de Kraker's method less accurate than the adaption of Hasselman and Kaplan's method and Beliveau and Soucy's method. The adaption of Hasselman and Kaplan's method and Beliveau and Soucy's method give the very best results and are recommended for the approximation of arbitrarily viscous damped substructured systems. One problem of Beliveau and Soucy's method is the fact that this method is not suited for overdamped systems. Further investigations can be conducted on the approximation accuracy of the presented methods for overdamped systems.

References

1. Craig, R.R., Kurdila, A.J.: Fundamentals of structural dynamics. Wiley, New York (2006)
2. Craig, R.R., Bampton, M.C.C.: Coupling of substructures for dynamic analyses. AIAA J. **6**(7), 1313–1319 (1968). https://doi.org/10.2514/3.4741
3. MacNeal, R.H.: A hybrid method of component mode synthesis. Comput. Struct. **1**(4), 581–601 (1971). https://doi.org/10.1016/00457949(71)900319
4. Rubin, S.: Improved component-mode representation for structural dynamic analysis. AIAA J. **13**(8), 995–1006 (1975). https://doi.org/10.2514/3.60497
5. Rixen, D.J.: A dual Craig-Bampton method for dynamic substructuring. J. Comput. Appl. Math. **168**(1–2), 383–391 (2004). https://doi.org/10.1016/j.cam.2003.12.014
6. Craig, R.R., Chang, C.-J.: On the use of attachment modes in substructure coupling for dynamic analysis. In: 18th Structural Dynamics and Materials Conference. Structures, Structural Dynamics, and Materials and Co-located Conferences, San Diego, pp. 89–99. American Institute of Aeronautics and Astronautics (1977). https://doi.org/10.2514/6.1977405
7. Herting, D.N.: A general purpose, multi-stage, component modal synthesis method. Finite Elem. Anal. Des. **1**(2), 153–164 (1985)
8. Park, K.C., Park, Y.H.: Partitioned component mode synthesis via a flexibility approach. AIAA J. **42**(6), 1236–1245 (2004). https://doi.org/10.2514/1.10423
9. Gruber, F.M., Rixen, D.J.: Evaluation of substructure reduction techniques with fixed and free interfaces. Strojniški vestnik – J Mech Eng **62**(7–8), 452–462 (2016). https://doi.org/10.5545/svjme.2016.3735
10. Craig, R.R., Chang, C.-J.: On the use of attachment modes in substructure coupling for dynamic analysis. In: 18th Structural Dynamics and Materials Conference. Structures, Structural Dynamics, and Materials and Co-located Conferences, San Diego, pp. 89–99. American Institute of Aeronautics and Astronautics (1977). https://doi.org/10.2514/6.1977405
11. Hasselman, T.K., Kaplan, A.: Dynamic analysis of large systems by complex mode synthesis. J. Dyn. Syst. Meas. Control. **96**(3), 327–333 (1974). https://doi.org/10.1115/1.3426810
12. Beliveau, J.-G., Soucy, Y.: Damping synthesis using complex substructure modes and a Hermitian system representation. AIAA J. **23**(12), 1952–1956 (1985). https://doi.org/10.2514/3.9201
13. de Kraker, A.: Generalization of the Craig-Bampton CMS procedure for general damping. Technical report. Technische Universiteit Eindhoven, Vakgroep Fundamentele Werktuigkunde, Rapportnr WFW 93.023 (1993)
14. Frazer, R.A., Duncan, W.J., Collar A.R.: Elementary Matrices and Some Applications to Dynamics and Differential Equations. Cambridge University Press, Cambridge (1938)
15. Hurty, W.C., Rubinstein, M.F.: Dynamics of Structures. Prentice-Hall, Englewood Cliffs (1964)
16. Lang, G.F.: Understanding modal vectors. In: Mains, M., Blough, J.R. (eds.) Topics in Modal Analysis & Testing. Proceedings of the 35th IMAC, A Conference and Exposition on Structural Dynamics 2017, vol. 10, pp. 55–68. Springer, Cham (2017). https://doi.org/10.1007/9783319548104_8

Chapter 4
Experimental Verification of a Recently Developed FRF Decoupling Method for Nonlinear Systems

Taner Kalaycıoğlu and H. Nevzat Özgüven

Abstract The FRF Decoupling Method for Nonlinear Systems (FDM-NS), recently proposed by the authors of this paper, is a technique based on predicting the dynamic behavior of a particular substructure of a coupled nonlinear structure from the knowledge of the measured FRFs of the coupled nonlinear structure and calculated or measured FRFs of the other substructure. The uncoupled substructure can be linear or nonlinear. The method is applicable to systems where the nonlinearity can be represented as a single nonlinear element. The method has been experimentally verified for a structure having a grounded nonlinear element. In this work, the applicability of the method to a structure having an internal nonlinearity is demonstrated. The test system used in this study is composed of two cantilever beams where their free ends are connected to each other with two identical thin beams which introduce an internal nonlinearity to the coupled structure. In this test, the FRFs of the coupled nonlinear assembly are measured in a frequency range for various different constant displacement levels of the nonlinear connection element. Tip point transverse FRFs of one of the cantilever beam, which is taken as the known subsystem, are also measured. By using the decoupling method proposed the modal parameters of the unknown nonlinear subsystem are calculated as a function of the relative displacement amplitude between ends of the nonlinear connection element, from which the dynamic response of the decoupled subsystem can be calculated for any harmonic excitation. In order to demonstrate the accuracy of the method, the decoupled system is connected to a cantilever beam with a different length, and firstly, the FRFs of the coupled new system are calculated for constant amplitude harmonic forcing. Then, the calculated FRF curves are compared with those which are directly measured.

Keywords Nonlinear decoupling · Nonlinear uncoupling · Nonlinear inverse substructuring · Nonlinear subsystem identification · Nonlinear substructure decoupling

4.1 Introduction

Most of the engineering structures are designed as an assembly of several parts and usually various alternatives are studied within design period. It is computationally expensive to perform FE analysis each time while reviewing these design alternatives. Hence, several structural coupling methods have been developed in order to reduce the computational effort in dynamic reanalysis of such systems [1–19]. Although there are several different coupling methods based on the linearity assumption in literature, engineering structures are usually nonlinear by nature. In the past three decades, structural coupling of nonlinear subsystems has been well investigated and various coupling methods have been developed considering the nonlinear effect [20–28].

Likewise, considerable effort has also been devoted to the reverse problem, i.e. structural decoupling of linear systems. This problem arises when the dynamic behavior of a subsystem cannot be measured due to difficulty in measuring and/or exciting it individually, yet it is possible to test the whole system as well as the other subsystems. The first investigation on decoupling problem was made by Okubo and Miyazaki [29] where they have proposed an uncoupling method to extract the dynamics of a bearing itself. Later, Gontier and Bensaibi [30] presented a time domain method for in situ identification of

T. Kalaycıoğlu (✉)
Department of Mechanical Engineering, Middle East Technical University, Ankara, Turkey

MGEO Business Sector, ASELSAN Inc., Ankara, Turkey
e-mail: tkalayci@aselsan.com.tr

H. N. Özgüven
Department of Mechanical Engineering, Middle East Technical University, Ankara, Turkey
e-mail: ozguven@metu.edu.tr

A. Linderholt et al. (eds.), *Dynamics of Coupled Structures, Volume 4*, Conference Proceedings of the Society for Experimental Mechanics Series, https://doi.org/10.1007/978-3-319-74654-8_4

joints through modal analysis of their assembled structure. Maia et al. [31] presented a decoupling methodology particularly for the dynamic characterization of joints without making any FRF measurement at the joints. Kalling et al. [32] also worked on the decoupling problem by performing state-space model identification along with a sensitivity analysis showing possible ill-conditioning due to inertia ratios at the interface. D'Ambrogio and Fregolent [33] proposed impedance and mobility based approaches that suffers from ill-conditioning problems due to the internal resonances of the known subsystem with fixed interface. Sjövall and Abrahamsson [34] proposed a subsystem identification method that is based on reconstruction of the interface forces acting between the unknown subsystem and its neighbor. A general framework for dynamic substructuring is provided in [15] and [16] in which the so called dual domain decomposition technique that allows retaining the full set of global degrees of freedom (DOFs) by ensuring equilibrium at the interface between substructures is introduced. When performing substructuring by using the dual domain decomposition, the coupling problem can be directly formulated from [16], whereas a similar formulation for the decoupling problem is developed and discussed in [35] for collocated approach where DOFs used to enforce equilibrium are the same as DOFs used to enforce compatibility, and in [36] and [37] for non-collocated approach where DOFs used to enforce equilibrium are not the same as DOFs used to enforce compatibility. Batista and Maia [38] proposed three different formulations based on the classical decoupling procedure of Jetmundsen et al. [5] considering the effects of including different sets of DOFs on the coupled system. They concluded that the formulation that needs measurements at the connection point of substructures gives the best results. D'Ambrogio and Fregolent [39] also proposed the so-called hybrid assembly approach which gives similar results as dual assembly approach [35] in terms of the predicted FRFs of an unknown subsystem when applied to an experimental test bed. Dual assembly approach [35] was afterwards used to predict the subsystem dynamics in machine tools [40, 41], wind turbines [42] and flexible space payloads [43].

Although the dynamic decoupling problem of linear structures is well investigated in literature, the first study on dynamic decoupling problem of nonlinear structures was given quite recently by the authors of this paper [44]. The present study is based on a real-life application of this recently developed method, which is called FDM-NS (FRF Decoupling Method for Nonlinear Systems). The method can predict the FRFs of an unknown subsystem, whether linear or nonlinear, from the measured FRFs of the coupled nonlinear system and the measured or calculated FRFs of the remaining known subsystem. This method can be used for any type of nonlinearity provided that nonlinearity can be modeled as a single nonlinear element and its location is known. In this paper, a nonlinear experimental test rig is used in order to demonstrate the validity of FDM-NS in real life practice. Note that the experimental test system used in this paper differs mainly from the one given in [44] in terms of the location of the nonlinear element and availability of its parameters.

4.2 Theory

In this section, the approaches used in FDM-NS are briefly summarized. The details of the theory of the method can be found in [44]. The decoupling problem can be studied in three different categories, based on the location of nonlinear element in the coupled system: The nonlinearity can be either in the unknown or in the known subsystem, or it can connect two subsystems.

4.2.1 Nonlinearity in the Unknown Subsystem

This is the case where the nonlinearity which can be modelled as a single nonlinear element is in the unknown subsystem and therefore its parameters are not available. The following approach is developed for this case after examining available nonlinear system identification techniques in literature, detailed reviews of which are presented by Kerschen et al. [45] and Noël et al. [46]. Firstly, the complete FRF matrix of the linear known subsystem for the coordinates of interest can be obtained either by using its available system parameters or experimentally. Also, sets of linear FRFs of the coupled system for the coordinates of interest can be measured via controlled response amplitude testing each set of which corresponds to a different displacement level of the nonlinear element. Then, sets of linear FRF curves for the unknown subsystem can be calculated by applying a decoupling method for linear systems (dual assembly approach [35] is used in this paper) to each set of coupled system FRFs measured. Finally, each FRF curve set can be identified through the modal identification technique developed by Richardson and Formenti [47]. Since each set of coupled system FRFs corresponds to a different displacement level of the nonlinear element, identified modal parameters of the unknown subsystem FRFs can be expressed as a function of the magnitude of the relative harmonic displacement between end coordinates of the nonlinear element [48]. As a result, unknown subsystem FRFs can be calculated for any level of excitation using the obtained modal parameter variations.

4.2.2 Nonlinearity in the Known Subsystem

This is the case where the nonlinearity which can be modelled as a single nonlinear element is in the known subsystem and therefore its parameters are available. Firstly, the point and transfer FRFs of the coupled system as well as of the known subsystem at coordinates that belong to the known subsystem should be obtained for a specific displacement level of the nonlinear element throughout the frequency range of interest [48]. This will ensure that the nonlinearity matrices [49], which are involved in the dynamic stiffness matrices of the coupled system and the known subsystem, will have the same values at each frequency point throughout the desired frequency range. While FRFs of the known subsystem can be calculated using its available parameters, those of the coupled system have to be measured by performing controlled displacement amplitude testing. Note that including the same nonlinearity matrix into the dynamic stiffness matrices of the coupled system and the known subsystem mathematically corresponds to adding an additional stiffness matrix to the known parts of both systems. Thus, the problem can be reduced into decoupling of linear systems. So, FRFs of the unknown subsystem at its connection DOFs can be calculated via direct application of a linear decoupling approach.

4.2.3 Nonlinearity at the Connection of Two Subsystems

This is the case where a nonlinear element connects two subsystems. Based on the existence of the nonlinear element parameters, this problem can be transformed into one of those defined in Sect. 4.2.1 or Sect. 4.2.2. If the parameters of the nonlinear connection element are not available, the nonlinear element will be a part of the unknown subsystem. Then a massless node can be defined at the end coordinate of the nonlinear element which will be rigidly connected to the known subsystem when coupled. So, the problem transforms into the one considered in Sect. 4.2.1. If the parameters of the nonlinear connection element are available, the problem can likewise be transformed into the one given in Sect. 4.2.2.

4.3 Experimental Application

In this section, FDM-NS is applied to a nonlinear experimental test system in order to validate it through a real-life application. This experiment is a practical implementation of the case where nonlinearity is at the connection of two subsystems. Firstly, FRFs of the coupled system are measured at several different constant harmonic displacement levels of the nonlinear connection element. Then, FRFs of the known subsystem are also measured via modal testing and FDM-NS is applied. Thus, the modal parameters of the nonlinear unknown subsystem are calculated as a function of the magnitude of the relative harmonic displacement between end coordinates of the nonlinear element. Afterwards, in order to study the validity of the dynamic properties of the uncoupled nonlinear subsystem, the nonlinear subsystem is coupled to another beam which is longer than the previous one, and its FRFs are measured again through modal testing. Finally, measured FRFs of the new subsystem are coupled with those of the decoupled subsystem by using the modal parameter variations obtained. Calculated FRFs of the coupled new system are compared with those obtained performing controlled force amplitude testing on it.

4.3.1 Experimental Setup

The setup consists of a long cantilever beam and a shorter one connected tip to tip with each other by means of two thin identical beams which introduce nonlinear stiffness to the overall system. Both cantilever beams are of St37 steel alloy while the thin connection beams are of 6061-T3 aluminum alloy. Each cantilever beam is riveted to two thin identical beams in order to have fixed connections. A view of the experimental setup is given in Fig. 4.1. The dimensions and technical details of the test system are also illustrated in the same figure.

It is very difficult to achieve perfect fixed boundary conditions in physical applications. Therefore, both cantilever beams were manufactured longer than their effective lengths of 353.5 and 244 mm, respectively, so that the excess lengths could be squeezed between fixture blocks. A list of equipment used in the experiment is given in Table 4.1.

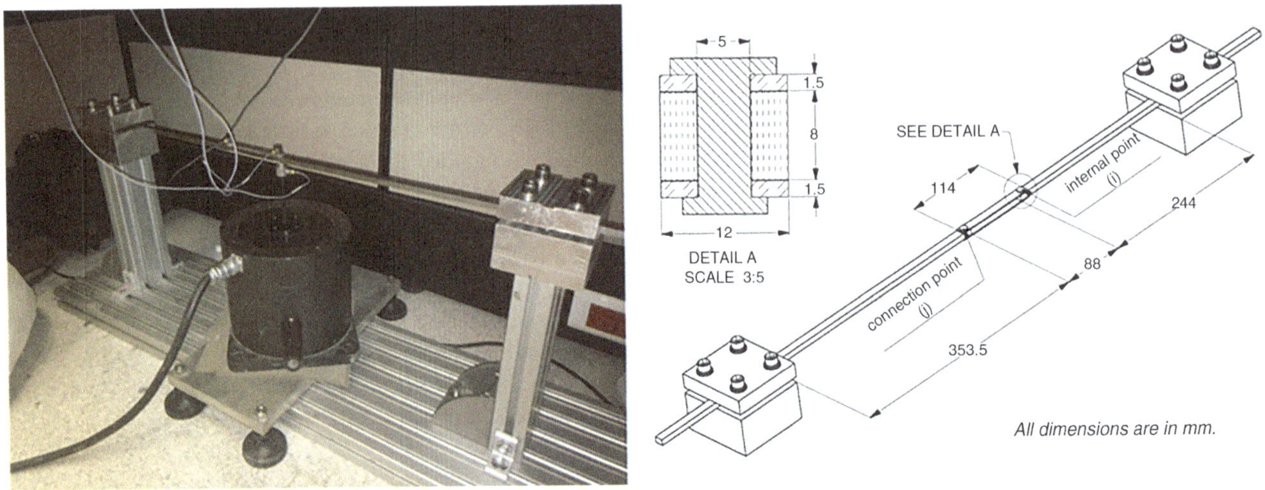

Fig. 4.1 View of the experimental setup (left) with its dimensions and technical details (right)

Table 4.1 List of modal test equipment

Data acquisition system	Brüel & Kjaer type 3560 C frontend
Shaker	Brüel & Kjaer type 4808
Force transducer	Brüel & Kjaer type 8230–002
Power amplifier	Brüel & Kjaer type 2712
Accelerometer	Brüel & Kjaer 4507B

4.3.2 The Preliminary Analyses

As explained in previous section, the experimental test system is mainly composed of two cantilever beams with different lengths. The longer one is taken as the linear known substructure. Two thin identical beams connect two cantilever beams to each other from their free ends and thus introduce nonlinearity into the coupled system. Due to unavailability of the type and parameter information regarding the nonlinear connection elements, short cantilever beam together with the two thin identical beams is considered as the unknown substructure. The aim in this application is to decouple this unknown substructure from the coupled cantilever beams. Preliminary modal analysis is performed in advance using a commercial FEA software to determine a proper frequency range for the FRF measurements, so that it includes the fundamental natural frequencies of the coupled structure and the known substructure. Modeling errors in the FEM of the analyzed structures are minimized by including the mass loading effects of the sensors in the analyses. The first natural frequency of the coupled cantilever beams is calculated as 72.2 Hz by performing modal analysis with ANSYS R15.0®. Results are illustrated in Fig. 4.2 along with the corresponding mode shape.

On the other hand, the fundamental natural frequency of the longer cantilever beam alone is obtained as 52.9 Hz by performing modal analysis with ANSYS R15.0®. The ANSYS result is presented in Fig. 4.3.

As a consequence, the frequency span to be used in the experiments is determined as 35–91 Hz which covers the wide neighborhood of pre-estimated fundamental resonances of the coupled cantilever beams and the long cantilever beam itself.

4.3.3 Experimental Work and Application of FDM-NS

As the initial step, point and transfer FRFs of the coupled cantilever beams in transverse direction ($\mathbf{H}_{jj}^{KU}$ and $\mathbf{H}_{ji}^{KU}$, respectively) are experimentally obtained by employing controlled displacement amplitude test procedure given in Fig. 4.4. An adaptive frequency resolution is employed during measurements so that the frequency resolution is further improved in the close neighborhood of the fundamental resonance.

In order to obtain relative transverse displacement between two ends of the thin identical beams, an additional measurement is taken from the tip of the short cantilever beam. A number of modal tests are conducted for different

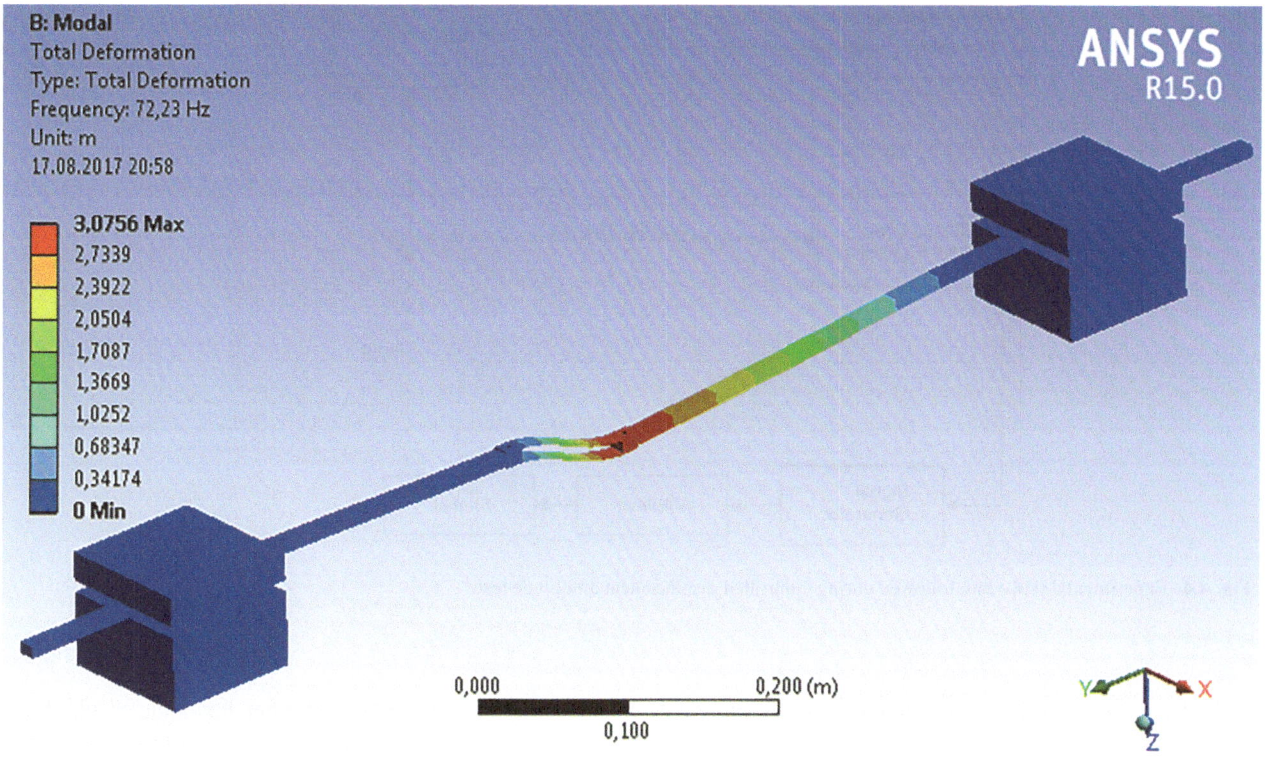

Fig. 4.2 Fundamental mode of the coupled cantilever beams

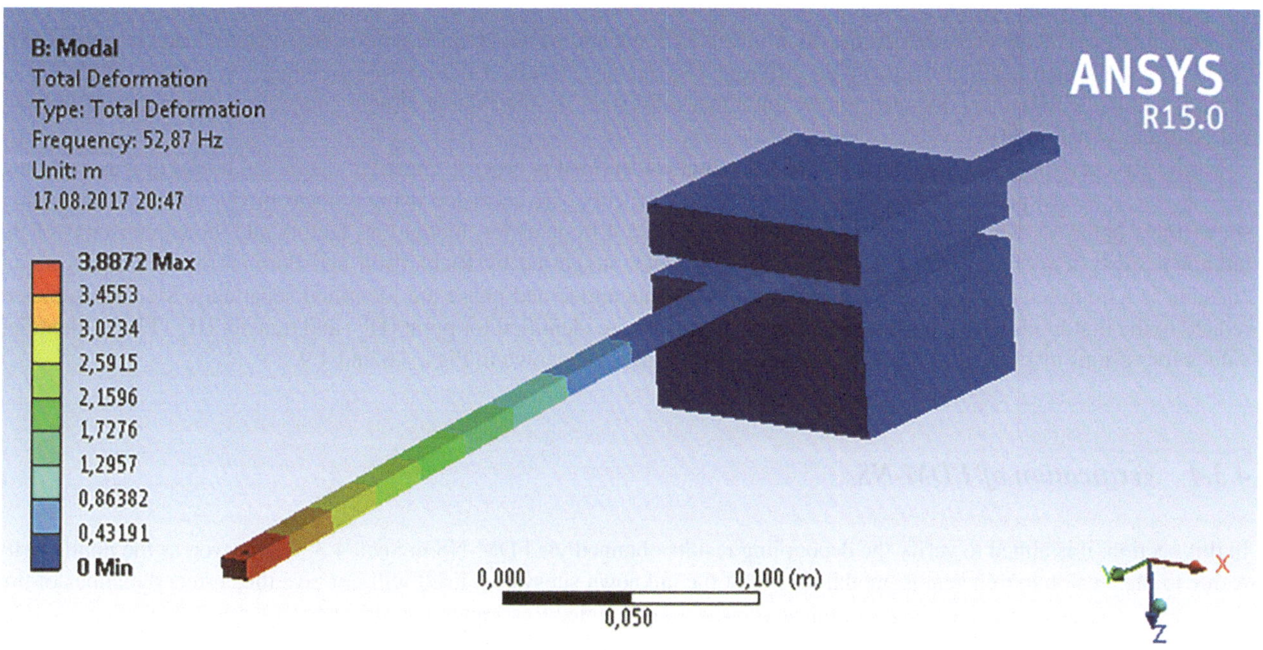

Fig. 4.3 Fundamental mode of the long cantilever beam alone

amplitudes of the relative harmonic displacement (0.1, 0.3, 0.5, 0.7 and 1.0 mm) between two ends of the nonlinear element. Measured data points are given along with FRF curves fitted to them in Fig. 4.5.

As the next step, tip point FRFs of the long cantilever beam are measured by conducting a modal test under transverse excitation. Results are shown in Fig. 4.6 in the form of a fitted FRF curve to experimental data.

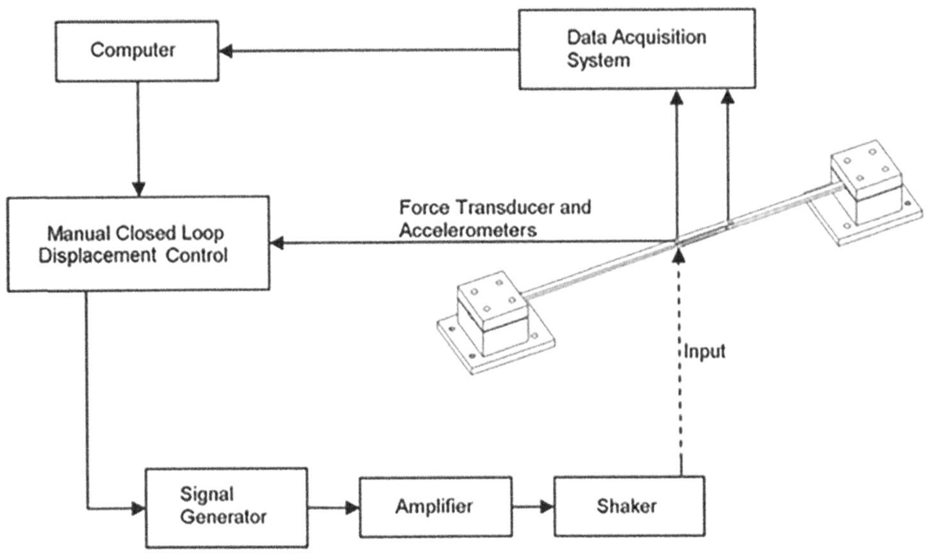

Fig. 4.4 Experimental procedure followed during controlled displacement amplitude test

FRFs of the unknown substructure at its coupling DOF are obtained for various different harmonic displacement amplitudes of the nonlinear element by applying FDM-NS. Dual assembly approach [35] is used during decoupling calculations. Calculated unknown substructure FRFs are shown in Fig. 4.7 together with those of the coupled structure given in Fig. 4.5.

Not surprisingly, decoupling of the nonlinear subsystem from coupled nonlinear system by using a single linear FRF curve for long cantilever beam and several equivalent linear FRF curves for the coupled nonlinear structure, leads to several FRF curves each of which again corresponds to different displacement levels of the nonlinear connection element. It should be emphasized that FRF curves obtained for the unknown subsystem includes the nonlinear connection dynamics which could not be directly measured.

Each FRF curve obtained for the unknown subsystem is identified in order to construct the modal model of it as explained in Sect. 4.2.1. Modal parameters corresponding to each FRF curve are extracted using a linear modal identification as they show linear behavior. Modal parameters in the form of natural frequencies, modal loss factors and modal constants of the unknown subsystem are identified by using the formulation suggested by Richardson and Formenti [47]. Since a modal constant is a complex quantity for a damped system, its magnitude and phase are identified separately. Modal parameters related to the first mode of the unknown nonlinear subsystem are identified for point ($\mathbf{H}_{jj}^{U}$) and transfer ($\mathbf{H}_{ji}^{U}$) FRFs. Identified values for various displacement levels of the nonlinear element are shown in Figs. 4.8 and 4.9.

4.3.4 Verification of FDM-NS

In this section, it is aimed to verify the decoupling results obtained via FDM-NS in Sect. 4.3.3. However, as the nonlinearity is due to the tension in thin beams, modal testing of the unknown subsystem itself will not give the correct dynamics of this subsystem. Therefore, the identified nonlinear subsystem is coupled to a longer cantilever beam and a coupled new system is obtained as shown in Fig. 4.10, and its dynamic response is predicted by using the decoupled subsystem dynamics, so that it can be compared with measured values for the verification of FDM-NS.

In this new test system, the decoupled subsystem is coupled to a longer cantilever beam of length 378 mm. Firstly, FRFs of 378 mm long cantilever beam alone are measured at its free tip by performing modal testing. The results obtained are given in Fig. 4.11 along with those obtained for 353.5 mm long cantilever beam used in Sect. 4.3.3.

As the next step, FRFs of the decoupled subsystem and those of the 378 mm long cantilever beam are coupled in order to obtain the response of the coupled new system. It should be noted that an iterative solution is required for the computation, as the modal parameters of the decoupled subsystem obtained in Sect. 4.3.3 are dependent on the displacement level of

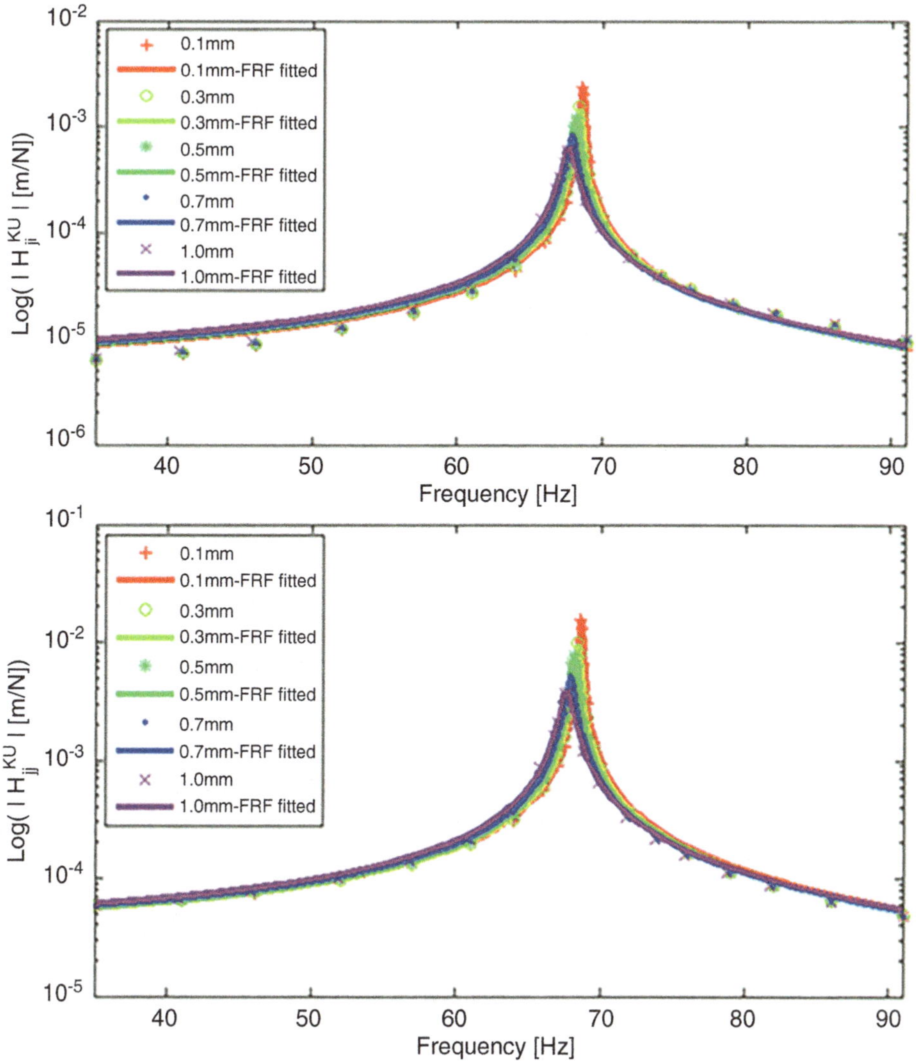

Fig. 4.5 Measured transfer FRFs (symbols) and fitted FRF curves (lines) of the coupled system at internal point (i) and at connection point (j) in transverse direction for various response levels of the nonlinear element

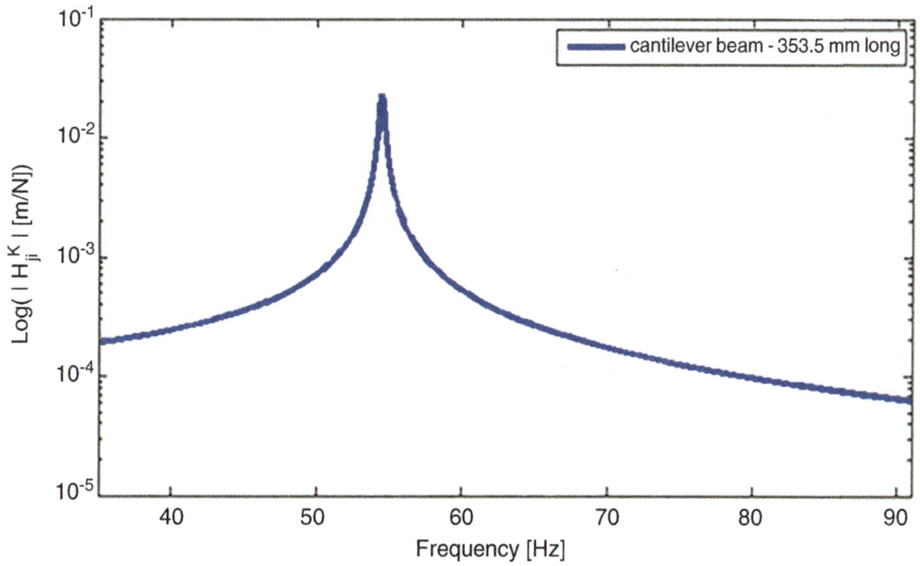

Fig. 4.6 FRF curve fitted to measured point FRFs of the known subsystem alone at its coupling DOF (j) in transverse direction

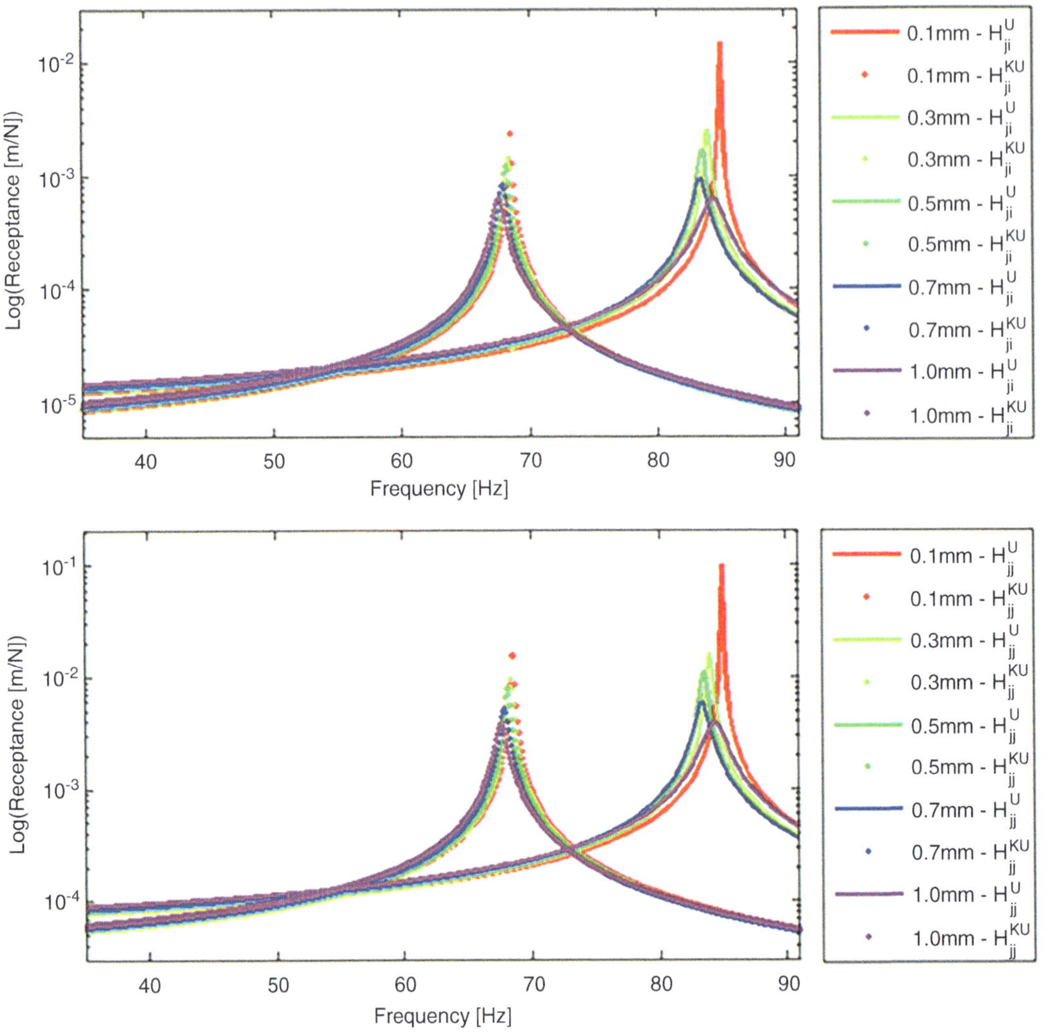

Fig. 4.7 Calculated point (above) and transfer (below) FRFs of the unknown subsystem (lines) and those of the coupled system (dots) under transverse excitation at connection point (j)

the nonlinear element. Thus, the response of the coupled new system is numerically obtained for a harmonic excitation of magnitude 0.4 N, applied transversely to the connection point j.

Finally, point FRFs of the new test system are measured again for a harmonic excitation of amplitude 0.4 N applied to the connection point j in transverse direction. In this test, the same procedure given in Fig. 4.4 is followed where harmonic forcing level is controlled by manually regulating the voltage output of the power amplifier at each frequency step this time. Measured and numerically calculated point FRFs of the coupled new system at its coupling DOF are compared in Fig. 4.12. A zoomed comparison around resonance is given in Fig. 4.13.

It can be observed from Figs. 4.12 and 4.13 that nonlinearity introduces softening stiffness, and measured response agrees quite well with those calculated through forward and backward frequency sweeping, particularly around frequency that jump occurs. Small discrepancy observed between the magnitudes of the resonances is believed to be due to the inaccuracies in extracting the loss factors from experimental FRFs. Because, unlike the case in linear systems, several controlled displacement amplitude tests are performed on the coupled nonlinear system, and therefore quite less number of frequency points had to be used in measuring FRFs due to technical issues. Although frequency resolution is further increased in the immediate vicinity of the resonance, it may not be good enough to catch the frequency where the peak amplitude occurs. As a result, the unknown subsystem FRFs obtained through decoupling using these measured FRFs may involve some errors which may be due to the misidentified loss factor values.

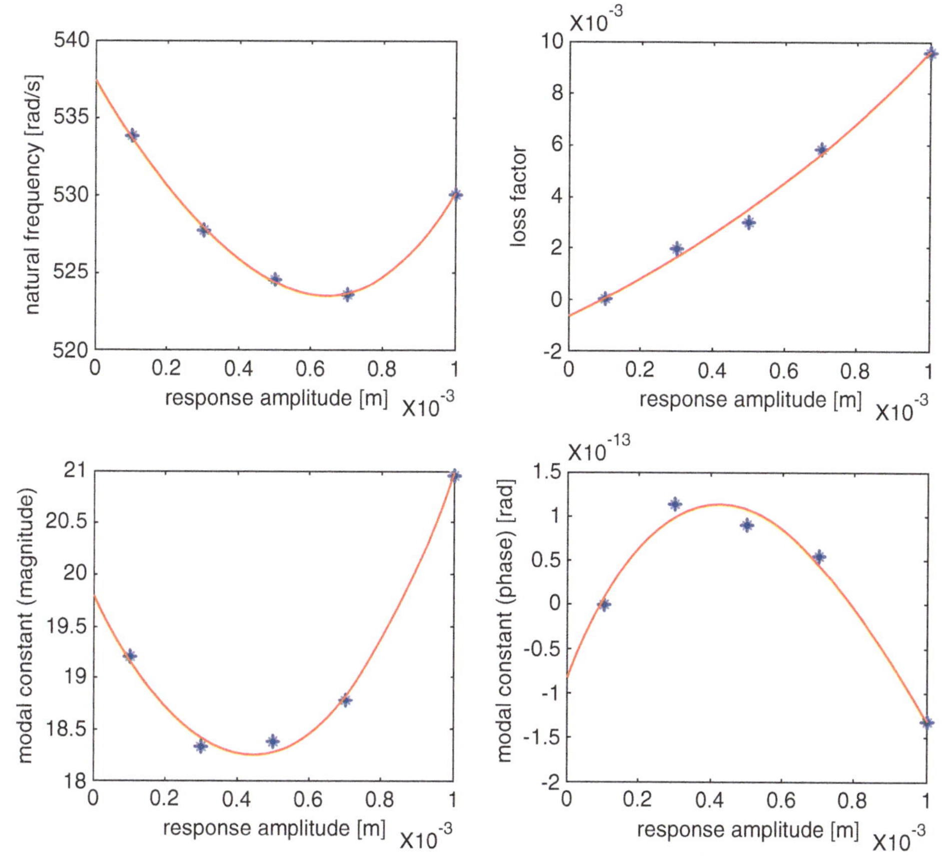

Fig. 4.8 Variation of the modal parameters of the unknown nonlinear subsystem at the first mode for $\mathbf{H}^{U}_{jj}$ with the relative response amplitude, $|X_j - X_i|$ (*, identified parameters & —, fitted curves)

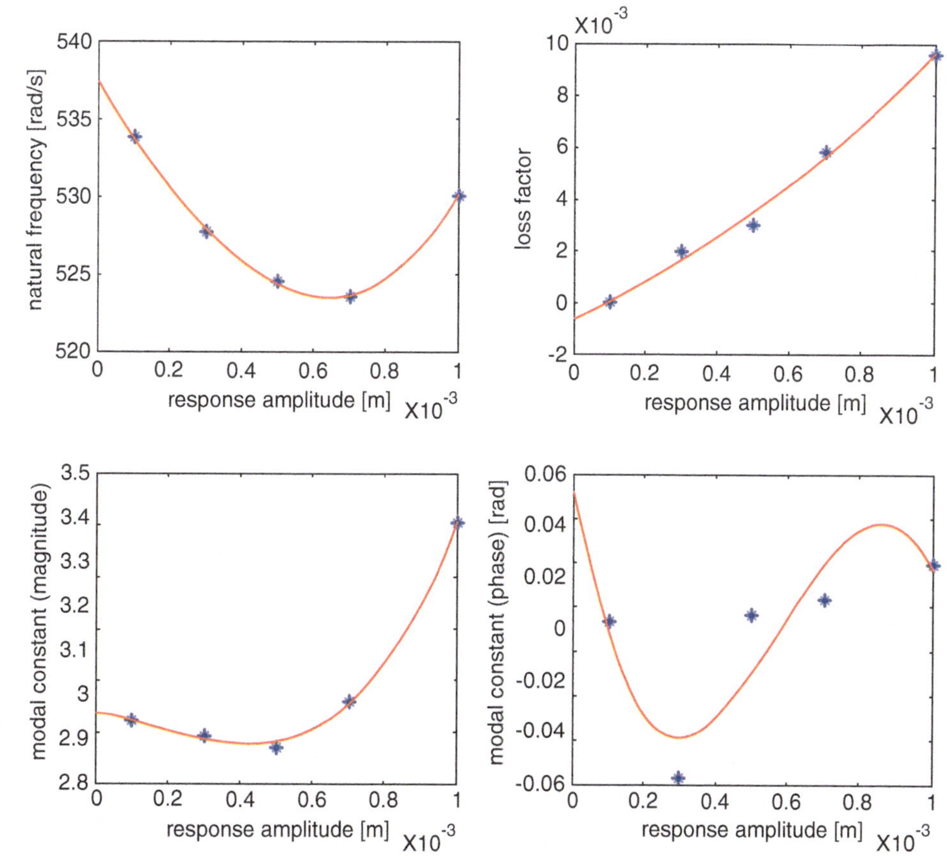

Fig. 4.9 Variation of the modal parameters of the unknown nonlinear subsystem at the first mode for $\mathbf{H}^{U}_{ji}$ with the relative response amplitude, $|X_j - X_i|$ (*, identified parameters & —, fitted curves)

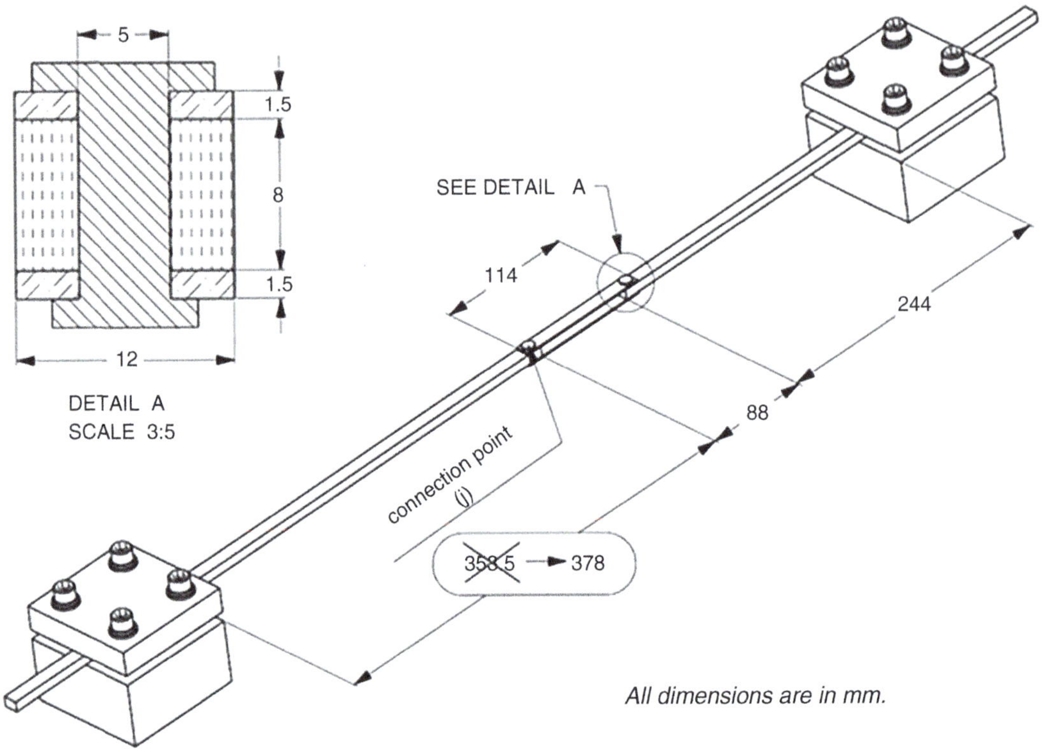

Fig. 4.10 Dimensions and technical details of the new test system

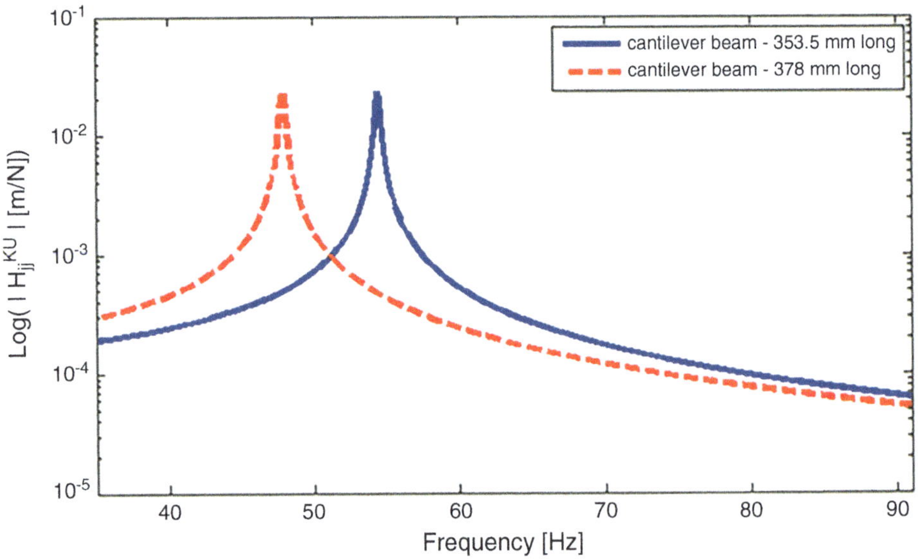

Fig. 4.11 FRF curves fitted to measured point FRFs of the known subsystems (both original and new) at their coupling DOF (j) in transverse direction

4.4 Discussion and Conclusions

In this work, the FRF Decoupling Method for Nonlinear Systems (FDM-NS) proposed recently [44] is applied to a test system in order to show its real-life applicability and validity. This method, which is believed to be the first method proposed for nonlinear decoupling, is capable of decoupling nonlinear systems having a nonlinearity of any type that can be modeled as a single element. Yet, the method allows the nonlinearity be either in the known or unknown subsystem, or connect both subsystems.

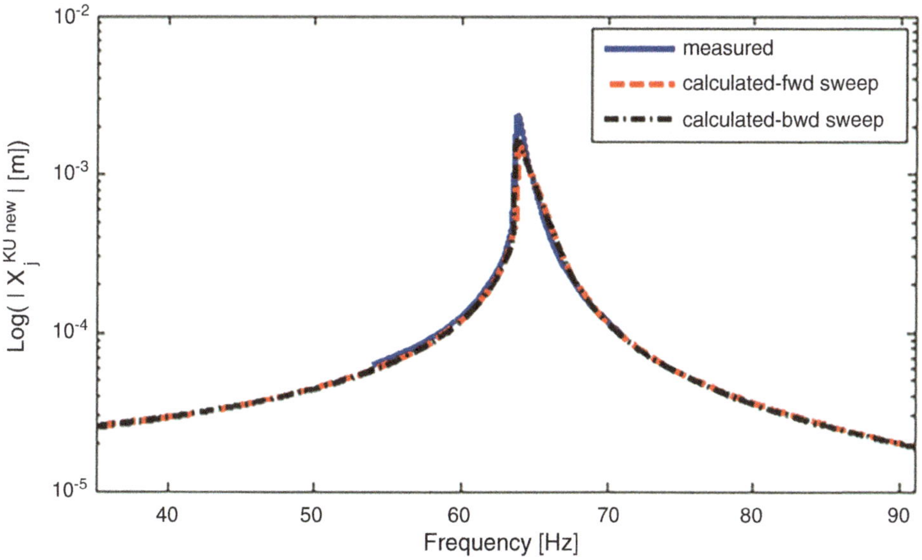

Fig. 4.12 Measured and calculated point FRFs of the coupled new system at its coupling DOF for harmonic forcing amplitude of 0.4 N in transverse direction

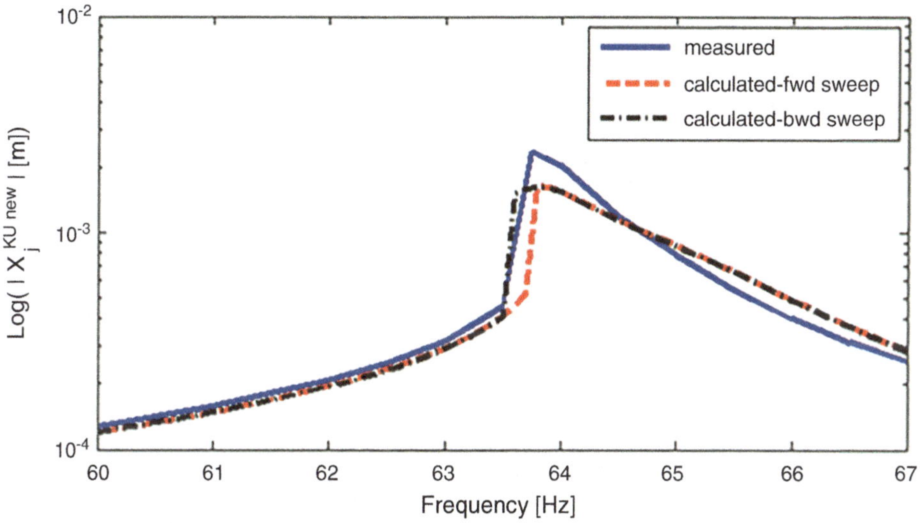

Fig. 4.13 Measured and calculated point FRFs of the coupled new system at its coupling DOF for harmonic forcing amplitude of 0.4 N in transverse direction – zoomed in the frequency range of 60–67 Hz

The proposed technique is applied to an experimental test system composed of two cantilever beams coupled with two thin identical beams. This study is an implementation of the case where two linear subsystems are coupled with an unknown nonlinear connection element. First, FRFs of the coupled system are measured through controlled displacement amplitude testing, while those of the known cantilever beam are obtained via classical modal testing. Then, modal parameter variations of the unknown subsystem are obtained as a function of the displacement level of the nonlinear element by applying FDM-NS. In order to verify these results, the experimental test system is modified such that the length of the known subsystem is increased whose FRFs are again obtained via classical modal testing. Then, the FRFs of the new system are calculated for a constant amplitude harmonic force by using the modal parameter variations of the unknown subsystem along with FRFs of the modified known subsystem. Finally, they are compared with those directly measured. Although the calculated results show slight deviations from the measured values around resonance, it is concluded that the agreement is quite well.

In this study, FDM-NS is verified for a practical case, where the nonlinear element couples two linear structures such that the unknown nonlinear subsystem cannot be tested alone. It can be concluded that FDM-NS proves itself as a promising tool in obtaining substructure dynamics of a nonlinear structure.

References

1. Craig, R., Bampton, M.: Coupling of substructures for dynamic analysis, American Institute of Aeronautics and Astronautics. Journal. **6**(7), 1313–1319 (1968)
2. Hallquist, J., Snyder, V.W.: Synthesis of two discrete vibratory systems using eigenvalue modification. AIAA J. **11**(2), 247–249 (1973)
3. Crowley, J.H., Rocklin, G.T., Klosterman, A.L., Vold, H.: Direct structural modification using frequency response functions. In: Proceedings of the 2nd International Modal Analysis Conference, Orlando, Florida, pp. 58–65 (1984)
4. İmregün, M., Robb, D.A., Ewins, D.J.: Structural modification and coupling dynamic analysis using FRF data. In: Proceedings of the 5th International Modal Analysis Conference, London, UK (1987)
5. Jetmundsen, B., Bielawa, R.L., Flannelly, W.G.: Generalized frequency domain substructure synthesis. J. Am. Helicopter Soc. **33**(1), 55–64 (1988)
6. Urgueira, A.P.V.: Dynamic analysis of coupled structures using experimental data (Ph.D. thesis), Department of Mechanical Engineering, Imperial College London, London, England, 1989 (http://www.imperial.ac.uk/dynamics/phd-theses/)
7. Sestieri, A., D'Ambrogio, W.: A modification method for vibration control of structures. Mech. Syst. Signal Process. **3**(3), 229–253 (1989)
8. Özgüven, H.N.: Structural modifications using frequency response functions. Mech. Syst. Signal Process. **4**(1), 53–63 (1990)
9. Ren, Y., Beards, C.F.: On substructure synthesis with FRF data. J. Sound Vib. **185**(5), 845–866 (1995)
10. Park, Y.H., Park, Y.S.: Structural modification based on measured frequency response functions: an exact eigenproperties reallocation. J. Sound Vib. **237**, 411–426 (2000)
11. Mottershead, J.E., Mares, C., Friswell, M.I.: An inverse method for the assignment of vibration nodes. Mech. Syst. Signal Process. **15**, 87–100 (2001)
12. Liu, W., Ewins, D.J.: Substructure synthesis via elastic media. J. Sound Vib. **257**(2), 361–379 (2002)
13. D'Ambrogio, W., Sestieri, A.: A unified approach to substructuring and structural modification problems. Shock. Vib. **11**(3–4), 295–310 (2004)
14. Kyprianou, A., Mottershead, J.E., Ouyang, H.: Assignment of natural frequencies by an added mass and one or more springs. Mech. Syst. Signal Process. **18**(2), 263–289 (2004)
15. De Klerk, D., Rixen, D.J., de Jong, J.: Frequency based substructuring (FBS) method reformulated according to the dual domain decomposition method. In: Proceedings of the 24th International Modal Analysis Conference (2006)
16. De Klerk, D., Rixen, D.J., Voormeeren, S.N.: General framework for dynamic substructuring: history, review, and classification of techniques. AIAA J. **46**(5), 1169–1181 (2008)
17. Hang, H., Shankar, K., Lai, J.C.S.: Prediction of the effects on dynamic response due to distributed structural modification with additional degrees of freedom. Mech. Syst. Signal Process. **22**(8), 1809–1825 (2008)
18. Mayes, R.L.: Tutorial on experimental dynamic substructuring using the transmission simulator method. In: Proceedings of the 30th International Modal Analysis Conference (2012)
19. Sayın, B., Ciğeroğlu, E.: A new structural modification method with additional degrees of freedom for dynamic analysis of large systems. In: Proceedings of the 31st International Modal Analysis Conference (2013)
20. Watanabe, K., Sato, H.: A modal analysis approach to nonlinear multi-degrees-of freedom system. J. Vib. Acoust. Stress. Reliab. Des. **110**, 410–411 (1988)
21. Ferreira, J.V., Ewins, D.J., Nonlinear receptance coupling approach based on describing functions. In: Proceedings of the 14th International Modal Analysis Conference, Dearborn, Michigan, USA, pp. 1034–1040 (1996)
22. Chong, Y.H., İmregün, M.: Coupling of non-linear substructures using variable modal parameters. Mech. Syst. Signal Process. **14**, 731–746 (2000)
23. Huang, S.: Dynamic analysis of assembled structures with nonlinearity (Ph.D. thesis), Department of Mechanical Engineering, Imperial College London, London, England (2007) (http://www.imperial.ac.uk/dynamics/phd-theses/)
24. Kuether, R.J., Allen, M.S.: Structural modification of nonlinear FEA subcomponents using nonlinear normal modes. In: Proceedings of the 31st International Modal Analysis Conference, Garden Grove, California, USA (2013)
25. Kalaycıoğlu, T., Özgüven, H.N.: Nonlinear structural modification and nonlinear coupling. Mech. Syst. Signal Process. **46**(2), 289–306 (2014)
26. Wenneker, F., Tiso, P.: A substructuring method for geometrically nonlinear structures. In: Proceedings of the 32nd International Modal Analysis Conference, Orlando, Florida, USA (2014)
27. Tepe, C., Ciğeroğlu, E.: Structural coupling of two-nonlinear structures. In: Proceedings of the 33rd International Modal Analysis Conference, Orlando, Florida, USA (2015)
28. Kuether, R.J., Allen, M.S.: Modal substructuring of geometrically nonlinear finite-element models. AIAA J. **54**(2), 691–702 (2016)
29. Okubo, N., Miyazaki, M.: Development of uncoupling technique and its application. In: Proceedings of 4th International Modal Analysis Conference (1986)
30. Gontier, C., Bensaibi, M.: Time domain identification of a substructure from in situ analysis of the whole structure. Mech. Syst. Signal Process. **9**(4), 379–396 (1995)
31. Maia, N.M.M., Silva, J.M.M., Ribeiro, A.M.R., Silva, P.L.C.: On the dynamic characterization of joints using uncoupling techniques. In: Proceedings of the 16th International Modal Analysis Conference. Santa Barbara, California, USA (1998)
32. Kalling, P., Abrahamsson, T., McKelvey, T.: Subsystem state-space model identification and its sensitivity to test variability. In: Sas, P., De Munck, M. (eds.) Proceedings of ISMA 2004 – International Conference on Noise and Vibration Engineering, Leuven, Belgium, pp. 2729–2744 (2004)
33. D'Ambrogio, W., Fregolent, A.: Promises and pitfalls of decoupling procedures. In: Proceedings of the 26th International Modal Analysis Conference (2008)
34. Sjövall, P., Abrahamsson, T.: Substructure system identification from coupled system test data. Mech. Syst. Signal Process. **22**(1), 15–33 (2008)
35. D'Ambrogio, W., Fregolent, A.: The role of interface DOFs in decoupling of substructures based on the dual domain decomposition. Mech. Syst. Signal Process. **24**(7), 2035–2048 (2010)

36. Voormeeren, S.N., Rixen, D.J.: A dual approach to substructure decoupling techniques. In: Proceedings of the 28th International Modal Analysis Conference (2010)
37. D'Ambrogio, W., Fregolent, A.: Direct decoupling of substructures using primal and dual formulation. In: Proceedings of the 29th International Modal Analysis Conference (2011)
38. Batista, F.C., Maia, N.M.M.: Uncoupling techniques for the dynamic characterization of sub-structures. In: Proceedings of the 29th International Modal Analysis Conference (2011)
39. D'Ambrogio, W., Fregolent, A.: Inverse dynamic substructuring using direct hybrid assembly in the frequency domain. Mech. Syst. Signal Process. **45**(2), 360–377 (2014)
40. Law, M., Rentzsch, H., Ihlenfeldt, S., Putz, M.: Application of substructure decoupling techniques to predict mobile machine tool dynamics: numerical investigations. Procedia CIRP. **46**, 537–540 (2016)
41. Law, M., Rentzsch, H., Ihlenfeldt, S.: Predicting mobile machine tool dynamics by experimental dynamic substructuring. Int. J. Mach. Tools Manuf. **108**, 127–134 (2016)
42. Brunetti, J., Culla, A., D'Ambrogio, W., Fregolent, A.: Experimental dynamic substructuring of the Ampair wind turbine test bed. In: Proceedings of the 32nd International Modal Analysis Conference (2014)
43. D'Ambrogio, W., Fregolent, A.: Predicting the dynamics of flexible space payloads under different boundary conditions through substructure decoupling. In: Proceedings of the 35th International Modal Analysis Conference (2017)
44. Kalaycıoğlu, T., Özgüven, H.N.: FRF decoupling of nonlinear systems. Mech. Syst. Signal Process. **102**, 230–244 (2018). https://doi.org/10.1016/j.ymssp.2017.09.029
45. Kerschen, G., Worden, K., Vakakis, A.F., Golinval, J.C.: Past, present and future of nonlinear system identification in structural dynamics. Mech. Syst. Signal Process. **20**(3), 505–592 (2006)
46. Noël, J.P., Kerschen, G.: 10 years of advances in nonlinear system identification in structural dynamics: a review. In: Proceedings of ISMA 2016 – International Conference on Noise and Vibration Engineering, Leuven, Belgium (2016)
47. Richardson, M.N., Formenti, D.L.: Parameter estimation from frequency response measurements using rational fraction polynomials. In: Proceedings of the 1st International Modal Analysis Conference (1982)
48. Arslan, Ö., Aykan, M., Özgüven, H.N.: Parametric identification of structural nonlinearities from measured frequency response data. Mech. Syst. Signal Process. **25**(4), 1112–1125 (2011)
49. Tanrıkulu, Ö., Kuran, B., Özgüven, H.N., İmregün, M.: Forced harmonic response analysis of non-linear structures using describing functions. AIAA J. **31**(7), 1313–1320 (1993)

Chapter 5
Automated Correction of Sensor Orientation in Experimental Dynamic Substructuring

M. Haeussler, S. Sendlbeck, and D. Rixen

Abstract Experimentally modeling the dynamics of substructures in a large assembly is of great benefit. This comes into play, if some components are too complex to model numerically, or certain effects, e.g. friction, within these parts cannot be modeled accurately. In engineering practice, this task also arises when building a product containing multiple supplier parts, with often no numerical models provided. When the dynamics of the substructures shall be assembled, it is well known that modeling the connection between them properly is essential for obtaining good results. For example, when coupling two pieces of a beam the interface between them must not be assumed as a point with only three translational degrees of freedom. Coupling only these, corresponds to a ball joint connection, which is clearly insufficient for coupling the pieces of a beam. For accurate results one needs to consider rotational degrees of freedom as well. One way of doing so is to apply a so called virtual point transformation. It is based on the assumption that a small region around a connection point is locally rigid. Measuring with sensors distributed around this connection point allows to calculate the rigid motions of the interface (including rotations), provided that their *position and orientation is known*. This paper focuses on the fact that in an actual experiment the sensor positions can only be known up to a certain measurement accuracy. We will show how a wrong estimate of sensor positions deteriorates the quality of the experimental model obtained from the virtual point transformation and thus its usefulness for dynamic substructuring. However, there are some measures for checking the quality of an experimental model. For instance if the assumption of rigidity on the interface is valid, then the sensors should move with negligible flexible motion between each other. The remaining residual in the transformation should thus be minimal. Another well known property from mechanics is the reciprocity principle. When exchanging the input and output on a linear structure one should get the same transfer function. In other words, the frequency response function (FRF) matrix should be symmetric. Those quality indicators can be used to formulate a cost function for an optimization, with the *unknown positions* as optimization variables. We will show how wrong position estimates of inputs and outputs can be removed to a large extend by the optimization. We will also show how removing these errors improves the quality of the experimental model and thus the results of dynamic substructuring.

Keywords Experimental dynamic substructuring · Virtual point transformation · Sensor orientation · Optimization

Nomenclature

$\mathbf{u}, \mathbf{f}$	measured displacements/forces	$(\star)_1$	pertaining to set of DoFs 1
$\mathbf{q}, \mathbf{m}$	generalized displacements/forces	$(\star)^{AB}$	coupled quantity of A and B
$\mathbf{Y}$	admittance matrix	$(\star)^A$	quantity pertaining to substructure A
$\boldsymbol{\mu}$	residual in displacement transformation		

DS	dynamic substructuring	FBS	frequency based substructuring
FRF	frequency response function	IDM	interface displacement mode
VPT	virtual point transformation	DoF	degree of freedom

M. Haeussler (✉) · D. Rixen
Faculty of Mechanical Engineering, Technical University of Munich, Garching, Germany
e-mail: michael.haeussler@tum.de

S. Sendlbeck
Technical University of Munich, Garching, Germany
e-mail: rixen@tum.de

A. Linderholt et al. (eds.), *Dynamics of Coupled Structures, Volume 4*, Conference Proceedings of the Society for Experimental Mechanics Series, https://doi.org/10.1007/978-3-319-74654-8_5

5.1 Introduction

The framework of dynamic substructuring (DS) allows to virtually assemble substructures into one larger product and predict its overall dynamics. This is a necessary tool for analyzing the dynamic properties in an early design stage, where a full prototype is not yet available. The framework provides different methods for enforcing the coupling conditions on the interface: equal displacements (*compatibility*) and force equilibrium (*actio est reactio*). The formulation of these conditions in different domains (physical, modal, frequency [1], state space [2] and impulse response [3]) can be derived straight forward. They can then be used for coupling substructure models that were obtained from different modeling approaches. It is for example desirable to couple two structures where one model comes from experiments and the other model is from a numerical simulation. One theoretically well suited method therefore is the frequency based substructuring (FBS) method. It was first published by Jetmundsen [4], although similar methods where proposed earlier [5], and reformulated so that fewer inversions are needed in the process [6]. However, the definition of the interface coupling in experimental DS remains challenging. On the one hand the measured degrees of freedom (DoFs) on either side of the interface are often non-matching due to practical instrumentation problems. For some interface geometries it is simply impossible to mount the sensors and apply the force inputs on the same geometric positions as in the later assembly (see e.g. substructure A and B in Fig. 5.1). They can thus not be coupled by simply enforcing compatibility in each measured DoF, as one would do in purely numerical analysis. Another problem is the right description of the interface dynamics, including e.g. rotational degrees of freedom which are hard to measure with conventional accelerometers and maybe even harder to excite. One way for solving this problem was recently proposed by measuring rotations with a rotational sensor and perform subsequent experimental modal analysis (EMA) to obtain also the FRFs for the moment excitations [7].

Another way for solving this problem is to use the so called virtual point transformation [8, 9], where interface motion is measured on several positions with conventional accelerometers. These measurements are then projected on so called interface displacement modes (IDMs). The IDMs are assumed the describe the dynamics of the interface nearly completely (see the generalized IDM DoFs $\mathbf{q}$ in Fig. 5.1 and [8, 9] for more information). Provided that the interface motion of both substructures was projected on the same IDMs, the enforcement of compatibility and actio est reactio is straight forward again. This approach comes with the advantage of not requiring an extra EMA step, which would be challenging for very complex structures like e.g. a car body, where hardly any clear modes are visible. However, this is also a disadvantage in the sense that the smoothing effect of the EMA on the FRFs is very well able to at least reduce some of the conditioning problems common in FBS (see e.g. [10–12]).

One example for this approach can be seen in Fig. 5.2. The complete structure is modeled with a Finite Elements, which serves as a reference solution (see Fig. 5.2a). It is then divided into two different pieces which are also modeled separately via Finite Elements (see Fig. 5.2b). All FRF matrices (of the reference structure and the two single substructures) are synthesized from the first 250 free vibration modes. The FRF matrices of the substructures are then assembled via first performing a virtual point transformation on the interface FRFs (see Fig. 5.2c for the simulated sensors and force inputs) and subsequently

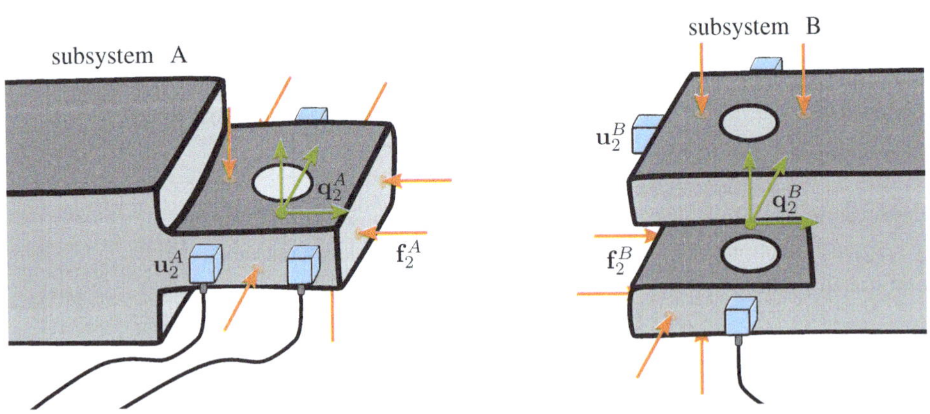

Fig. 5.1 Assembly problem in experimental dynamic substructuring. Sensors $\mathbf{u}_2$ and force inputs $\mathbf{f}_2$ on either side of the interface are non-matching

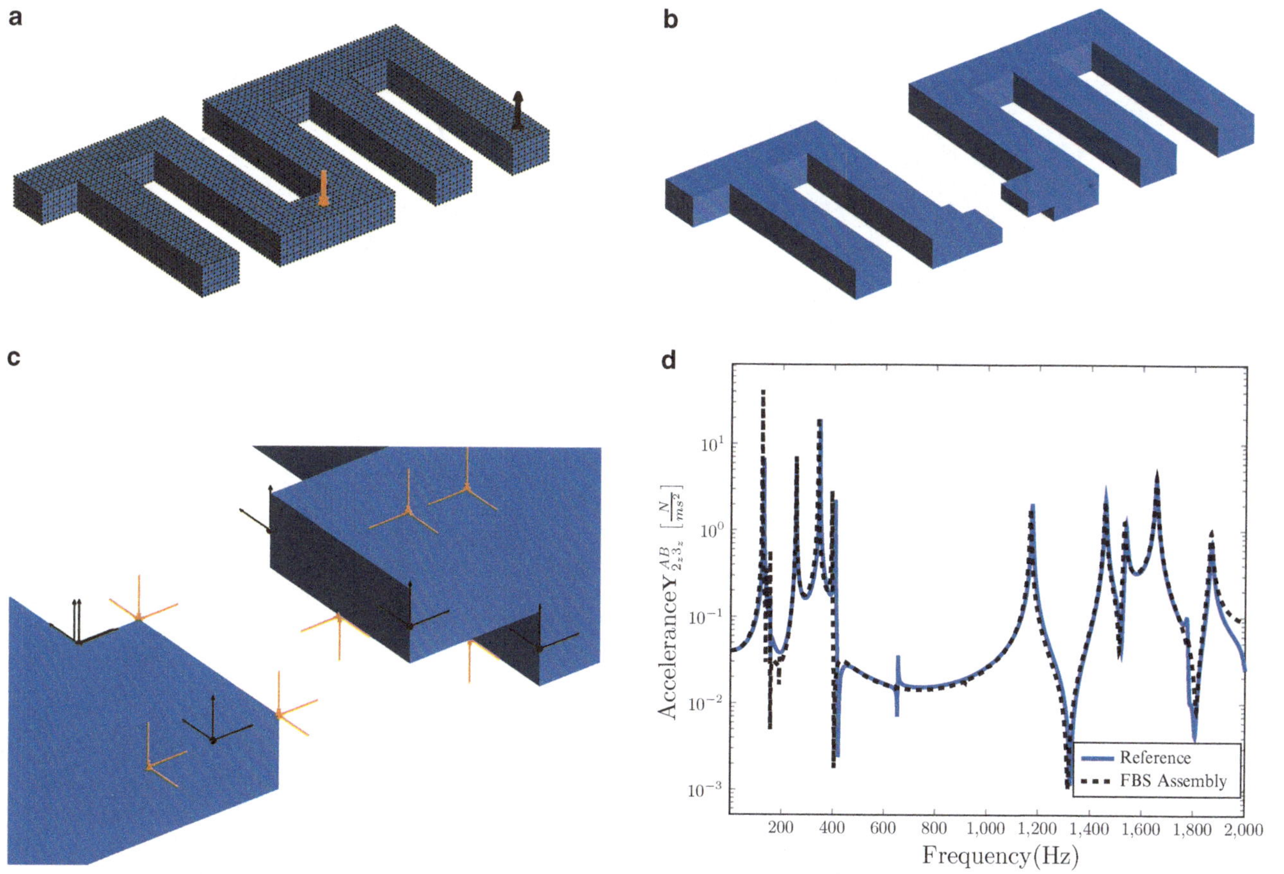

Fig. 5.2 Overview of example structure. (**a**) Input (orange arrow) and output (black arrow) for reference FRF on example structure. The finite element nodes of the reference model are indicated as black dots. (**b**) Example structure divided into two substructures. Left and right part will be referred to as substructure A and substructure B respectively. FEM nodes of substructures not shown. (**c**) Zoom on interface inputs (orange) and outputs (black) which are used for synthesis of FRFs and subsequent virtual point transformation. (**d**) Comparison of the reference FRF for the full model shown in (**a**) and the virtual point transformed and FBS assembled models of the substructures in (**b**, **c**)

coupling these transformed FRFs with FBS. The reference FRF between an input on the interface and an output on an external DoF of substructure B (see Fig. 5.2a) is compared to the result from the FBS assembly in Fig. 5.2d.[1]

One can observe that the FBS results provide a good estimate for the FRF of the reference model. However, the quality of the FBS results relies of course on the quality of the single substructure models. Here the virtual point transformation was applied as a preprocessing step to the 'measured' (in this example simulated of course) FRFs on both substructures. This preprocessing step is what makes the substructures', originally incompatible FRFs on the interface DoFs, compatible and allows to subsequently couple them via FBS (see Figs. 5.1 and 5.2c). In the example shown, we have known all positions and orientations of the force inputs and sensors perfectly. However, we cannot be so self confident with our positioning and orientation estimates for the sensors and force inputs in a real measurement. Assume for example we have the same 'measurement' data for both substructures as for the results shown in Fig. 5.2. I.e. the untransformed FRF matrices are the same as before. The only thing different is, that now we did not have good estimates for the positions and orientations of the sensors and force inputs. This is shown on the example of substructure A in Fig. 5.3a. The untransformed FRFs are all the same, so the real sensor and force input positions are still the ones shown in solid lines. However, due to measurement inaccuracies, we have assumed that our measurement positions were actually at the positions shown with the dashed lines in Fig. 5.3a.[2] When we perform a virtual point transformation with these wrong estimates for positions and orientations, we

[1]The transfer from an interface DoF to a final receiver DoF on substructure B is purposely chosen. Especially these FRFs are important, if the FBS results shall be used together with component based transfer path analysis (TPA), where the source activity is completely modeled via equivalent forces on the interface between source and receiver [13].

[2]In reality it is probably the other way around, in the sense that we have estimated the positions of e.g. the sensors in nearly perfect alignment to edges and corners, because that is how we tried to glue them on the structure. But in reality the real positions were slightly off. However,

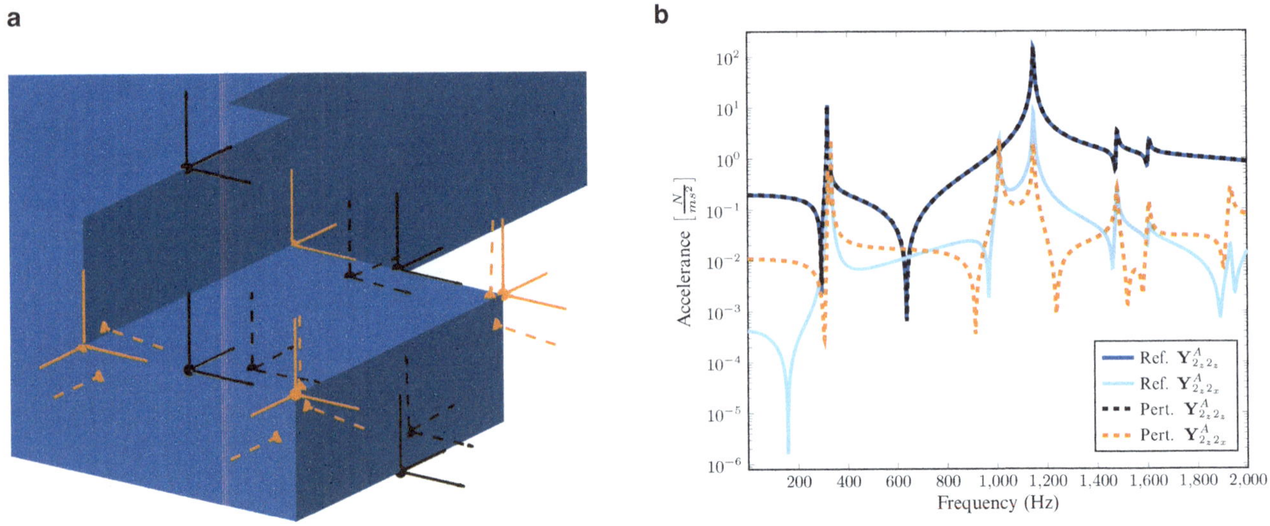

Fig. 5.3 Effects of wrong position estimates on the resulting FRFs after the virtual point transformation. (**a**) Perfect positions of sensors (black) and force inputs (orange) shown in solid lines. The actually 'assumed' and perturbed positions are shown as dashed lines. (**b**) Comparison of the virtual point transformed FRFs with the optimal/reference (Ref.) positions and wrong/perturbed (Pert.) position estimates

will necessarily get different results for the virtual point transformed FRFs which shall be used for coupling. If we project all sensor outputs and force inputs on only rigid motions (i.e. translations, rotations, forces and moments) we get a different result for the FRFs of these virtual point degrees of freedom. The driving point FRFs (i.e. input DoF equals output DoF) of the virtual point transformed FRFs do not change significantly, as can be seen for the $z - z$ FRF in Fig. 5.3b. However the non-diagonal entries of the transformed FRF matrix are strongly affected, as can be seen for the example of the $x - z$ cross FRF shown in Fig. 5.3b. This result is not entirely surprising, since the orientation errors (all below 5°) and positioning errors (all below 0.2 of the distance to the virtual point) were still small enough so that a small angle approximation for the driving point FRFs should be valid. Similar results were already found in [14, p. 63ff.].

5.1.1 Outline of the Paper

The intent of this paper is the following: We want to remove the positioning and orientation errors from our wrong estimates. Therefore we will briefly recapitulate the theory for FBS and the virtual point transformation. We then move on to show some physically necessary properties that every virtual point transformed FRF matrix should posses. From these properties we are defining some cost functions that take any given FRF matrix as an argument, and return a minimal value for a perfectly physical FRF. With this cost function we formulate an optimization problem that uses the unknown positions and orientations as design variables and tries to update them in a way that the 'physicalness' of the transformed FRF Matrix is optimal. We then show the performance of this approach on some examples:

- A 2D rigid plate,
- a 3D rigid block,
- the simulated flexible structure shown in the introduction,
- measurements on an example for teaching vibration testing.

The examples are followed by a discussion about the cost function. Especially its ability to correct wrong positions and orientations must be considered under the knowledge about the nullspace of the cost function. I.e. some combinations of position updates do not result in a change of the cost function and are therefore not able to be found by the optimization. It seems that this knowledge is essential for obtaining good results. The conclusion provides some general recommendations for the use and capability of this method and provides some outlook on further work.

---

for illustrative purposes in this paper we will show it this way around, i.e. real positions aligned with edges and corners and assumed positions unaligned and skew. This way it is easier to visually inspect the success of the optimized positions compared to the assumed positions.

5.2 Examples

5.2.1 Flexible Body

As mentioned in the introduction it is possible to formulate an optimization problem with the unknown positions and orientations of the sensors and force inputs. The cost function for the optimization is composed of the following properties that the final virtual point transformed matrix shall possess:

- *Rigidness of the interface*: The virtual point transformation can be seen as a filter, which is filtering out all motion in the sensors that does not comply with rigid motion of the interface (see e.g. [14, p.49ff.]). However if this filtering effect is very large, it may well be that it is due to positioning errors in the assumed sensor and hammer positions. This part of the cost function returns a scalar value between 0 and 1. Zero means that no filtering is applied at all, i.e. the interface indeed behaves like a rigid body and one means that all motion is non-rigid, i.e. filtered out.
- *Symmetry of the FRF Matrix*: As the FRF matrix in the virtual point is a square driving point matrix (i.e. the inputs and outputs are the same generalized IDM DoFs) it should be symmetric (often this property is also denoted as *reciprocity* [14]). This part of the cost function also delivers a scalar value between 0 and 1, where zero means again perfect symmetry and one the opposite.
- *Passivity of the FRF Matrix*: The scalar product of the input forces on the FRF matrix and the resulting velocities on the interface shall have a positive real part. If this is violated it would physically mean that the substructure is not passive, i.e. producing instead of dissipating energy. The requirement of passivity can be formulated as a requirement on the FRF matrix. This part of the cost function is also yielding a relative measure between 0 and 1, where zero means perfect passivity at all frequencies and one means strong violation of passivity at all frequencies.

The capabilities for finding the original positions and, maybe even more important, finding the right FRF matrix in the virtual point is demonstrated in Fig. 5.4. It can be seen that the above mentioned physical requirements on the virtual point transformed FRF matrix are actually sufficient for greatly improving the position estimates and thus adjusting the FRF matrix to its actual value.

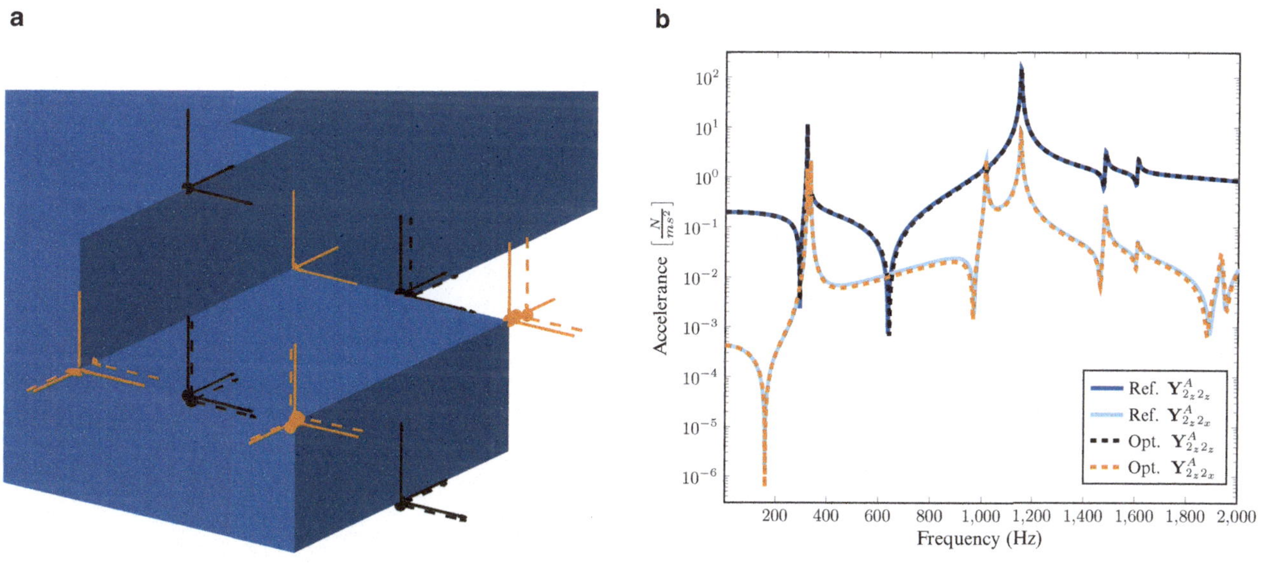

Fig. 5.4 Effects of optimized position estimates on the resulting FRFs after the virtual point transformation. (**a**) Perfect positions of sensors (black) and force inputs (orange) shown in solid lines. The optimized position estimates are shown as dashed lines. (**b**) Comparison of the virtual point transformed FRFs with the optimal/reference (Ref.) positions and optimized position estimates (Opt.)

Acknowledgements We want to thank the BMW Group and Vibes technology for financial and technical support respectively.

References

1. de Klerk, D., Rixen, D.J., Voormeeren, S.N.: General framework for dynamic substructuring: history review and classification of techniques. AIAA J. **46**(5), 1169–1181 (2008). ISSN: 0001-1452. https://doi.org/10.2514/1.33274
2. Sjövall, P., Abrahamsson, T.: Component system identification and state-space model synthesis. Mech. Syst. Signal Process. **21**(7), 2697–2714 (2007). ISSN:08883270. https://doi.org/10.1016/j.ymssp.2007.03.002
3. Van der Valk, P.L.C., Rixen, D.J.: An impulse based substructuring method for coupling impulse response functions and finite element models. Comput. Methods Appl. Mech. Eng. **275**, 113–137 (2014). ISSN:00457825. https://doi.org/10.1016/j.cma.2014.03.003
4. Jetmundsen, B., Bielawa, R.L., Flannelly, W.G.: Generalized frequency domain substructure synthesis. J. Am. Helicopter Soc. **33**(1), 55–64 (1988)
5. Crowley, J.R., Klosterman, A.: Direct structural modification using frequency response functions. In: IMAC II – International Modal Analysis Conference (1984)
6. Klerk, D.D., Rixen, D., Jong, J.D.: The frequency based substructuring (FBS) method reformulated according to the dual domain decomposition method. In: Sem: IMAC Xxiv (2006). ISSN:21915644
7. Drozg, A., Čepon, G., Boltežar, M.: Full-degrees-of-freedom frequency based substructuring. Mech. Syst. Signal Process. **98**, 570–579 (2018). ISSN:10961216. https://doi.org/10.1016/j.ymssp.2017.04.051
8. de Klerk, D., et al.: Solving the RDoF problem in experimental dynamic substructuring. In: Proceedings of the 26th International Modal Analysis Conference – IMAC (2008). ISSN:9781605600666
9. van der Seijs, M.V., et al.: An improved methodology for the virtual point transformation of measured frequency response functions in dynamic substructuring. In: 4th ECCOMAS Thematic Conference on Computational Methods in Structural Dynamics and Earthquake Engineering, June 2013. https://doi.org/10.13140/RG.2.1.2715.3126
10. Mayes, R.L., Stasiunas, E.C.: Lightly damped experimental substructures for combining with analytical substructures. In: 25th International Modal Analysis Conference (IMAC XXV), Orlando (2007)
11. Allen, M.S., Mayes, R.L.: Comparison of FRF and modal methods for combining experimental and analytical substructures. In: 25th International Modal Analysis Conference (IMAC XXV) (2007)
12. Rixen, D.: How measurement inaccuracies induce spurious peaks in frequency based substructuring. In: IMAC XXVI – International Modal Analysis Conference (2008). ISSN:0-912053-98-4
13. van der Seijs, M.V., de Klerk, D., Rixen, D.J.: General framework for transfer path analysis: history, theory and classification of techniques. Mech. Syst. Signal Process. **68–69**, 217–244 (2016). ISSN:10961216. https://doi.org/10.1016/j.ymssp.2015.08.004
14. van der Seijs, M.V.: Experimental dynamic substructuring, analysis and design strategies for vehicle development. PhD thesis, Delft University of Technology (2016)

Chapter 6
Real-Time Hybrid Substructuring Shake Table Test of a Seismically Excited Base Isolated Building

Muammer Avci and Richard Christenson

Abstract This paper present a real-time hybrid substructuring (RTHS) shake table test to evaluate the seismic performance of a base isolated building. Significant experimental research has been conducted on base isolators and dampers toward developing high fidelity numerical models. Shake table testing where the building superstructure is tested while the isolation layer is numerically modeled can allow for a range of isolation strategies to be examined for a single shake table experiment. Further, recent concerns in base isolation for long period, long duration earthquakes necessitate adding damping at the isolation layer which can allow higher frequency energy to be transmitted into the superstructure and can result in damage to structural and nonstructural components that can be difficult to numerically model and accurately predict. As such, physical testing of the superstructure while numerically modeling the isolation layer may be desired. The RTHS approach has been previously proposed for base isolated buildings, however, to date it has not been conducted on base isolated structure isolated at the ground level and where the isolation layer itself is numerically simulated. This configuration provides multiple challenges associated with higher physical substructure frequencies and a low numerical to physical mass ratio. This paper demonstrates a base isolated RTHS test with a scaled idealized 4-story superstructure building model placed directly onto a shake table and the isolation layer simulated in MATLAB/Simulink using a dSpace real-time controller.

Keywords Real-time hybrid simulation · Base isolation

6.1 Introduction

Earthquakes pose a serious threat to human lives and can causes damage to building structures. Seismic protective systems can protect structures. Base isolation is one of the most widely used and accepted types of seismic protective systems. Seismic base isolation reduces the damage by shifting the resonant frequency of the structure below the energy content of the ground motion [1–2]. This isolation both reduces the inter-story drifts of the superstructure and reduces the floor accelerations throughout the building. Significant research has been conducted on base isolators and high fidelity numerical models are available for base isolator devices. Long period long duration earthquakes can result in large displacements across the isolation layer and result in detrimental pounding effects. One solution is to increase the damping across the isolation layer, however, that reduces the effectiveness of the isolation, and raises concerns of damage to nonstructural components from larger amplitude and higher frequency vibration of the superstructure. This nonstructural damage can be difficult to numerically model and predict and physical testing may be desired to evaluate performance. Further, by numerically modeling the isolation layer it can be easier and less costly to examine a wide range of isolator devices.

Real-time hybrid substructuring (RTHS) provides the capability to isolate and physically test the critical components of a structure at the design phase of the system by including the dynamic interaction using a numerical representation of the remainder structural system. Early research in RTHS showed this method as a useful tool in earthquake and structural engineering [3–8]. Recent advances in RTHS have been made possible by increased computer power, digital signal

M. Avci
Grad Student University of Connecticut, Storrs, CT, USA

R. Christenson (✉)
Department of Civil and Environmental Engineering, University of Connecticut, Storrs, CT, USA
e-mail: rchriste@engr.uconn.edu

A. Linderholt et al. (eds.), *Dynamics of Coupled Structures, Volume 4*, Conference Proceedings of the Society for Experimental Mechanics Series, https://doi.org/10.1007/978-3-319-74654-8_6

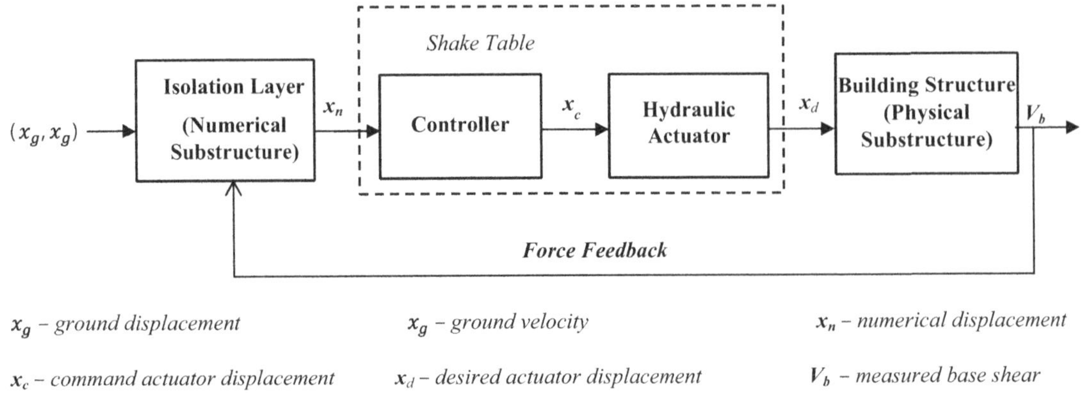

Fig. 6.1 Block diagram for real-time hybrid substructuring of a base isolated structure

processing hardware/software, and hydraulic control hardware. RTHS allows a structural dynamic system to be partitioned into physical and numerical substructures. The substructure of interest is physically tested, while the substructure that is better understood is simulated in real-time using analytical or numerical models. Considering the availability of numerical models for many of the isolators and dampers considered for the isolation layer and the uncertainty of the damage accrued in the superstructure when isolation is balanced with the threat of long-period long-duration earthquakes, this paper considered a RTHS configuration where the physical substructure is the building superstructure and the numerical superstructure is the isolation layer. This choice of substructuring will also facilitate the exploration of different isolation layer configurations. Figure 6.1 shows a block diagram of a RTHS closed loop test for a base isolated structure.

6.2 RTHS Shake Table Testing

Shake table testing, which has been used extensively, is an experimental technique to identify the seismic behavior of a structural system [9]. In this testing, the specimens on the shake table are subjected to excitations representative of actual earthquakes and the results are often considered more representative of the behavior during an actual earthquake. By combining experimental testing and numerical simulation RTHS is an attractive alternative method to the traditional shake table testing to provide efficient and cost effective methods of considering larger systems than can be tested on the shake Table [10]. For example, Ashasi-Sorkhabi et al. [11] conducted a study implementing RHTS with a shake table where a tuned liquid damper (TLD) was physically tested as connected to a shake table and the building structure below the TLD numerically modeled. Further, Zhang, et al. [12] demonstrated RTHS for a building with a mid-height isolation layer. In that research the upper portion of the structure is physically tested along with the isolation layer. The bottom portion of the building structure is numerically modeled. In the prior studies of RTHS shake table testing the substructure configuration whereby the building is physical and the isolation layer is numerical has not been conducted. This configuration will be examined here.

In order to conduct RTHS for a base isolated structure, an alternative model is desired where the isolation layer and building structure are modeled separately and interfaced appropriately. The derivation of numerical and substructure models necessary for RTHS are provided here. This system consists of a single degree of freedom (SDOF) numerical substructure corresponding to the lower DOF (which is the isolation layer in this study) and a 4DOF physical substructure above that represents a 4-story superstructure building model. In this way, the numerical model is a feedback system where the inputs for the SDOF isolation layer are the ground displacement, ground velocity and the base shear force coming as an output from the 4DOF building structure.

6.3 Numerical Substructure

A SDOF model which consist of base mass, m_1, isolator stiffness, k_1, and isolator damping, c_1, is used as a numerical model of the base isolation layer. This model can be extended to provide more complex and nonlinear realizations. Frequency of the base is 0.56 *Hz* and damping ratio is set to 10%, ζ (*zeta*) = 0.1 . The second order ordinary differential equation (ODE) for the isolation layer is

$$m_1\ddot{x}_1^a + c_1\left(\dot{x}_1^a - \dot{x}_g\right) + k_1\left(x_1^a - x_g\right) - V_b = 0 \tag{6.1}$$

$$\ddot{x}_1^a = -\frac{k_1}{m_1}x_1^a - \frac{c_1}{m_1}\dot{x}_1^a + \frac{k_1}{m_1}x_g + \frac{c_1}{m_1}\dot{x}_g + \frac{V_b}{m_1} \tag{6.2}$$

where m_1, k_1 and c_1 are the mass, stiffness and damping of the isolation layer, respectively. The equations of motion for the isolation layer can be written in state space form as

$$\dot{Z}_n = A_n \bullet Z_n + B_n \bullet U_n \tag{6.3}$$

$$Y_n = C_n \bullet Z_n + D_n \bullet U_n \tag{6.4}$$

where $A_n = \begin{bmatrix} 0 & 1 \\ -\frac{k_1}{m_1} & -\frac{c_1}{m_1} \end{bmatrix}$, $Z_n = \begin{bmatrix} x_1^a \\ \dot{x}_1^a \end{bmatrix}$, $B_n = \begin{bmatrix} 0 & 0 & 0 \\ \frac{k_1}{m_1} & \frac{c_1}{m_1} & \frac{1}{m_1} \end{bmatrix}$, $U_n = \begin{bmatrix} x_g \\ \dot{x}_g \\ V_b \end{bmatrix}$, $C_n = \begin{bmatrix} 1 & 0 \\ 0 & 1 \end{bmatrix}$ and $D_n = \begin{bmatrix} 0 & 0 & 0 \\ 0 & 0 & 0 \end{bmatrix}$.

The outputs of the base are the absolute displacement and velocity, which are in turn the inputs for the building superstructure. Further, the base shear which is measured from the building structure is feed back into the isolation layer as an input.

6.4 Physical Substrucure

A scaled and idealized 4-story superstructure building is considered in this study as the physical substructure. Four 0.5 in diameter steel threaded rods are used as columns, and the floor height is set to 6 in. The floors consist of 24 in square by 1 in thick steel plates held in place with nuts and washers. Stiffness calculated by using material properties of the threaded rods which is $K = (4 * 12 * E * I)/h^3$ where $I = \frac{1}{4} * pi * \left(\frac{d}{2}\right)^4$ is the moment of inertia (d is the diameter of the rods) and h is the story height. The mass of the each floor is 0.4222 lbs-s2/in, (81.5 lb). A picture of the building structure is shown in Fig. 6.2.

A state space model of the building model can be developed to compare the full numerical results with the physically tested structure. The ODE for the building model is

$$M_2 * \ddot{X}_2^a + C_2 * \left(\dot{X}_2^a - \Gamma * \dot{X}_1^a\right) + K_2 * \left(X_2^a - \Gamma * X_1^a\right) = 0 \tag{6.5}$$

$$\ddot{X}_2^a = -M_2^{-1} \bullet C_2 * \left(\dot{X}_2^a - \Gamma * \dot{X}_1^a\right) - M_2^{-1} \bullet K_2 * \left(X_2^a - \Gamma * X_1^a\right) \tag{6.6}$$

where, M_2, K_2 and C_2 are the mass, stiffness and damping matrix of the building structure, respectively.

The outputs of the isolation layer correspond to the inputs for the (4DOF) building model. The SS representation is given by

$$\dot{Z}_p = A_p \bullet Z_p + B_p \bullet U_p \tag{6.7}$$

$$Y_p = C_p \bullet Z_p + D_p \bullet U_p \tag{6.8}$$

Fig. 6.2 The four story structure physically tested

where $A_p = \begin{bmatrix} \mathbf{0} & I \\ -M_2^{-1} \bullet K_2 & -M_2^{-1} \bullet C_2 \end{bmatrix} Z_p = \begin{bmatrix} X_1{}^a \\ \dot{X}_1^a \end{bmatrix}$, $B_p = \begin{bmatrix} \mathbf{0} & \mathbf{0} \\ M_2^{-1} \bullet K_2 \bullet \Gamma & M_2^{-1} \bullet C \bullet \Gamma \end{bmatrix} U_p \begin{bmatrix} x_1{}^a \\ \dot{x}_1^a \end{bmatrix}$, $C_p = [K_2 \; C_2] \; D_p = [-K_2 \bullet \Gamma \; -K_2 \bullet \Gamma]$.

Figure 6.3 illustrates the comparison of the transfer function of the numerical model and physically tested building structure.

6.5 Transfer System

The transfer system used in this RTHS test is a 5 ft. by 5 ft. uniaxial shake Table. A picture of the shake table with structure mounted is shown in Fig. 6.4.

Since RTHS is a feedback system, the frequency-dependent dynamics of the shake table transfer system in the feedback loop can lead to inaccuracy in the actuator tracking and potential instability during closed-loop testing. The effect of time delay in the actuator dynamics on RTHS testing using hydraulic actuators was initially considered by Horiuchi et al., 1996 [6]. Figure 6.5 provides the measured transfer function of the servo-hydraulic actuator for the uniaxial shake table, which shows approximately 20 msec. of apparent time delay.

There are various approaches to mitigate the effect of time delays and lags. One of the most effective is to compensate the dynamic of the servo-hydraulic system for accurate displacement tracking. An inverse compensation method is used to reduce the time delay to ensure the system stability. Different types of compensation techniques have been developed for compensating the actuator time delay to enhance the close-loop stability and performance of RTHS. The techniques vary from polynomial extrapolation in Horiuchi et al., 1999 [13] and inverse compensation in Chen and Ricles, 2009 [14] to decrease the actuator delay as well as adaptive techniques in Chen et al., 2010 [15] and Chae et al., 2013 [16]. Carrion and Spencer, 2007 [17] used a controls approach to develop model-based feedforward-feedback control for actuator dynamics compensation. The ultimate objective of these compensation techniques is to ensure effective displacement tracking of the

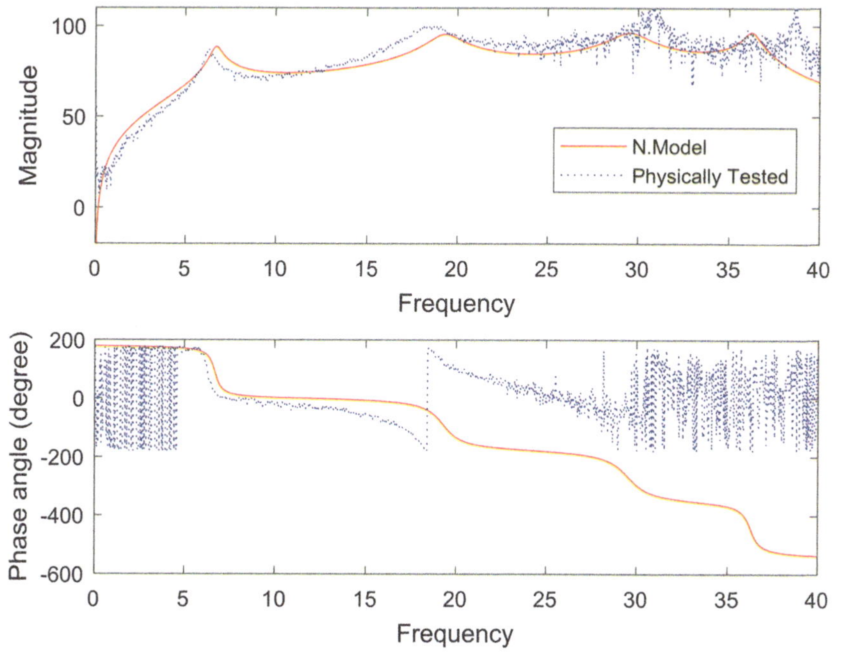

Fig. 6.3 Building model transfer function for numerical model and physical substructure

Fig. 6.4 Shake table with the physical structure

actuator transfer system, decreasing both magnitude and phase error over the RTHS control band. This research utilizes model based feedforward inverse compensation. Figure 6.5 shows the differences between compensated and uncompensated transfer function of the uniaxial shake table that is in the University of Connecticut Structural Lab.

It can be seen from Fig. 6.5 that the apparent time delay is reduced from 20 msec to 5 msec which provides a significant improvement for system stability.

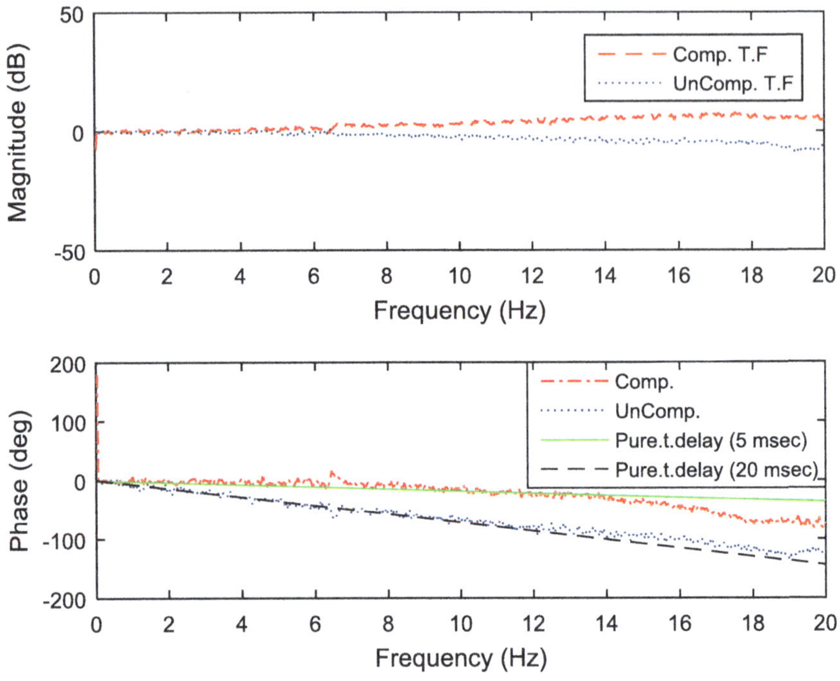

Fig. 6.5 Compensated versus uncompensated transfer function of the shake table transfer system

6.6 Results

Tests were conducted at the University of Connecticut Shock and Vibration Laboratory. The four story shear frame structure is mounted on the shake table as the building model and physical substructure. Promising results have been obtained from the RTHS test and the some of the result presented as follows. In Fig. 6.5 both full simulation of the system and the RTHS results of the full system has been show. Fully simulated and RTHS results match very well for both isolated and non-isolated cases. Figure 6.6 also illustrates the effect of isolation layer on the transfer function of the top floor acceleration. The magnitude of the transfer function of the top floor acceleration is observed to be decreased about 30 dB when the structure is isolated. Effect of the isolation layer can be seen clearly in the RTHS results.

Figure 6.6 show the transfer function of the top floor acceleration up to 20 Hz which is the bandwidth of the ground acceleration used in the RTHS test, however, Fig. 6.7 shows the results up to 40 Hz which captures all four lateral modes of the superstructure.

6.7 Conclusions

In this paper the RTHS approach has been proposed for base isolated buildings isolated at the ground level and where the isolation layer numerically simulated and the physical structure tested on the shake table. The isolation layer is numerically modeled and the absolute displacement and velocity used instead of acceleration as an input for the system which is a new approach to conduct the RTHS. The RTHS test is conducted at the University of Connecticut Shock and Vibration Laboratory using a uniaxial shake table. The results of RHTS are very promising that in the same RHTS experimental setup different kinds of isolation layers can be tested.

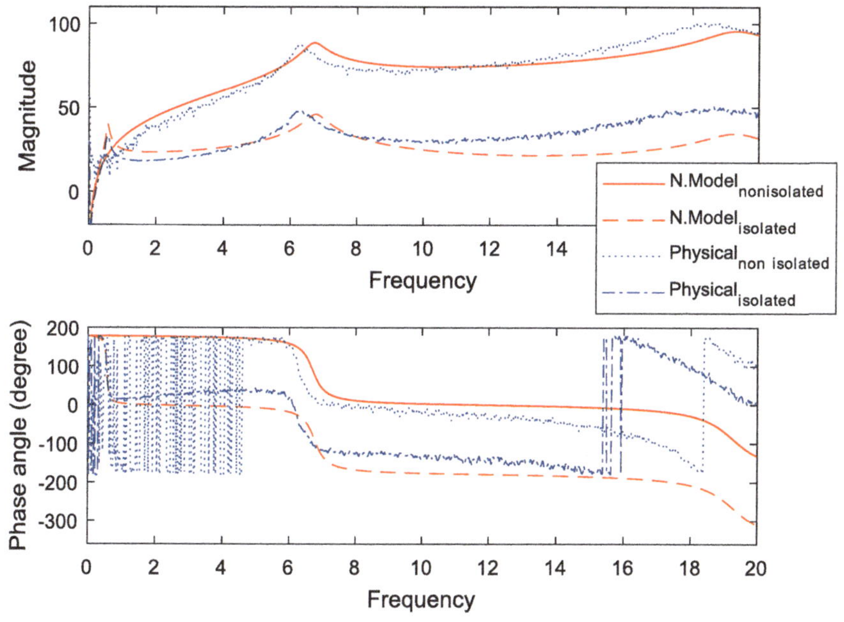

Fig. 6.6 Transfer functions of the top floor acceleration up to 20 Hz

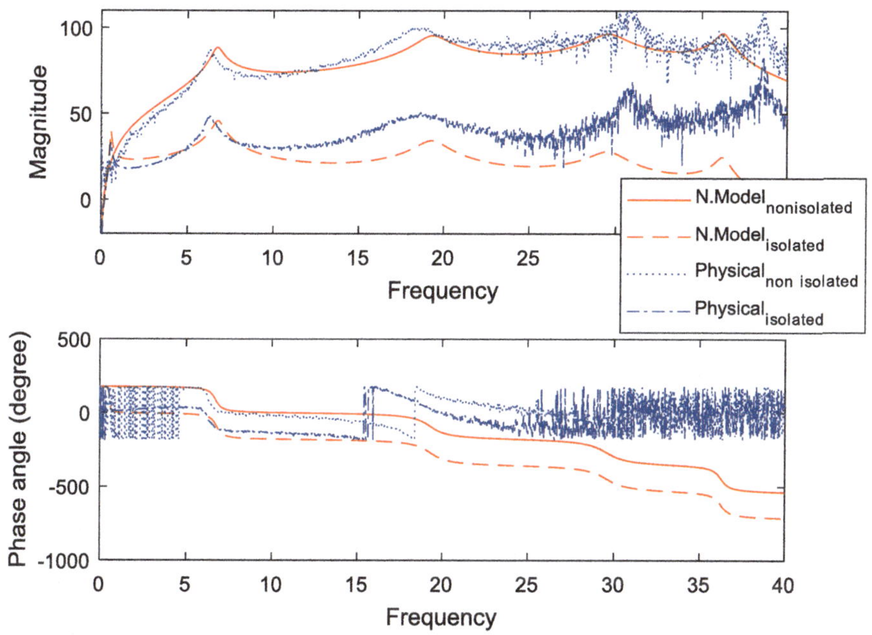

Fig. 6.7 Transfer functions of the top floor acceleration up to 40 Hz

References

1. Skinner, R.I., Robinson, W.H., McVerry, G.H.: An Introduction to Seismic Isolation. Wiley, Chichester (1993)
2. Naeim, F., Kelly, J.M.: Design of Seismic Isolated Structures: from Theory to Practice. Wiley, Chichester (1999)
3. Nakashima, M., Masaoka, N.: Real time on-line test for MDOF systems. Earthq. Eng. Struct. Dyn. **28**, 393–420 (1999)
4. Darby, P., Blakeborough, A., Williams, M.S.: Real time substructure test using hydraulic actuator. J. Eng. Mech. **125**(10), 1133–1139 (1999)
5. Dimig, J., Shield, C., French, C., Bailey, F., Clark, A.: Effective force testing: a method of Siesmic simulation for structural testing. J. Struct. Eng. **125**(9), 1028–1037 (1999)
6. Horiuchi, T., Inoue, M., Konno, T., Namita, Y.: Real time hybrid experimental system with actuator delay compensation and it's applications to a piping system with energy absorber. Earthq. Eng. Struct. Dyn. **28**(10), 1121–1141 (1999)

7. Christenson, R., Lin, Y.Z.: Real time hybrid simulation of a siesmically excited structure with large scale magneto Rheological Fluid Dampers. In: Hybrid Simulation Theory, Implementation and Applications
8. Jiang, Z., Kim, S.J., Plude, S., Christenson, R.: Real time hybrid simulation of a complex bridge model with MR dampers using the convolution integral method. J. Smart Mater. Struct. **22**(10), 31–59 (2013)
9. Roth, S., Cheney, D.: Directory of International Earth-Quake Engineering Research Facilities. SRI International Center for Science, Technology and Educational Development Policy Division (2001)
10. Nakashima, M., Kato, H., Takaoka, E.: Development of real-time pseudo dynamic testing. Earthq. Eng. Struct. Dyn. **21**(1), 79–92 (1992)
11. Ashasi-Sorkhabi, A., Malekghasemi, H., Mercan, O.: Implementation and verification of real-time hybrid simulation (RTHS) using a shake table for research and education. J. Vib. Control. **21**(8), 1459–1472 (2015)
12. Zhang, R., Phillips, B.M., Taniguchi, S., Ikenaga, M., Ikago, K.: Shake Table Real-Time Hybrid Simulation Techniques for the Performance Evaluation of Buildings with Inter-Story Isolation. Monitoring, Structural Control and Health (2017)
13. Horiuchi, T., M. Nakagawa, M. Sugano, Konno, T.: Development of a real-time hybrid experimental system with actuator delay compensation. In: 11th World Conference on Earthquake Engineering, Paper No. 660 (1996)
14. Chen, C., Ricles, J.: Analysis of actuator delay compensation methods for real-time testing. Eng. Struct. **31**, 2643–2655 (2009)
15. Chen, C., Ricles, J.: Tracking error-based Servohydraulic actuator adaptive compensation for real-time hybrid simulation. J. Struct. Eng. **136**(4), 432–440 (2010)
16. Chae, Y., Kazemibidokhti, K., Ricles, J.M.: Adaptive time series compensator for delay compensation of servo-hydraulic actuator systems for real-time hybrid simulation. Earthq. Eng. Struct. Dyn. **42**(11), 1697–1715 (2013)
17. Carrion, J.E., Spencer, B.: Model-based strategies for real-time hybrid testing." NSEL Report No NSEL-006 (2007)

Chapter 7
A Comparison of Two Reduction Techniques for Forced Response of Shrouded Blades with Contact Interfaces

Fahimeh Mashayekhi, Stefano Zucca, and Ali Salehzadeh Nobari

Abstract Two reduction methods for dynamic analysis of structures with local nonlinearity are compared. Dual and primal formulation have the same projection basis including flexibility residual attachment modes and free interface modes, but there are significant differences in their implementation. Both methods can be applied to nonlinear forced response analysis of turbine blades with contact interfaces in shroud. In this study, the shroud contact elements are employed using the adequate description of friction and 3D tangential coupled contact forces considering the effect of normal load variation. In order to examine and compare the accuracy of the two formulations, a rod and a simplified shrouded turbine blade was considered as case studies.

Keywords Reduced order models · Primal formulation · Dual formulation · Component mode synthesis · Shroud contact · Forced response

Nomenclature

[M],[C], and [K]	Mass, damping, and stiffness matrix
X, F	Displacement and Force vector
[Φ]	Matrix of free interface dynamic modes
[Ψ], $[\Psi]_{ar}$	Matrix of attachment modes and residual attachment modes
ω_p and Φ_p	p_{th} eigenvalue and eigenvectors of free interface body
$*_N$ and $*_L$	Pertaining to nonlinear(interface) and linear(inner) dofs of substructure
R	Reduction matrix
h	Harmonic Component Number
α	Mass proportional damping
$*_C$ and $*_E$	Pertaining to Contact and excitation
$\bar{X}^h$	Vector of Fourier coefficient of h_{th} harmonic of X
ω	Circular frequency
$*_{\underline{n}}$	Pertaining to n_{th} contact nodes
$[\tilde{A}]$	Pertaining to characteristic matrix [A] of system with interfaces added stiffness in fully suck condition
$[k_{c,n}]$	nth sub matrix regarding the added stiffness to nth contact node
k_t, k_n	Tangential and normal contact stiffness parameters
α	Proportional damping constant
N_0	Static normal preload on contact area

7.1 Introduction

Bladed disks are subject to not only very high static load but also great periodic excitations which make the study of blades dynamic behavior inevitable. Resulted high amplitude of structure in forced response may create stresses that lead to High Cycle Fatigue (HCF). This vibration must be prevented to avoid failure. Accurate prediction of dynamic stress needs detailed

F. Mashayekhi (✉) · A. S. Nobari
Department of Aerospace Engineering, Amirkabir University of Technology, Tehran, Iran

S. Zucca
Department of Mechanical and Aerospace Engineering, Politecnico di Torino, Torino, Italy

A. Linderholt et al. (eds.), *Dynamics of Coupled Structures, Volume 4*, Conference Proceedings of the Society for Experimental Mechanics Series, https://doi.org/10.1007/978-3-319-74654-8_7

finite element model of bladed disk that are very large and usually are not computationally affordable. On the other hand, frictional interfaces like blade roots, contact between shrouds, and underplatform dampers lead to nonlinear behavior of blades while their magnificent influences on resonant vibration and also on computational time cannot be ignored.

Despite growing high-performance computers, designers require tools to speed up the nonlinear dynamic analysis of bladed disk to achieve better understanding of structure behavior. Many studies like paper published by Petrov [1] have shown that Multi Harmonic Balance Method (MHBM) is an efficient method to increase computational efficiency by solving the governing equation in frequency domain, although, the number of unknown grows proportionally with the number of retained harmonics.

MHBM is an effective but not enough tool because of bladed disks huge number of degree of freedoms (dof). The best way of decreasing computational costs is to reduce the dimension of problem using reduction techniques. Component Mode Synthesis (CMS) methods are commonly used for model order reduction of structure with contact. In order to this, structure is initially divided to substructures, then modal properties of each part are extracted, and finally the whole structure matrices are assembled to derive the response. In CMS methods, each substructure displacement vector is partitioned into inner and interface dofs, where latter are retained as master dofs in the reduced model [2]. This feature makes them applicable for the dynamic analysis of structures with intermittent contacts, such as bladed disks [3], because the computed interface displacements can be used to contact force calculation.

In reduced order modeling, a set of generalized coordinates are used to project the governing equations to a subspace. The Guyan reduction model [4] relies on the static condensation approach, while in Component Mode Synthesis approaches, described by Tran [5], mode shapes computed with special boundary conditions at interface dofs are also employed.

One of the most exploited method is the Craig-Bampton [2] method, in which the reduction basis is made of constraint and fixed-interface modes, respectively computed as the static deformed shapes due to unit displacements at the interface nodes and as linear normal modes of the body with zero displacements at the interface nodes.

Another approach is the so-called Rubin (or Rubin-Martinez) method [6], in which the reduction basis is made of attachment and free interface modes, respectively computed as the static deformed shapes due to unit forces at the interface nodes and as linear normal modes of the body with free displacements at the interface nodes.

In Zucca [7], it was shown that the dual formulation of the Rubin method, used by Rixen [8] to develop the so-called Dual Craig-Bampton method, can be successfully applied to the forced response analysis of floating and non-floating structures with contact interfaces.

Both the Rubin and the Dual-Craig Bampton method use the same reduction basis, although the final ROMs used for the nonlinear forced response by MHBM are different.

In this paper, the Rubin method, hence referred to as the primal approach and the dual approach used by Zucca [7] are compared to each other in case of turbine blades with friction contact at shrouds.

Two formulations are used for the dual approach, as introduced by Zucca [7], in the first one the reduction basis is obtained with open contacts, while in the second one fully-stuck contacts are used.

7.2 Methodology

The governing equation of motion of a discrete dynamic system in time domain is as following:

$$M\ddot{X} + C\dot{X} + KX = F_E + F_C$$

$$F_C = \begin{bmatrix} F_{C,N} \\ 0 \end{bmatrix}, X = \begin{bmatrix} X_N \\ X_L \end{bmatrix} \tag{7.1}$$

In this equation, M, C, and K are mass, damping and stiffness matrices of the system, respectively. X is the vector of displacement and the dot script representing the time derivative. F_E and F_C correspond to external and contact forces. It should be noticed that the force and displacement vectors are divided into linear (L) and nonlinear (N) partitions related to inner and interface dofs of the system. In this study, proportional damping is also considered as $C = \alpha M$ where α is a constant.

7.2.1 *Harmonic Balance Method*

Local nonlinearity in contact area makes the solution in time domain very time consuming. Considering that the steady state part of the solution is of interest, the equations are re-written in the frequency domain using MHBM. In this way, the differential equations are turned to a set of nonlinear algebraic equations by approximating the harmonic components by a Fourier series truncated at H_{th} term and imposing the equilibrium of the Fourier coefficients of each retained harmonic. The governing Eq. 7.1 in frequency domain will be

$$\left[-(h\omega)^2 M + ih\omega C + K\right] \bar{X}^{(h)} = \bar{F}_E^{(h)} + \bar{F}_C^{(h)} \tag{7.2}$$

where $\bar{Y}^{(h)}$ is the vector of Fourier coefficient of h_{th} harmonic component of Y.

7.2.2 *Reduced Order Modeling*

Finite element models of turbine blades for industrial applications are so large that even their linear dynamic analysis needs huge computational time and memory. This issue makes the use of reduced order models necessary to perform nonlinear or sensitivity analysis. Component Mode Synthesis (CMS) methods are frequently used in nonlinear analysis of turbine blades with friction contacts. CMS methods can be classified according to reduction basis into free or fixed interface elastic modes. A common type of projection basis, only valid for non-floating assemblies, consists of free interface modes (Φ) and residual flexibility attachment modes (Ψ_{ar}) when the dofs of the body in contact is expressed as Eq. 7.3 for each harmonic component after using MHBM.

$$\bar{X}^{(h)} = \Psi_{ar} \bar{F}_N^{(h)} + \Phi \bar{q}^{(h)},$$

$$\Psi_{ar} = \begin{bmatrix} \Psi_{ar,NN} & \Psi_{ar,NL}(5) \\ \Psi_{ar,LN} & \Psi_{ar,LL} \end{bmatrix}, \Phi = \begin{bmatrix} \Phi_N \\ \Phi_L \end{bmatrix} \tag{7.3}$$

when q is the normal modes contribution factors (modal dofs). Free interface modes are also determined by modal analysis of the system with free boundary condition at the interface dofs. While attachment modes (Ψ) are static displacement of the component under unit force at each contact dofs (i.e. columns of the flexibility matrix). Because of the similar elastic information in these modes, the residual flexibility attachment modes are calculated as Eq. 7.4:

$$\Psi_{ar} = \Psi - \sum_{p=1}^{m} \frac{\Phi_p \Phi_p^T}{\omega_p^2} \tag{7.4}$$

In the following, two approaches based on this displacement shapes are described.

7.2.2.1 Primal Approach

In the primal approach called Rubin's [9], Martinez's [10], or Rubin-Martinez's [11] in different references, reduced matrices of the system are generated and then the ROM is used to compute the nonlinear forced response of the system.

Based on Eq. 7.3, the reduction matrix (R) is calculated as

$$R = \begin{bmatrix} I & 0 \\ \Psi_{ar,LN}(\Psi_{ar,NN})^{-1} & \Phi_L + \Psi_{ar,LN}(\Psi_{ar,NN})^{-1}\Phi_N \end{bmatrix}, \tag{7.5}$$

The reduced matrices and vectors are defined as:

$$A_r = R^T A R,$$

$$y = R^T Y, \tag{7.6}$$

where A is a structural matrix (M, C and K) and Y is force or displacement vector.

The size of the model is reduced with respect to the original one because only a subset of free interface modes is retained in the projection basis.

After transformation, the governing equation in the frequency domain become

$$\left[-(h\omega)^2 M_r + ih\omega C_r + K_r\right] \bar{x}^{(h)} = \bar{f}_E^{(h)} + \bar{f}_C^{(h)} \tag{7.7}$$

where $\bar{x}^{(h)} = \left[\bar{X}_N^{(h)}; \bar{q}^{(h)}\right]$ is vector of dofs of the reduced system, including interface dofs and modal coordinates.

7.2.2.2 Dual Approach

The dual approach uses the same reduction basis as the primal one. However, the reduced mass and stiffness matrices are not calculated anymore and Eq. 7.3 is used directly for forced response analysis, by replacing $\bar{q}^{(h)}$ by the response of the linear system in modal coordinate as below [7]:

$$\bar{X}_N^{(h)} = \Psi_{ar,NN}\left(\bar{F}_{C,N}^{(h)} + \bar{F}_{E,N}^{(h)}\right) + \Phi_N \operatorname{diag}\left[\frac{1}{\omega_p^2 - (h\omega)^2 + i\alpha_p h\omega}\right]\left(\Phi_N^T \bar{F}_{C,N}^{(h)} + \Phi_E^T \bar{F}_E^{(h)}\right) \tag{7.8}$$

This equation can be solved by a nonlinear iterative solver when the reduction basis is determined.

Another formulation for the dual approach is possible, as introduced by Zucca [7] for the case of floating structures. According to this formulation the final set of equations is:

$$\bar{X}_N^{(h)} = \tilde{\Psi}_{ar,NN}\left(\tilde{\bar{F}}_{C,N}^{(h)} + \bar{F}_{E,N}^{(h)}\right) + \tilde{\Phi}_N \operatorname{diag}\left[\frac{1}{\tilde{\omega}_p^2 - (h\omega)^2 + i\alpha_p h\omega}\right]\left(\tilde{\Phi}_N^T \tilde{\bar{F}}_{C,N}^{(h)} + \tilde{\Phi}_E^T \bar{F}_E^{(h)}\right) \tag{7.9}$$

where $\tilde{\Psi}_{ar}$ and $\tilde{\Phi}$ are the vector of residual flexibility attachment modes and free interface modes computed with fully stuck contacts, while the vector $\tilde{\bar{F}}_{C,N}^{(h)}$ is modified contact force vector and

$$\tilde{\bar{F}}_{C,n}^{(h)}\left(\bar{X}_{N,n}^{(h)}\right) = \bar{F}_{C,n}^{(h)}\left(\bar{X}_{N,n}^{(h)}\right) + [k_{c,n}]\,\bar{X}_{N,n}^{(h)} \tag{7.10}$$

represent the modified contact force at the n_{th} contact node, and $[k_{c,\,n}]$ is the matrix of contact stiffness corresponding to the n_{th} contact node.

7.2.3 Contact Force

Contact forces which make the dynamic behavior of structure nonlinear because of their dependency on displacement amplitude are determined using Alternating Frequency Time (AFT) method and considering an appropriate contact model.

In AFT method (Fig. 7.1), the approximated displacement Fourier coefficients of nonlinear dofs in each iteration are transferred to time domain using Inverse Fast Fourier Transform (IFFT) in order to calculate the interface forces using the defined contact model as the core of AFT method. In this paper, the 3D contact model represented by Afzal [12] is used. This model considers a coupled 2D tangential motion in the plane of interface and a normal motion perpendicular to the tangential plane. Finally, Fourier transform are applied on calculated forces in time domain. This procedure is repeated in each step of iterative nonlinear solver.

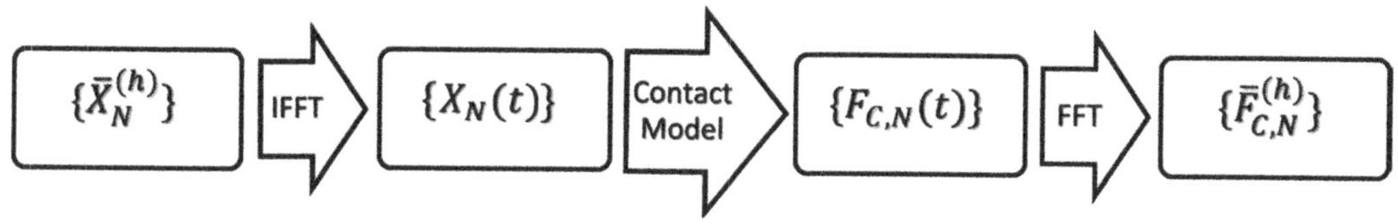

Fig. 7.1 The AFT method flowchart

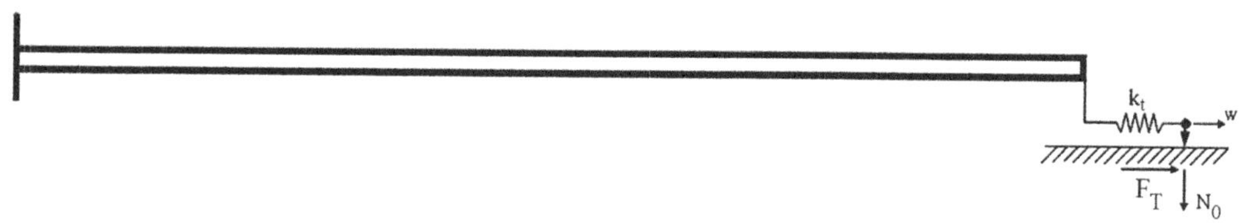

Fig. 7.2 Rod model with outer friction contact

In order to reduce the computational time of the Jacobian matrix, which is typically the most computationally expensive step, the analytical formulation of the Jacobian matrix formulated by Afzal [12] is implemented.

7.3 Results

To compare the efficiency of primal and dual approaches, they are implemented for the forced response analysis of two test cases:

- a rod with a friction contact at its tip
- a shrouded blade.

7.3.1 Rod

The forced response of the rod is first investigated. It is 20 dofs system with axial motion, clamped at one end node and with friction contact at the other end node (Fig. 7.2). Because of the 1D kinematics the normal load N0 is constant during vibration. The nominal contact stiffness is $k_t = 1e3$ N/mm and the friction coefficient $\mu = 0.5$.

Modulus of elasticity and density of the rod is as $E = 2e5$Mpa and $\rho = 7800 kg/m^3$. The rod has the length of 200 mm and cross section area of 10 mm^2.

A preliminary analysis has been performed in order to plot the backbone curve (Fig. 7.3) of the structure, where the maximum vibration amplitude is plotted versus the static normal pre-load N0.

Thanks to the backbone curve, four different levels of pre-load have been identified as representative of different contact conditions:

- N0 = 0.5 N (dominant slip conditions)
- N0 = 2 N (optimum value)
- N0 = 10 N (dominant stuck conditions, also called microslip conditions in case of multiple contact points)
- N0 = 60 N (fully stuck contact)

The three different approaches:

- Primal formulation (in the picture referred to as Primal)
- Dual formulation with open contacts (in the pictures referred to as Dual-O)
- Dual formulation with stuck contacts (in the pictures referred to as Dual-S)

are implemented for all the 4 contact conditions and their performance investigated by increasing the number of free-interface modes included in the reduction basis.

Results are given in terms of resonance peak and frequency of the corresponding maximum vibration amplitude.

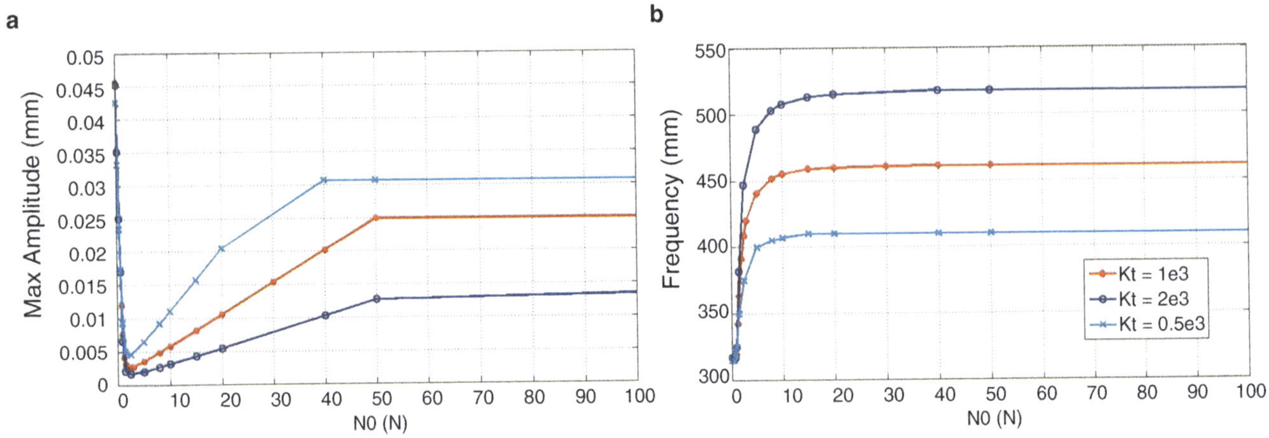

Fig. 7.3 Maximum amplitude and resonance frequency backbone curves of the rod

Results in Fig. 7.4, which refer to the resonance frequency, show that Primal and Dual-O, which include the mode shapes of the structure with open contacts in the reduction basis, better represent the dynamics of the system for low values of static pre-load (N0 = 0.5 N and N0 = 2 N). On the contrary, if the dual approach is chosen, the Dual-S formulation becomes more accurate as the system behavior tends towards the full-stick conditions, since the mode shapes used in the Dual-S reduction basis are those of the fully-stuck system.

Nevertheless, with the nominal value of contact stiffness, the Primal approach proves to be always the best reduction strategy, with the exception of the full-stick case, since in that case the Dual-S formulation corresponds to the exact solution.

It is well known, that in structures with contact interfaces, the operating deformed shape of the structure depends on the contact conditions. Since the contact elasticity is modelled here by the contact stiffness k_t, a second investigation is performed for different values of k_t.

As shown in Fig. 7.5, the performance of the Primal approach gets worse as the value k_t and the number of free interface modes necessary to accurately predict the system dynamics increases. This behavior is confirmed by Fig. 7.6, where the MAC between the 1st open mode and the first full-stick mode of the system is shown for different values of kt. As the value of k_t increases, the MAC decreases, since the full-stick mode becomes more and more different from the open mode, due to the extra-constraint at the contact node.

The effect of k_t on the accuracy of the Dual-S approach in full-stick conditions is not investigated, since the Dual-S approach is exact in full-stick.

Finally, in order to compare Primal and Dual-S performances, dominant stuck conditions are investigated for different values of contact stiffness k_t. In Fig. 7.7 it is observed that, apart from some numerical instability observed for the Dual-S, whose reason will be further investigated, the Dual-S approach gives better results than the Primal one.

7.3.2 Shrouded Blade

The second test case is a simplified shrouded blade model, shown in Fig. 7.8. This FE model is generated in ANSYS and is made of 2448 nodes and 745 elements. Unit point force is applied on two nodes and internal structural damping is introduced as mass proportional damping where α is equal to 50. Blade is composed of steel with Young's modulus E = 2E5 MPa, density $\rho = 7800 \text{kg/m}^3$ and a Poisson's ratio of $\nu = 0.3$. Contact friction coefficient is also assumed as $\mu = 0.5$.

Referring to the backbone curve shown in Fig. 7.9, corresponding to k_t = 1e6 N/m, it is preliminary observed that shrouded blades are usually designed to operate in the microslip regime (i.e. in the range included between the optimum pre-load value and full-stick conditions). Therefore, our analysis focused on N0 = 1 N, within that range of static pre-loads.

In Fig. 7.10, the convergence of the response at N0 = 1 N (in the microslip region of the backbone curve in Fig. 7.9) is shown. It is evident (Fig. 7.10b) that the Dual-O approach gives poorer results compared to Dual-S and Primal approaches, consistently with the results of the 1st test case.

Also (Fig. 7.10a) Dual-S looks more accurate than Primal formulations in frequency plots, while in terms of vibration amplitude the two methods look equivalent to each other.

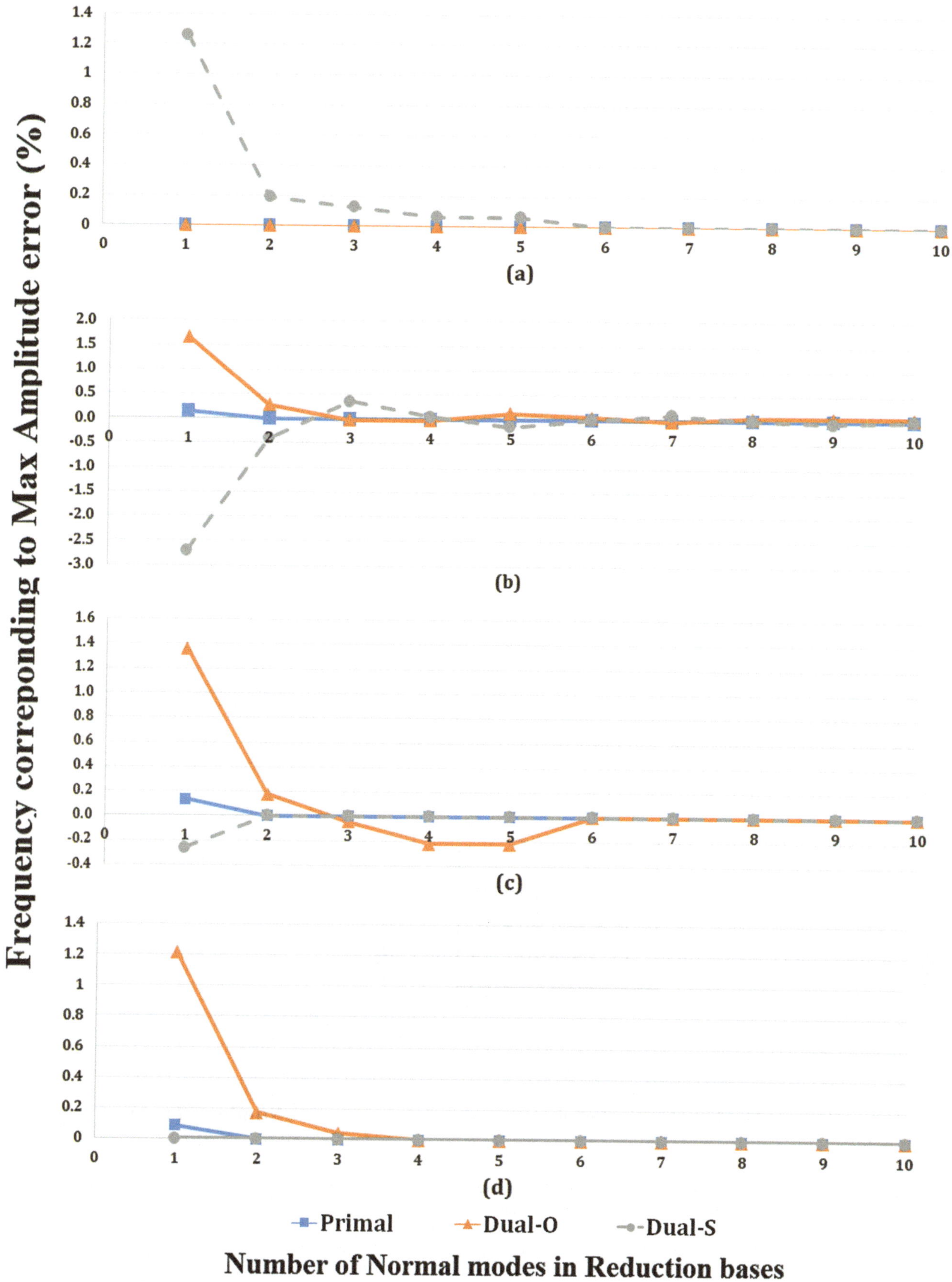

Fig. 7.4 Resonance frequency results: N0 = 0.5 N (**a**), N0 = 2 N (**b**), N0 = 10 N (**c**), and N0 = 10 N (**d**)

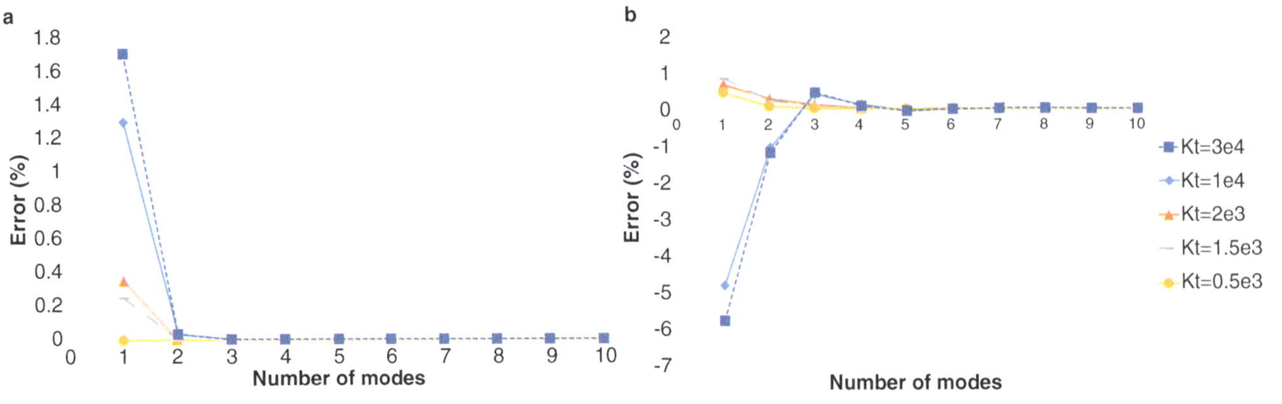

Fig. 7.5 (**a**) Frequency and, (**b**) Max amplitude results for Primal approach with different contact stiffness in full stick condition

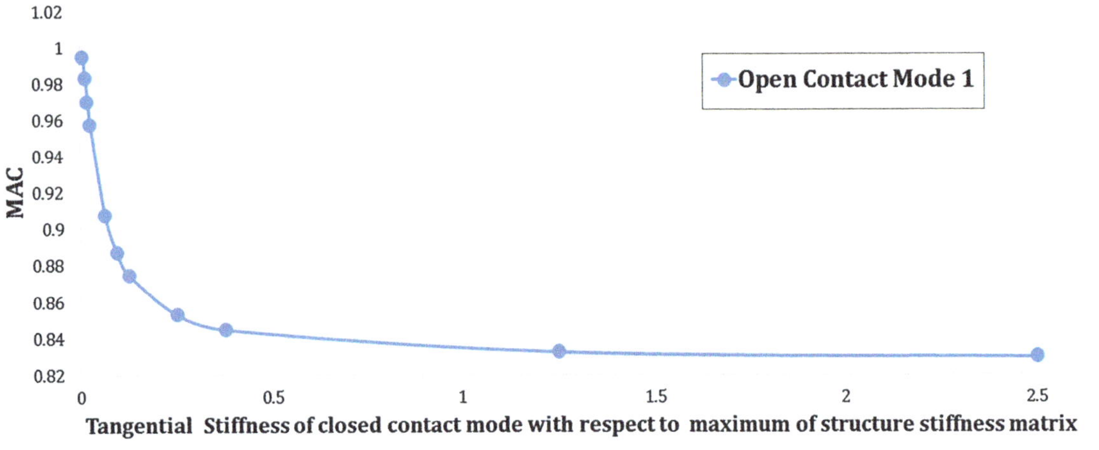

Fig. 7.6 MAC of fully stuck mode shapes with different value of tangential stiffness

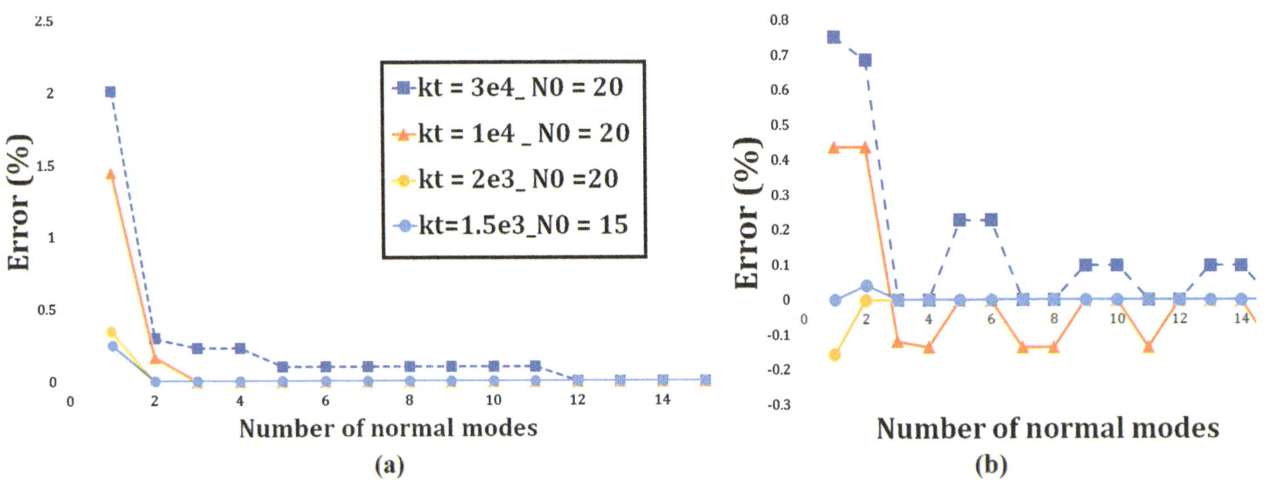

Fig. 7.7 Convergence of resonance frequency for Primal (left) and Dual-S (right) in case of stick-dominated contact conditions for different values of k_t

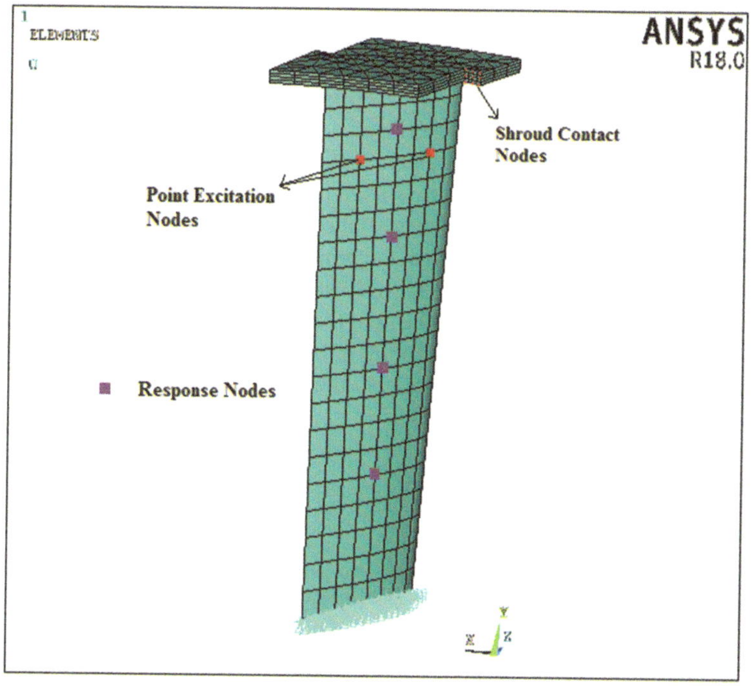

Fig. 7.8 FEM model of simplified shrouded blade

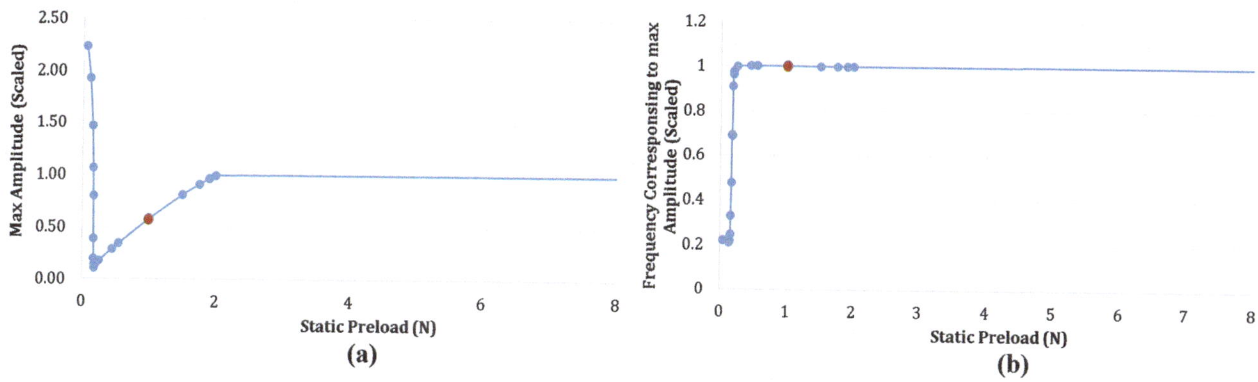

Fig. 7.9 Normalized backbone curves of the shrouded blade

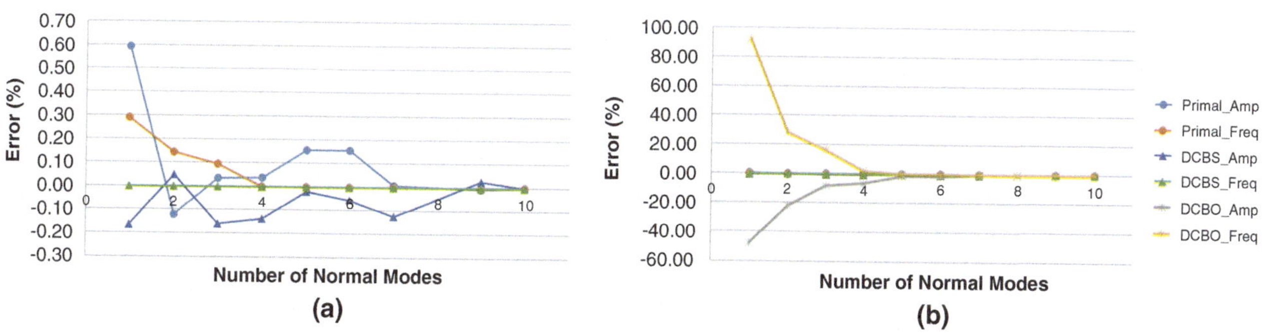

Fig. 7.10 Results of shrouded blade when $k_t = 1e6$ and $N_0 = 1$ obtained by different reduction technidues (a) without and (b) with Dual-O results

7.4 Conclusion

In this paper, two different reduction techniques, referred to as primal and dual approach, are compared to each other in case of forced response analysis of structures with contact interfaces.

The primal formulation and the dual approach with open contacts use the same projection basis (i.e. residual flexibility attachment modes and free interface modes). Results have shown that the primal formulation, which requires the calculation of the reduced order matrices gives more accurate results in all the contact conditions investigated in the paper.

In order to increase the accuracy of the dual formulation, the projection basis has been changed by using residual flexibility attachment modes and free interface modes computed with fully-stuck interfaces.

In this way, it has been observed that the accuracy of the dual formulation in case of microslip (also called dominant stuck conditions in the paper), which corresponds to the design point of blade shrouds, is comparable to the accuracy of the Primal formulation, although Dual-S ROMs predicted more accurately the resonance frequency.

References

1. Petrov, E.: A method for use of cyclic symmetry properties in analysis of nonlinear multiharmonic vibrations of bladed discs. In ASME Turbo Expo 2003, collocated with the 2003 International Joint Power Generation Conference, pp. 235–245. (2003)
2. Craig Jr., R.: A brief tutorial on substructure analysis and testing,# 325. Proc. IMAC-XVIII Conf. Struct. Dyn. **4062**, 899 (2000)
2. Charleux, D., Gibert, C., Thouverez, F., Dupeux, J.: Numerical and experimental study of friction damping blade attachments of rotating bladed disks. Int. J. Rotating Mach. **2006**, 1–13 (2006)
4. Guyan, R.J.: Reduction of stiffness and mass matrices. AIAA J. **3**(2), 380 (1965)
5. Tran, D.-M.: Component mode synthesis methods using partial interface modes: application to tuned and mistuned structures with cyclic symmetry. Comput. Struct. **87**(17), 1141–1153 (2009)
6. Rubin, S.: Improved component-mode representation for structural dynamic analysis. AIAA J. **13**(8), 995–1006 (1975)
7. Zucca, S.: On the dual Craig–Bampton method for the forced response of structures with contact interfaces. Nonlinear Dynamics. **87**(4), 2445–2455 (2017)
8. Rixen, D.J.: A dual Craig–Bampton method for dynamic substructuring. J. Comput. Appl. Math. **168**(1), 383–391 (2004)
8. Van der Valk, P.: Model reduction & interface modeling in dynamic substructuring: application to a multi-megawatt wind turbine, Master's thesis. TU Delft, The Netherlands, pp. 40–43 (2010)
10. Martinez, D.R., Carrie, T.G., Gregory, D.L., Miller, A.K.: Combined experimental/analytical modeling using component mode synthesis. Comput. Struct. **78**, 583–590 (1984)
11. Craig, R.R.: Coupling of substructures for dynamic analyses: an overview, structures, structural dynamics and material conference, (2000)
12. Afzal, M., Arteaga, I.L., Kari, L.: An analytical calculation of the Jacobian matrix for 3D friction contact model applied to turbine blade shroud contact. Comput. Struct. **177**, 204–217 (2016)

Chapter 8
Experimental-Analytical Substructuring of a Complicated Jointed Structure Using Nonlinear Modal Models

Daniel R. Roettgen, Benjamin R. Pacini, Randall L. Mayes, and Tyler F. Schoenherr

Abstract This work extends recent methods to calculate dynamic substructuring predictions of a weakly nonlinear structure using nonlinear pseudo-modal models. In previous works, constitutive joint models (such as the modal Iwan element) were used to capture the nonlinearity of each subcomponent on a mode-by-mode basis. This work uses simpler polynomial stiffness and damping elements to capture nonlinear dynamics from more diverse jointed connections including large continuous interfaces. The proposed method requires that the modes of the system remain distinct and uncoupled in the amplitude range of interest. A windowed sinusoidal loading is used to excite each experimental subcomponent mode in order to identify the nonlinear pseudo-modal models. This allows for a higher modal amplitude to be achieved when fitting these models and extends the applicable amplitude range of this method. Once subcomponent modal models have been experimentally extracted for each mode, the Transmission Simulator method is implemented to assemble the subcomponent models into a nonlinear assembled prediction. Numerical integration methods are used to evaluate this prediction compared to a truth test of the nonlinear assembly.

Keywords Nonlinear system identification · Nonlinear simulation · Structural dynamics · Modal model · Dynamic substructuring

Nomenclature

ω	Frequency in radians per second
f_n	Frequency in hertz
ζ	Modal damping ratio
ϕ	Mode shape matrix
$\ddot{q}$, $\dot{q}$, and q	Modal acceleration, velocity, and position
F	Physical force
$\ddot{x}$, $\dot{x}$, and x	Physical acceleration, velocity, and position
$\ddot{\eta}$, $\dot{\eta}$, and η	Generalized synthesized acceleration, velocity, and position
B	Physical Boolean constraints matrix
$\tilde{B}$	Modal Boolean constraints matrix
L	Synthetization matrix
$F_{j,i}$	Nonlinear modal forces for the ith subcomponent
ψ	Synthesized transformation matrix
$\ddot{p}$, $\dot{p}$, and p	Generalized uncoupled acceleration, velocity, and position
c_i	Damping coefficient for the ith term
k_i	Stiffness coefficient for the ith term
$\overline{U}$	Known restoring force vector
P	Modal response matrix
DOF	Degree-of-freedom
TS	Transmission simulator

Sandia National Laboratories is a multimission laboratory managed and operated by National Technology and Engineering Solutions of Sandia, LLC., a wholly owned subsidiary of Honeywell International, Inc., for the U.S. Department of Energy's National Nuclear Security Administration under contract DE-NA-0003525.

D. R. Roettgen · B. R. Pacini · R. L. Mayes (✉) · T. F. Schoenherr
Structural Dynamics Department, Sandia National Laboratories, Albuquerque, NM, USA
e-mail: drroett@sandia.gov; brpacin@sandia.gov; randy.mayes@awe.co.uk; tfschoe@sandia.gov

A. Linderholt et al. (eds.), *Dynamics of Coupled Structures, Volume 4*, Conference Proceedings of the Society for Experimental Mechanics Series, https://doi.org/10.1007/978-3-319-74654-8_8

SMAC	Synthesize Modes and Correlate algorithm
MAC	Modal assurance criterion
FRF	Frequency response function
FFT	Fast Fourier transform
A	Subscript for subcomponent A
B	Subscript for subcomponent B
nl	Subscript for nonlinear
$+$	Superscript indicating the Moore-Penrose pseudo-inverse of a matrix

8.1 Introduction and Motivation

Dynamic substructuring is a methodology by which the dynamics of multiple components can be assembled to predict the dynamic response of an assembled structure. While there are many methods to complete dynamic substructuring, most common methods synthesize two or more subcomponents through a series of connection degrees of freedom. In some cases, an experimental model is combined with a numerical (finite element) model to predict the assembled response. This is a special case of dynamic substructuring known as experimental-analytical substructuring which is often useful when a model of a subcomponent is difficult to describe using traditional finite element approaches.

Physically, bolted joints are a common means of assembling two structures. These joints introduce nonlinearity into a system, normally composed of linear components, due to frictional interfaces at the joining surfaces. These nonlinear surface interactions are difficult to model but component models describing these dynamics are obtainable using measured data. These nonlinear models can then be combined with a finite element model to formulate a prediction for a nonlinear assembled structure.

Many of these bolted structures are weakly nonlinear in nature, meaning the modes of the nonlinear system remain uncoupled in the amplitude range of interest and the mode shapes do not change with amplitude. These weakly nonlinear structures often exhibit small changes in modal frequency with potentially large changes in modal damping with respect to amplitude of vibration. Linear behavior can be characterized at low vibration levels as the structure exhibits increasingly nonlinear behavior as vibration amplitude is increased.

Previous works [1] have implemented nonlinear pseudo-modal models to describe subcomponent motion and have shown that these nonlinear modal models can be used to predict the dynamics of a structural system. These modal models can take many forms including the modal Iwan model [2, 3], the Palmov model [4], and polynomial stiffness and damping models [5]. In this work, polynomial stiffness and damping models, developed using a Restoring Force Surface technique, are utilized to describe the nonlinear dynamics of two different subcomponent models. These subcomponent models are then assembled using the Transmission Simulator method to predict the response of the assembled structure at both linear and nonlinear vibration amplitude levels.

This paper is organized as follows. In Sect. 8.2 a brief overview of the structure of interest and dynamic substructuring theory is presented. This describes how the nonlinear subcomponent models are built up to form an assembled prediction. Section 8.3 describes linear testing and linear substructuring predictions, while Sect. 8.4 overviews the excitation method and nonlinear modal model fitting from high-level testing. Section 8.5 presents the substructuring prediction results. Finally, Sect. 8.6 contains closing remarks and comments on this methodology.

8.2 Nonlinear Dynamic Substructuring

8.2.1 Subcomponent Definition

The objective of this example is to demonstrate nonlinear modal substructuring capabilities on a complicated jointed structure. A finite element model of the structure of interest is shown in Fig. 8.1 and the physical hardware is shown in Fig. 8.2. This structure consists of 5 main parts: beam, plate, cylinder, plug, and tail. The beam is bolted to the plate which is attached to the cylinder using eight bolts. The aft plug threads into the cylinder and has a flange that seats on the aft of the cylinder. Finally, the tail is attached to the plug by two bolts.

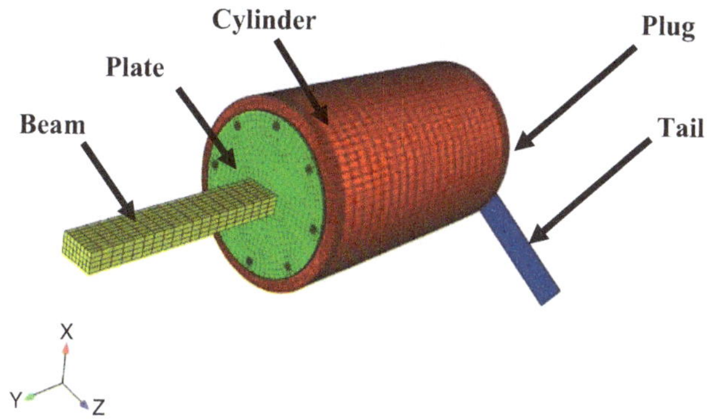

Fig. 8.1 Truth assembly solid model and structure identification

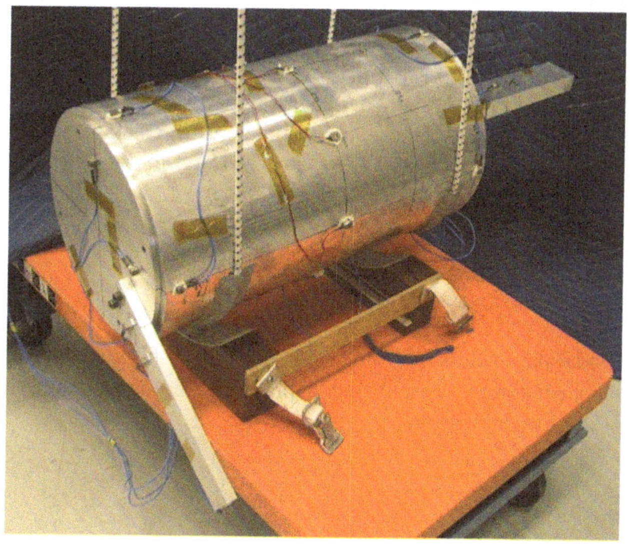

Fig. 8.2 Truth assembly test hardware

To complete substructuring predictions for this assembly, two subcomponent assemblies were tested. The first, subcomponent *A*, consists of the beam, plate, cylinder, and plug, with no aft tail. This structure is similar to the structure tested in [5] and [6]. The second subcomponent, *B*, consists of the front plate, cylinder, plug and aft tail. The tail is configured such that it should provide a different type of nonlinearity than the front beam.

This work uses the Transmission Simulator method to connect the two experimental subcomponent models [7]. Subcomponents *A* and *B* share many measurement locations to use as connection degrees-of-freedom (DOFs). Connecting at these DOFs with no correction would produce a combined structure with artificially high mass and stiffness since the effects of the plate, cylinder, and plug are included twice as they are present in both experimental models. To remedy this, a finite element model of the plate, cylinder, and plug components was used as a Transmission Simulator (TS) which is used to remove the effects of one set of these components from the predictions. Figure 8.3 shows the substructuring schematic using these three subcomponent models (*A*, and Transmission Simulator) to create a prediction of the truth assembly.

8.2.2 Dynamic Substructuring Theory

In this work, dynamic substructuring predictions are formulated using the Transmission Simulator method [7]. This method provides a rich modal basis creating more accurate boundary conditions at connection points between two subsystems by mass-loading the interface between two subcomponents. In this work, linear substructuring is performed first, then it is extended to include nonlinear effects. This substructuring methodology was implemented previously on a beam structure using modal Iwan elements [8] for just a few modes. Each subcomponent is written as a set of uncoupled modal equations of

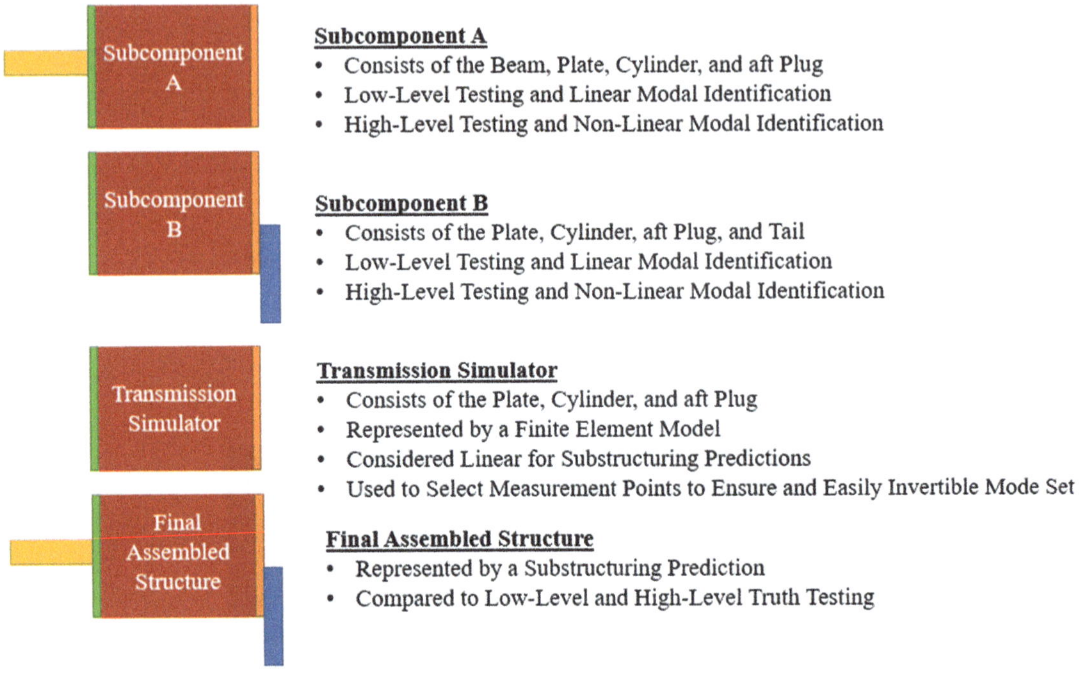

Fig. 8.3 Transmission simulator substructure schematic

motion, as shown in Eq. (8.1). Modal parameters ω, ζ, and ϕ are the natural frequency, damping ratio, and mode shapes of a subcomponent model. The modal acceleration, velocity and displacement are represented by $\ddot{q}$, $\dot{q}$, and q and F is the external force applied to a subcomponent in physical coordinates.

$$\begin{bmatrix} I_A & 0 & 0 \\ 0 & I_B & 0 \\ 0 & 0 & -I_{TS} \end{bmatrix} \begin{Bmatrix} \ddot{q}_A \\ \ddot{q}_B \\ \ddot{q}_{TS} \end{Bmatrix} + \begin{bmatrix} \begin{bmatrix} \ddots & & \\ & 2\zeta_A\omega_A & \\ & & \ddots \end{bmatrix} & 0 & 0 \\ 0 & \begin{bmatrix} \ddots & & \\ & 2\zeta_B\omega_B & \\ & & \ddots \end{bmatrix} & 0 \\ 0 & 0 & -\begin{bmatrix} \ddots & & \\ & 2\zeta_{TS}\omega_{TS} & \\ & & \ddots \end{bmatrix} \end{bmatrix} \begin{Bmatrix} \dot{q}_A \\ \dot{q}_B \\ \dot{q}_{TS} \end{Bmatrix}$$

$$+ \begin{bmatrix} \begin{bmatrix} \ddots & & \\ & \omega_A^2 & \\ & & \ddots \end{bmatrix} & 0 & 0 \\ 0 & \begin{bmatrix} \ddots & & \\ & \omega_B^2 & \\ & & \ddots \end{bmatrix} & 0 \\ 0 & 0 & -\begin{bmatrix} \ddots & & \\ & \omega_{TS}^2 & \\ & & \ddots \end{bmatrix} \end{bmatrix} \begin{Bmatrix} q_A \\ q_B \\ q_{TS} \end{Bmatrix} = \begin{Bmatrix} \phi_A^T F_A \\ \phi_B^T F_B \\ \phi_{TS}^T F_{TS} \end{Bmatrix} \tag{8.1}$$

In Eq. (8.1), the equations of motion for each subcomponent are uncoupled, to couple these subcomponents constraints must be enforced on these equations of motion. The constrains are first written in the form of Eq. (8.2). Here, x represents physical displacements of each substructure, and $\tilde{B}$ is a Boolean matrix such that the motion of shared DOFs between multiple subcomponents is equal. For example, $x_{A,1} - x_{B,1} = 0$, could be one row of this equation where the motion of DOF 1 on both subcomponents is equal.

$$\acute{B}\begin{Bmatrix} x_A \\ x_B \\ x_{TS} \end{Bmatrix} = 0 \tag{8.2}$$

Using a modal approximation, this constraint equation can be transformed into modal coordinates as shown in Eq. (8.3). Note, the constraints have been softened using the pseudo-inverse of the TS shapes as it would be difficult to enforce them strictly when using measured data.

$$\begin{bmatrix} \phi_{TS}^{+} & 0 \\ 0 & \phi_{TS}^{+} \end{bmatrix}\begin{bmatrix} \phi_A & 0 & -\phi_{TS} \\ 0 & \phi_B & -\phi_{TS} \end{bmatrix}\begin{Bmatrix} q_A \\ q_B \\ q_{TS} \end{Bmatrix} = \tilde{B}\begin{Bmatrix} q_A \\ q_B \\ q_{TS} \end{Bmatrix} = 0 \tag{8.3}$$

A coordinate transformation $q = L\eta$ is used to enforce these constraints on the uncoupled equations of motion, where L is the transformation matrix from uncoupled modal coordinate q, to synthesized general coordinate η. Rewriting the constraints in terms of the new generalized coordinate is shown in Eq. (8.4).

$$\tilde{B}L\eta = 0 \tag{8.4}$$

Because $\overline{B}$ is known and because η is arbitrary, L must reside in the nullspace of $\overline{B}$ for this equation to hold true. Thus, L can be defined as in Eq. (8.5).

$$L = null\left(\tilde{B}\right) \tag{8.5}$$

Using the transformation matrix L, the uncoupled equations of motion can be synthesized into a new set of equations of motion which describe the dynamics of the fully coupled structure, Eq. (8.6). The eigen value problem can be solved for these new equations of motion to provide predicted, linear natural frequencies and modes shapes of the assembled structure.

$$L^T\begin{bmatrix} I_A & 0 & 0 \\ 0 & I_B & 0 \\ 0 & 0 & -I_{TS} \end{bmatrix} L\begin{Bmatrix} \ddot{\eta}_A \\ \ddot{\eta}_B \\ \ddot{\eta}_{TS} \end{Bmatrix} + L^T\begin{bmatrix} \begin{bmatrix} \ddots & & \\ & 2\zeta_A\omega_A & \\ & & \ddots \end{bmatrix} & 0 & 0 \\ 0 & \begin{bmatrix} \ddots & & \\ & 2\zeta_B\omega_B & \\ & & \ddots \end{bmatrix} & 0 \\ 0 & 0 & -\begin{bmatrix} \ddots & & \\ & 2\zeta_{TS}\omega_{TS} & \\ & & \ddots \end{bmatrix} \end{bmatrix} L\begin{Bmatrix} \dot{\eta}_a \\ \dot{\eta}_B \\ \dot{\eta}_{TS} \end{Bmatrix}$$

$$+ L^T\begin{bmatrix} \begin{bmatrix} \ddots & & \\ & \omega_A^2 & \\ & & \ddots \end{bmatrix} & 0 & 0 \\ 0 & \begin{bmatrix} \ddots & & \\ & \omega_B^2 & \\ & & \ddots \end{bmatrix} & 0 \\ 0 & 0 & -\begin{bmatrix} \ddots & & \\ & \omega_{TS}^2 & \\ & & \ddots \end{bmatrix} \end{bmatrix} L\begin{Bmatrix} \eta_A \\ \eta_B \\ \eta_{TS} \end{Bmatrix} = L^T\begin{Bmatrix} \phi_A^T F_A \\ \phi_B^T F_B \\ \phi_{TS}^T F_{TS} \end{Bmatrix} \tag{8.6}$$

This work uses the nonlinear pseudo-modal model approach to add nonlinearity into subcomponent models, see Sect. 8.4.1. Therefore, nonlinear modal forcing elements, F_{nl}, can be added to the original uncoupled equations of motion, as seen in Eq. (8.7), for each subcomponent. The definition of these nonlinear modal forcing elements is discussed in Sect. 8.4 of this paper but, as can be seen in Eq. (8.7), each subcomponent can contain nonlinear modal forcing elements.

$$\begin{bmatrix} I_A & 0 & 0 \\ 0 & I_B & 0 \\ 0 & 0 & -I_{TS} \end{bmatrix} \begin{Bmatrix} \ddot{q}_A \\ \ddot{q}_B \\ \ddot{q}_{TS} \end{Bmatrix} + \begin{bmatrix} \begin{bmatrix} \ddots & & \\ & 2\zeta_A\omega_A & \\ & & \ddots \end{bmatrix} & 0 & 0 \\ 0 & \begin{bmatrix} \ddots & & \\ & 2\zeta_B\omega_B & \\ & & \ddots \end{bmatrix} & 0 \\ 0 & 0 & -\begin{bmatrix} \ddots & & \\ & 2\zeta_{TS}\omega_{TS} & \\ & & \ddots \end{bmatrix} \end{bmatrix} \begin{Bmatrix} \dot{q}_A \\ \dot{q}_B \\ \dot{q}_{TS} \end{Bmatrix}$$

$$+ \begin{bmatrix} \begin{bmatrix} \ddots & & \\ & \omega_A^2 & \\ & & \ddots \end{bmatrix} & 0 & 0 \\ 0 & \begin{bmatrix} \ddots & & \\ & \omega_B^2 & \\ & & \ddots \end{bmatrix} & 0 \\ 0 & 0 & -\begin{bmatrix} \ddots & & \\ & \omega_{TS}^2 & \\ & & \ddots \end{bmatrix} \end{bmatrix} \begin{Bmatrix} q_A \\ q_B \\ q_{TS} \end{Bmatrix} = \begin{Bmatrix} \phi_A^T F_A \\ \phi_B^T F_B \\ \phi_{TS}^T F_{TS} \end{Bmatrix} + \begin{Bmatrix} F_{nl,A} \\ F_{nl,B} \\ F_{j,TS} \end{Bmatrix} \tag{8.7}$$

Because the transformation matrix L was derived solely from compatibility it can be applied to these nonlinear forces as it was to the external forces in the original equations of motion. To numerically integrate this prediction, the matrices are diagonalized by solving the eigen value problem and completing one final coordinate transformation ($\eta = \psi p$). Integrating these equations of motion and comparing the results to those of a truth test, the capabilities of this nonlinear substructuring technique can be demonstrated.

$$\psi^T L^T \begin{bmatrix} I_A & 0 & 0 \\ 0 & I_B & 0 \\ 0 & 0 & -I_{TS} \end{bmatrix} L\psi^T \begin{Bmatrix} \ddot{p}_A \\ \ddot{p}_B \\ p_{TS} \end{Bmatrix} + \psi^T L^T \begin{bmatrix} \begin{bmatrix} \ddots & & \\ & 2\zeta_A\omega_A & \\ & & \ddots \end{bmatrix} & 0 & 0 \\ 0 & \begin{bmatrix} \ddots & & \\ & 2\zeta_B\omega_B & \\ & & \ddots \end{bmatrix} & 0 \\ 0 & 0 & -\begin{bmatrix} \ddots & & \\ & 2\zeta_{TS}\omega_{TS} & \\ & & \ddots \end{bmatrix} \end{bmatrix} L\psi^T \begin{Bmatrix} \dot{p}_A \\ \dot{p}_B \\ \dot{p}_{TS} \end{Bmatrix}$$

$$+ \psi^T L^T \begin{bmatrix} \begin{bmatrix} \ddots & & \\ & \omega_A^2 & \\ & & \ddots \end{bmatrix} & 0 & 0 \\ 0 & \begin{bmatrix} \ddots & & \\ & \omega_B^2 & \\ & & \ddots \end{bmatrix} & 0 \\ 0 & 0 & -\begin{bmatrix} \ddots & & \\ & \omega_{TS}^2 & \\ & & \ddots \end{bmatrix} \end{bmatrix} L\psi^T \begin{Bmatrix} p_A \\ p_B \\ p_{TS} \end{Bmatrix} = \psi^T L^T \begin{Bmatrix} \phi_A^T F_A \\ \phi_B^T F_B \\ \phi_{TS}^T F_{TS} \end{Bmatrix} + \psi^T L^T \begin{Bmatrix} F_{nl,A} \\ F_{nl,B} \\ F_{nl,TS} \end{Bmatrix} \tag{8.8}$$

8.3 Linear Testing and Predictions

This section describes the modal tests performed to extract the linear modal models of each subcomponent and linear substructuring as described in the previous section. This information will then be used in Sect. 8.4 in the identification of subcomponent nonlinear modal models and nonlinear prediction of the final assembled structure.

8.3.1 Test Procedure and Extraction of Linear Modal Models

Low excitation level impact and shaker tests were performed on subcomponents *A* and *B*. Linear natural frequencies, damping values, and mode shapes were obtained using the Synthesize Modes and Correlate (SMAC) [9] program developed by Mayes and Hensley. The results for substructures *A* and *B* are shown in Tables 8.1 and 8.2 respectively.

A finite element model was used to find the natural frequencies and mode shapes of the transmission simulator. Material damping of 0.05% critical damping was assumed for the rigid and elastic modes of the TS. The finite element mesh used for this transmission simulator model shown in Fig. 8.4.

Table 8.1 Linear Modal Parameters Subcomponent A

Mode	f_n [Hz]	ζ [%cr]	Shape description
7	129.2	0.29	1st bend of beam in soft direction[a]
8	171.1	0.17	1st bend of beam in stiff direction[a]
9	552.0	0.24	Axial mode[a]
10	867.8	1.12	(2,0) ovaling of cylinder
11	883.3	0.93	(2,0) ovaling of cylinder[a]
12	980.8	0.44	(3,0) ovaling of cylinder
13	987.7	0.46	(3,0) ovaling of cylinder
14	1031.8	0.09	2nd bend of beam in soft direction[a]

[a]Marks modes that are designated as nonlinear in Sect. 8.4.2

Table 8.2 Linear Modal Parameters Subcomponent B

Mode	f_n[Hz]	ζ[%cr]	Shape description
7	232.5	0.17	1st bend of tail in soft direction
8	608.9	0.42	1st bend of tail in stiff direction
9	882.5	1.14	(2,0) ovaling of cylinder
10	892.4	0.81	(2,0) ovaling of cylinder
11	981.8	0.45	(3,0) ovaling of cylinder
12	990.0	0.39	(3,0) ovaling of cylinder
13	1171.0	1.08	Axial mode
14	1284.5	0.48	2nd bend of tail in soft direction

Marks modes that are designated as nonlinear in Sect. 8.4.2

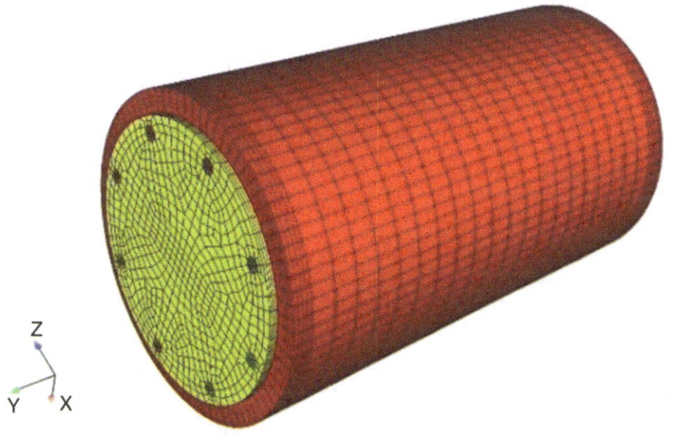

Fig. 8.4 Finite Element Model for Transmission Simulator

8.3.2 Linear Substructuring Predictions

Linear substructuring predictions were calculated following Eq. (8.6). Note that many different combinations of modes could be used to form $\overline{B}$ and create a prediction. Substructuring heuristics recommend including modes to twice the frequency of interest. Thus, to obtain quality predictions to 600 Hertz, modes up to 1200 Hz should be included in the subcomponent models. This results in 14 modes for both substructures a and b and only 10 modes for the TS. Using Eq. (8.6), 18 predicted modes for the assembled structure were found.

To verify the quality of our predictions, a truth test was conducted on the fully assembled structure. Low-level excitation was used in minimize its nonlinear response and ensure the data were comparable to that obtained from the subcomponent modal tests. SMAC was used to extract natural frequencies, damping, and mode shapes from the truth test. These are compared to the predicted values in Table 8.3.

The target modes of interest (7–11) have small frequency errors with a maximum frequency error of 1.32% and a maximum damping error of 6.94%. Modal assurance criterion (MAC) values for these modes are all 0.99 or higher. The prediction in the ovaling modes is less precise with frequency errors as high as 10% and damping errors up to 45%. A plot of the MAC values is shown in Fig. 8.5.

Figure 8.6 shows a comparison of a predicted and measured drive point frequency response function for an axial input. The modes under 800 Hz correlate well to the measured response, and will be the primary focus of the nonlinear investigation of this structure.

While the linear prediction provides a quality representation of the structure's response to low-level input, it poorly predicts the response under high-level loadings. Figure 8.7 shows the drive point response of the linear substructuring prediction to a high-level chirp overlaid with the corresponding measured data to the same excitation. The amplitude of several modes are over-predicted, indicating an error in the damping estimated in the linear substructuring process. In addition, the frequency of several resonances has a slight error. Including the nonlinear elements in the subcomponent models should provide a better substructuring prediction and better match the truth data. Section 8.4 discusses the nonlinear modal modeling approach to incorporate nonlinear elements into the subcomponent models which will then be used in Sect. 8.5 to predict the response of the assembled structure.

Table 8.3 Linear Substructuring Predictions[a]

Mode	Predicted f [Hz]	Measured f [Hz]	% Diff.	Predicted ζ [%cr]	Measured ζ [%cr]	% Diff.	MAC	Shape Description
7	129.62	128.22	1.09%	0.295	0.276	6.94%	1.00	1st bend of beam in soft direction
8	171.47	169.46	1.19%	0.171	0.176	-2.90%	0.99	1st bend of beam in stiff direction
9	232.56	233.37	-0.35%	0.168	0.176	-4.47%	1.00	1st bend of tail in soft direction
10	552.01	551.39	0.11%	0.241	0.245	-1.43%	0.99	Axial mode
11	608.08	616.25	-1.32%	0.415	0.426	-2.56%	1.00	1st bend of tail in stiff direction
12	798.44	895.22	-10.81%	1.358	0.931	45.90%	0.99	(2,0) ovaling of cylinder
13	868.99	898.00	-3.23%	1.127	0.993	13.51%	0.97	(2,0) ovaling of cylinder
14	966.39	982.79	-1.67%	0.409	0.458	-10.56%	0.98	(3,0) ovaling of cylinder
15	978.91	990.68	-1.19%	0.356	0.423	-15.71%	0.99	(3,0) ovaling of cylinder
16	1032.08	1031.75	0.03%	0.091	0.113	-19.05%	1.00	2nd bend of beam in soft direction
17	1170.47	1273.25	-8.07%	1.075	0.951	13.08%	0.51	Aft plate dishing mode
18	1282.73	1316.42	-2.56%	0.481	0.347	38.65%	0.95	2nd bend of tail in soft direction

[a]Rigid body modes not shown

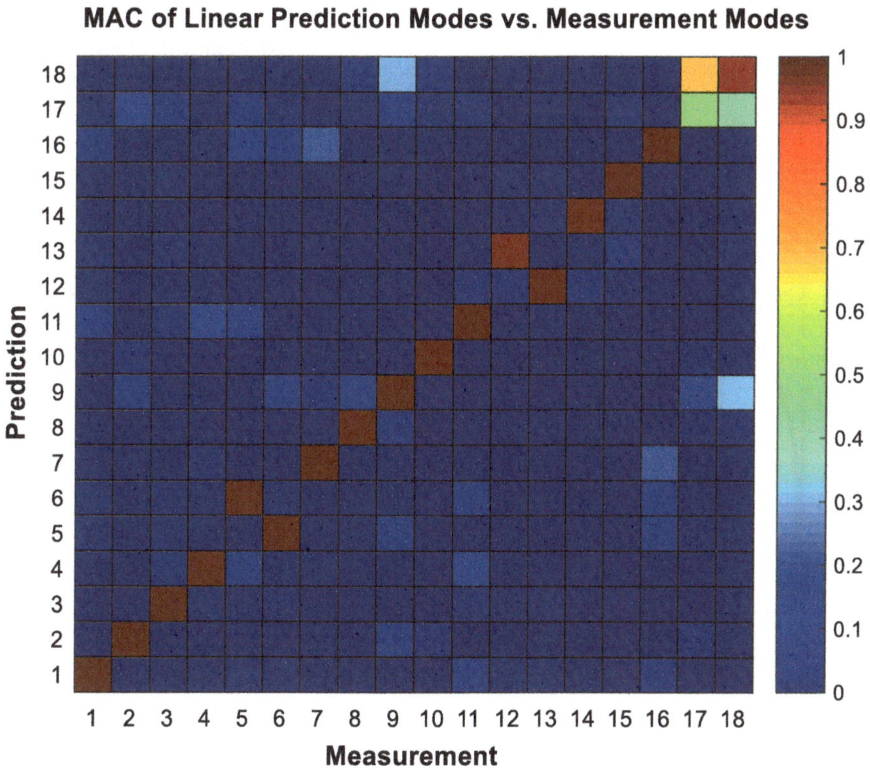

Fig. 8.5 Modal assurance criterion between predicted and measured linear mode shapes

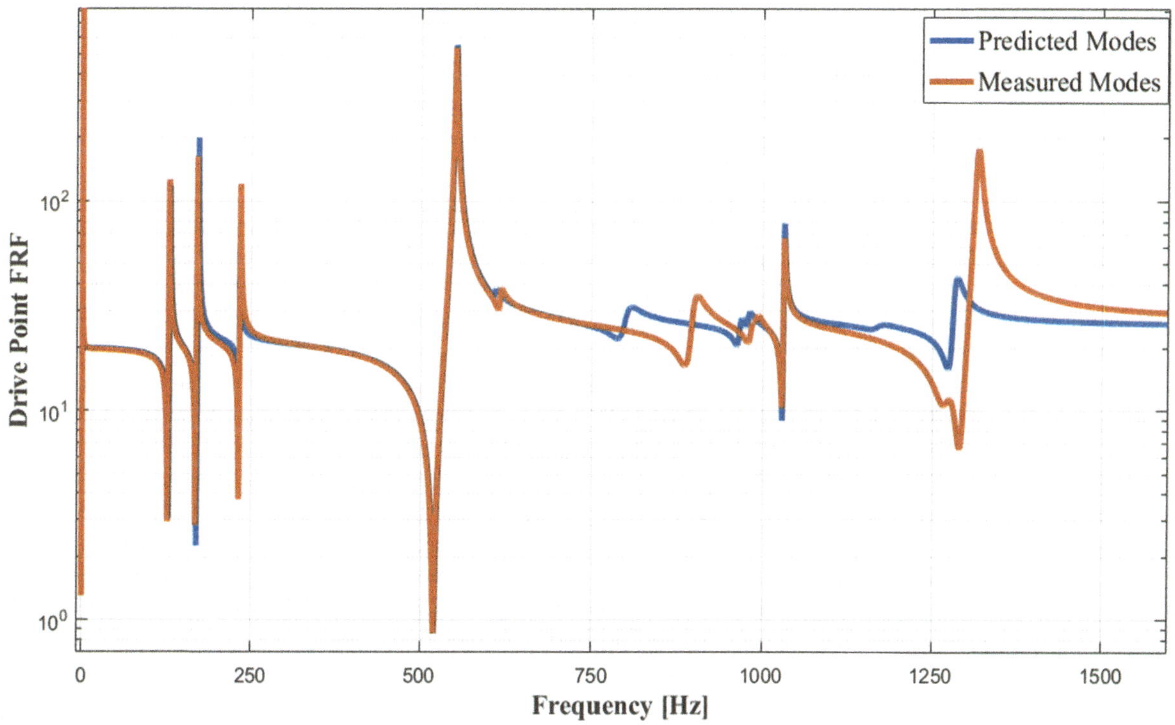

Fig. 8.6 Drive point frequency response function based on predicted and measured modes

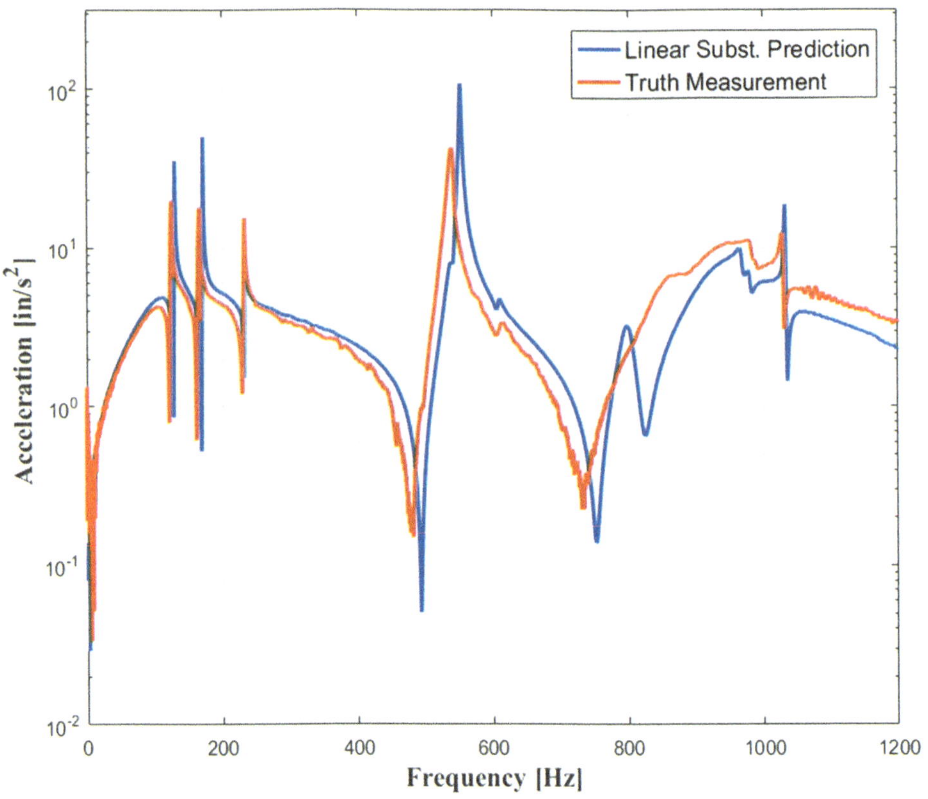

Fig. 8.7 Spectrum of drive point acceleration due to a high-level chirp

8.4 Nonlinear Modal Model

8.4.1 Nonlinear Modal Modeling Approach

This section provides an overview for the pseudo-modal model used to describe nonlinear subcomponent modes. This model was previously discussed in [5, 10] but is repeated here for convenience. In a linear structure, each resonance of the system is represented as single modal degree-of-freedom. This consists of a modal mass connected to ground with a spring and damper, see Fig. 8.8. This mode is assumed to not couple with any other modes of the system and the shape of the mode does not change with vibration amplitude.

A nonlinear modal model retains these assumptions and has each modal mass connected to ground by a spring and damper, but with the addition of some nonlinear forcing element. This nonlinear element can take many forms but for this work it takes the form of a cubic polynomial spring (k_{nl}) and damper (c_{nl}) as shown in Fig. 8.9.

The Restoring Force Surface method [11] was used to identify and define parameters for these nonlinear spring and damper elements. A classical linear modal model equation of motion can be written as in Eq. (8.9), where $\ddot{q}$, $\dot{q}$, and q are the modal acceleration, velocity, and displacement c_0 and k_0 are the linear modal damping and stiffness, and $\phi^T F$ is the modal force being applied to the system. Note, the sensors placed on the part measure physical acceleration, $\ddot{x}$, which must be transformed to the modal domain. This is accomplished using the SMAC modal filter as described in [5] and [12]. Thus, the many measurement channels from an experiment are transformed into a targeted single degree-of-freedom modal acceleration measurement.

$$\ddot{q} + c_0\dot{q} + k_0 q = \phi^T F \tag{8.9}$$

The nonlinear elements can be added to this equation of motion as seen in Eq. (8.10).

$$\ddot{q} + c_0\dot{q} + k_0 q + c_1\dot{q}\,|\dot{q}| + c_2\dot{q}^3 + k_1 q\,|q| + k_2 q^3 = \phi^T F \tag{8.10}$$

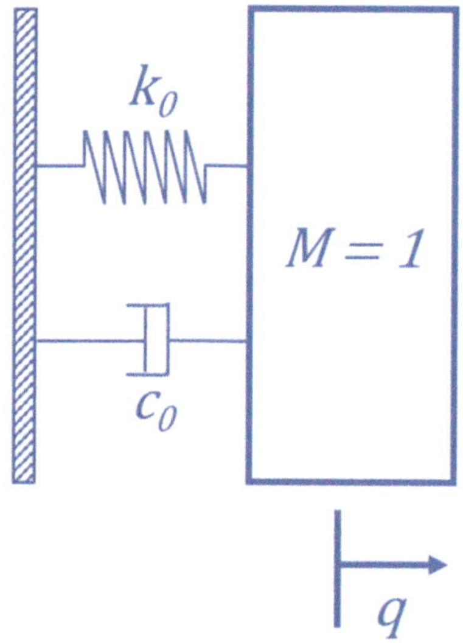

Fig. 8.8 Schematic of single degree-of-freedom linear modal model

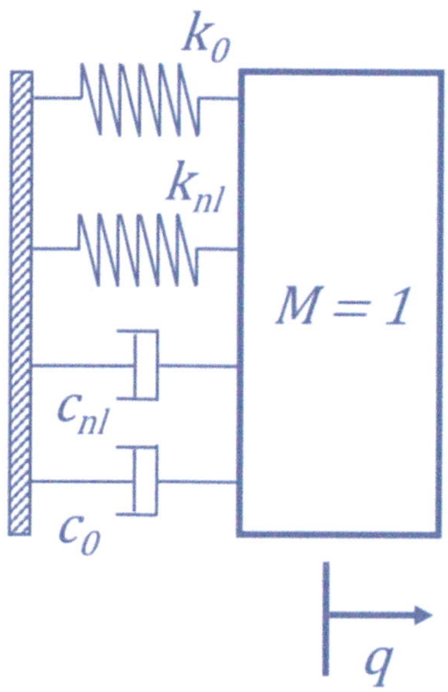

Fig. 8.9 Schematic of single degree-of-freedom nonlinear pseudo-modal model

To simulate the form of Eq. (8.7) during the substructuring process, these new nonlinear terms can be moved to the external forcing side of the equation and represented by F_{nl}.

$$F_{nl} = -c_1\dot{q}\,|\dot{q}| - c_2\dot{q}^3 - k_1 q\,|q| - k_2 q^3 \tag{8.11}$$

The mode shapes and linear coefficients (c_0 and k_0) were extracted from the low-level linear tests of the subcomponents. The procedure from [5] was used to obtain coefficients c_1, c_2, k_1, and k_2 which involved gathering terms with known or measured coefficients on one side of the equations and terms with unknown coefficients on the other.

$$\phi^T F - \ddot{q} - c_0\dot{q} - k_0 q = \left[\dot{q}\,|\dot{q}| \quad \dot{q}^3 \quad q\,|q| \quad q^3\right]\begin{bmatrix} c_1 \\ c_2 \\ k_1 \\ k_2 \end{bmatrix} \rightarrow \overline{U} = P\begin{bmatrix} c_1 \\ c_2 \\ k_1 \\ k_2 \end{bmatrix} \tag{8.12}$$

Note, the best parameters were found by transforming these quantities to the frequency domain, and stacking the real and imaginary parts of the spectra signals, as described in [5]. A least-squares estimate of these unknown coefficients can be obtains by pre-multiplying $\overline{U}$ by the pseudo-inverse of P

$$\begin{bmatrix} c_1 \\ c_2 \\ k_1 \\ k_2 \end{bmatrix} = P^{+}\overline{U} \tag{8.13}$$

This process was followed for each nonlinear mode in the system within the frequency band of interest.

8.4.2 Nonlinear Modal Model Identification

Nonlinear parameters for each mode for each substructure were extracted from experimental data in which windowed sinusoidal inputs were used to maximize the response of individual modes as described in [10]. An example excitation waveform is shown in Fig. 8.10 was created by applying a triangular window to a pure sine wave. The frequency and width of the main peak in the frequency domain are independently selectable and were determined by the frequency of sinewave and rise time of the triangular window, respectively. This focuses the input energy into a single target mode which drove the modal response to a large amplitude while also producing a "pure" response of just that mode which increases the accuracy of the extracted nonlinear parameters.

This input was used to excite each nonlinear mode for both subcomponents. The data from each experiment were then modally filtered and nonlinear parameters extracted for each targeted nonlinear mode for each substructure per the process described in the previous section. To verify their accuracy, the extracted nonlinear parameters for each mode were used along with their linear counterparts to simulate the response to the experimental force input (this was transformed to a modal force for the simulation). This response was then compared to the corresponding modally filtered measured data from the experiment. For brevity, the results for the first bending mode of subcomponent a are shown, see Fig. 8.11. These figures show good agreement between the nonlinear simulation and the measured response. Additionally, the nonlinear model shows a large improvement in both frequency and damping over those of the linear model. Table 8.4 lists the extracted nonlinear parameters for each nonlinear mode for both subcomponents.

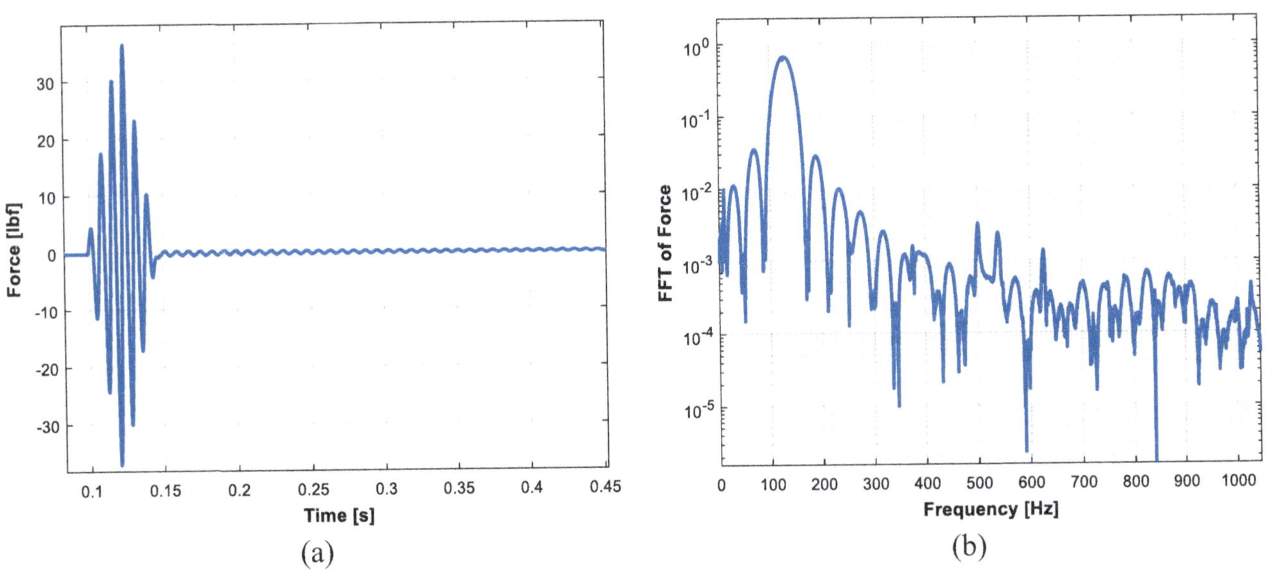

Fig. 8.10 Example waveform in the (**a**) time domain and (**b**) frequency domain used to excite targeted nonlinear modes of a substructure in order to maximize the its response

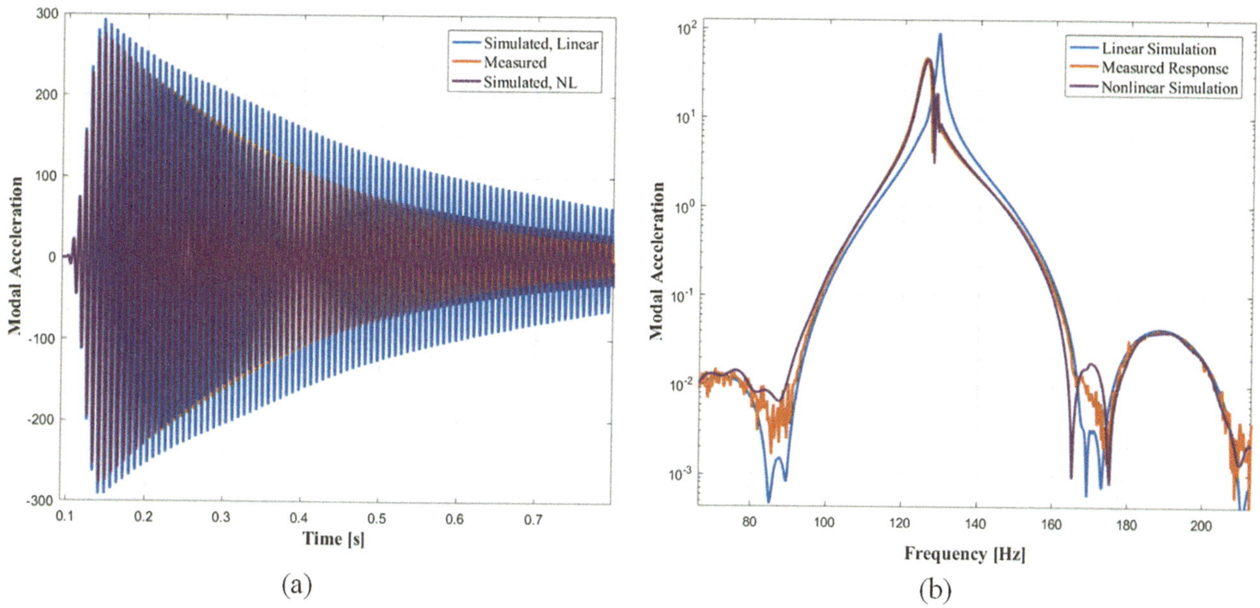

Fig. 8.11 Linear and nonlinear simulation results compared to measured modally filtered data for the first bending mode of subcomponent *a*; (**a**) time domain (**b**) frequency domain

Table 8.4 Nonlinear Pseudo-Modal Model Parameters

Subcomponent	Modal index	c_1	c_2	k_1	k_2
A	7	22.68	−62.07	−2.46E + 08	3.89E + 11
A	8	40.35	−98.70	−4.29E + 08	6.19E + 11
A	9	220.53	−508.18	−1.73E + 10	1.18E + 14
A	11	1.65E + 04	−4.78E + 04	−3.35E + 11	1.35E + 15
A	14	600.32	−4.00E + 03	−2.64E + 10	1.16E + 15
B	7	99.95	−581.08	−3.02E + 08	1.36E + 12
B	8	1.58E + 03	−1.82E + 04	−3.02E + 08	2.15E + 15
B	10	1.36E + 04	−3.42E + 03	−6.95E + 11	4.10E + 16

8.5 Nonlinear Substructuring Predictions

To evaluate the capability of these subcomponent nonlinear pseudo-modal models to predict the assembled structure's response, a dynamic substructuring prediction was completed for a high-level chirp using Eq. (8.8), see Fig. 8.12. Four representations of the drive point response spectrum are shown: substructuring prediction using linear subcomponent models (blue); the response of the assembled structure measured during the truth experiment (orange); linear simulation results from the modal model extracted from a low-level modal test of the assembly (green); and the nonlinear substructuring prediction using nonlinear subcomponent pseudo-modal models (purple). For clarity, two enhanced views are shown in Figs. 8.13 and 8.14.

The first enhanced view in Fig. 8.13 contains the 7th, 8th, and 9th modes. Note that comparing the linear substructuring prediction (blue) to the synthesized linear response (green) gives an indication of the error associated with the linear substructuring. This provides an upper bound on the accuracy achievable with the substructuring predictions and should be considered when comparing the nonlinear substructuring prediction (purple) to the measured truth test results (orange). In the first two modes shown, the nonlinear substructuring prediction shows an obvious improvement over the linear model, capturing both the frequency softening and increase in damping. For mode 9, the linear model did slightly better than the nonlinear model. This mode was not well excited from the drive point, so it is possible that mode 9 was a poor fit of the nonlinear modal model or that the tail bending mode should possibly be treated as linear. This poor fit could be attributed to the small drive point amplitude obtained for this mode, or if a cubic nonlinearity is not a quality model form for this model.

The second enhanced view shows the 10th and 11th modes. The 10th mode is the axial mode of the system. For this mode, the linear measurement and linear prediction align very well. The nonlinear response shows an appropriate increase

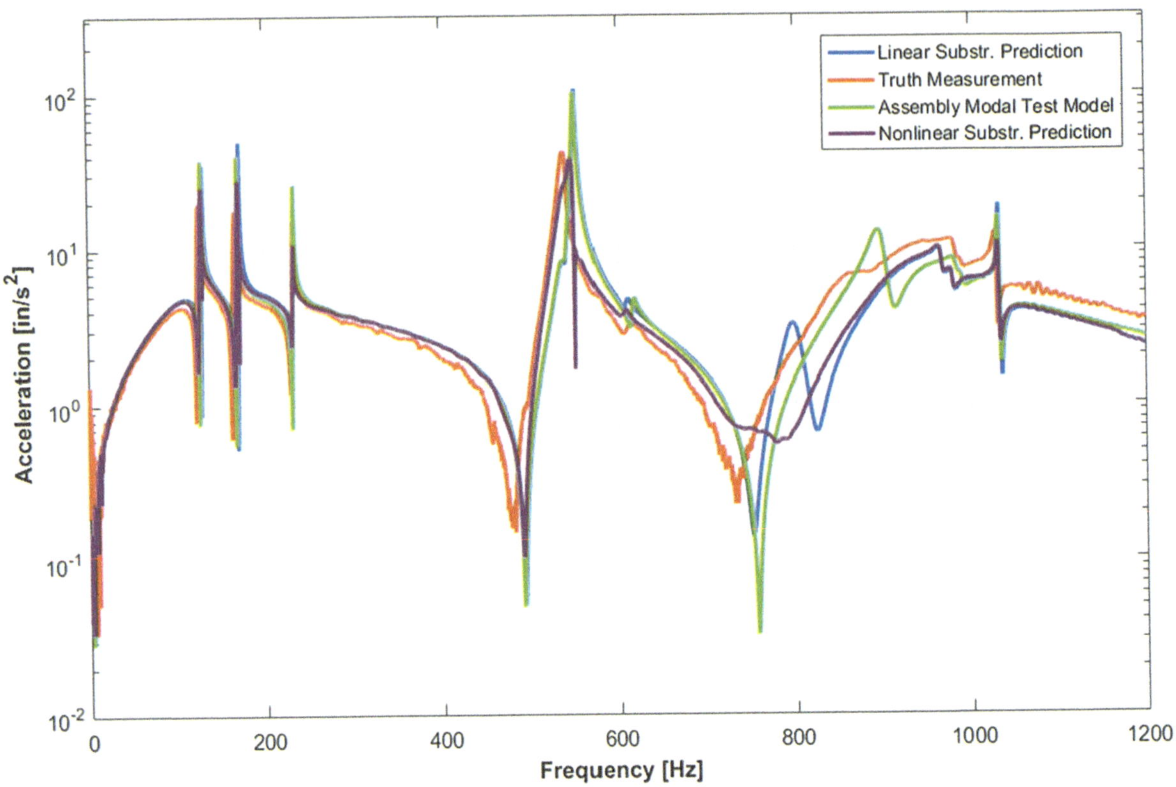

Fig. 8.12 FFT of drive point acceleration due to a high-level chirp using nonlinear modal models

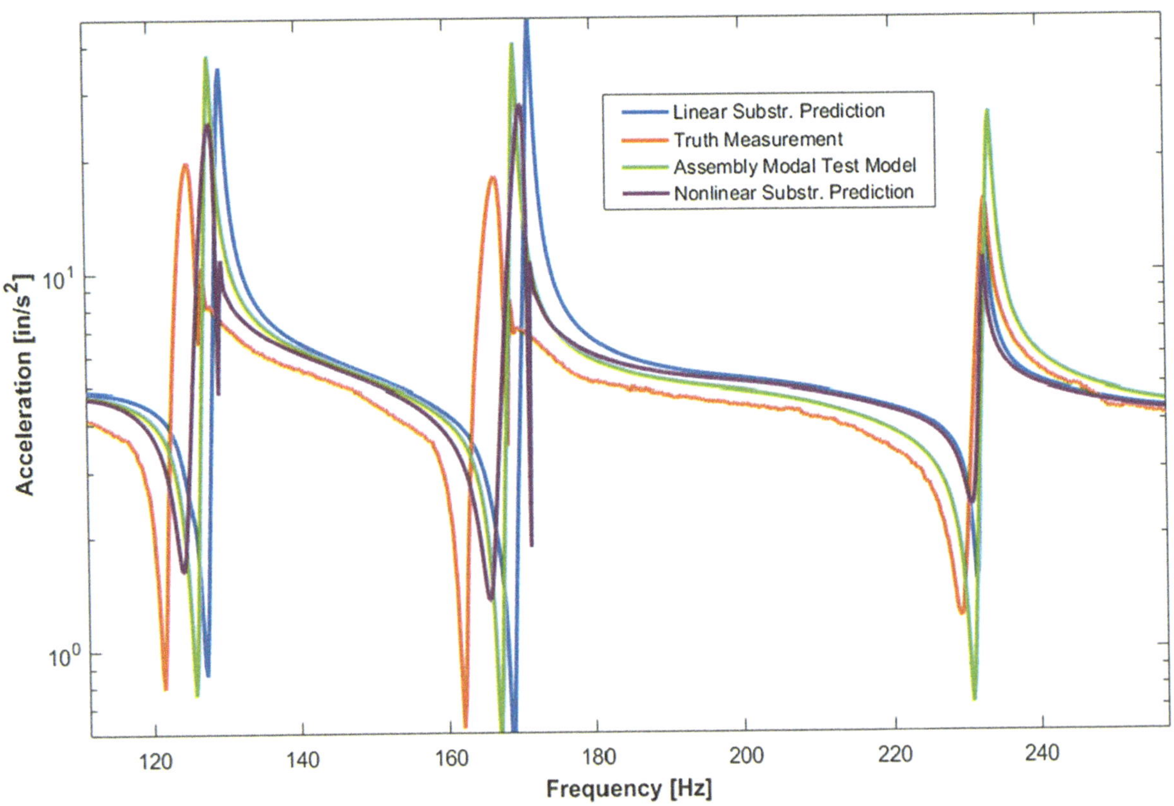

Fig. 8.13 Spectrum of drive point acceleration due to a high-level chirp using nonlinear modal models, modes 7 through 9

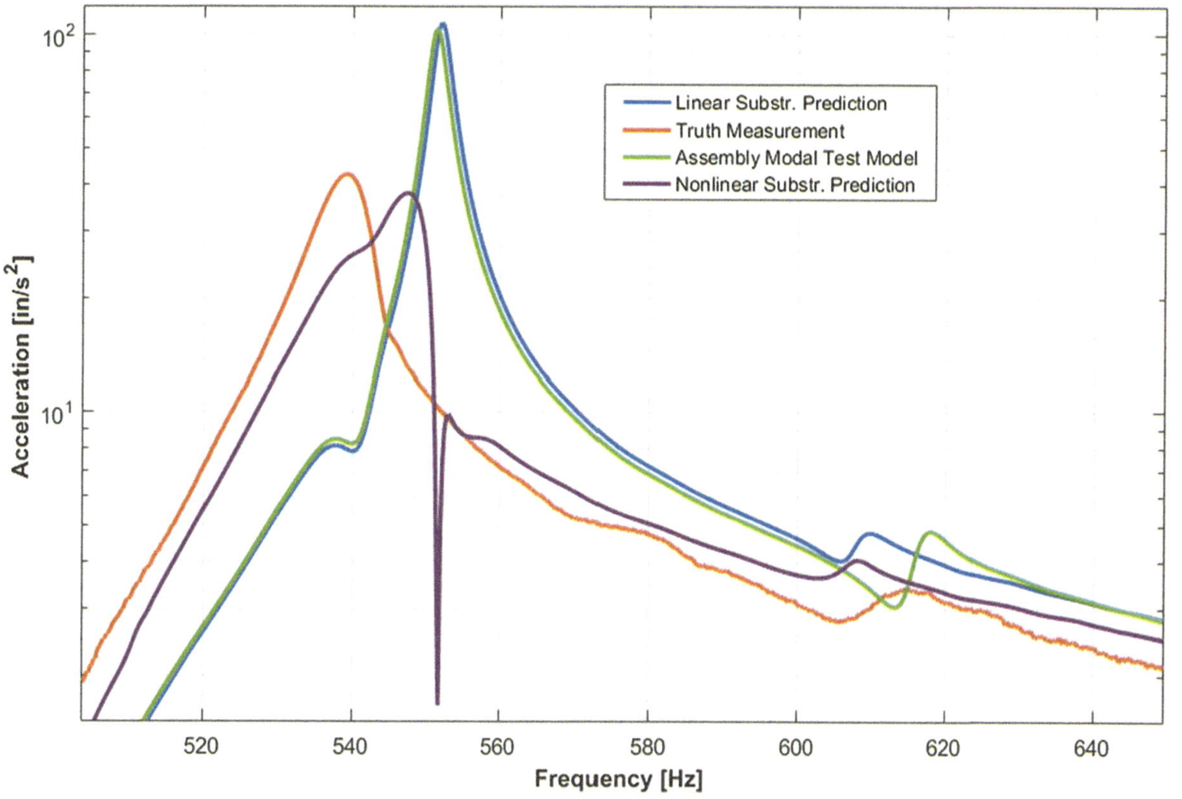

Fig. 8.14 FFT of drive point acceleration due to a high-level chirp using nonlinear modal models, modes 10 and 11

in damping but fails to accurately capture the frequency shift. The 11th mode is the circumferential bending of the tail which is not well excited from the current drive point.

8.6 Conclusions and Remarks

This work provided a nonlinear modal substructuring study using pseudo-modal models to describe the subcomponent modes for two experimental models. The nonlinear pseudo-modal models consisted of polynomial cubic springs and dampers. The parameters for these models were obtained using a windowed sinusoidal excitation in order to maximize amplitude of each individual mode. The linear substructuring predictions of the experimental system provided a good response when compared to a linear low-level test but overestimated response amplitude and failed to capture frequency shifts when the full system was tested at high loading levels.

The addition of uncoupled nonlinear modal forces to the subcomponents in traditional component mode synthesis provided a more accurate answer than using standard linear models; generally, the nonlinear method better captured the damping changes and frequency shifts of the assembled structure associated with high level inputs. Though not perfect, the truth test and nonlinear substructuring prediction comparison shows that using a cubic polynomial can provide an effective means of introducing measurable weak nonlinearities into substructuring predictions. Thus this method provides the benefits of substructuring while also adequately predicting the frequency and damping shifts of the assembled structure due to nonlinearity using nonlinear models of the subcomponents.

The nonlinear modal forces from each subcomponent were successfully spread to the correct assembly modes through the linear synthetization matrix, L. While successful in predicting nonlinearities, this method does bring some limitations. Using shaker inputs as proposed in [10], higher modal amplitudes can be obtained for subcomponent models. However, the nonlinear polynomial terms tend toward positive or negative infinity outside of the modal amplitude range excited in the subcomponent experiments. It is difficult to know whether these modal models are extrapolating once they are spread to the modes of the assembled system. In the high-level chirp truth test it is possible that the modal amplitude of difficult to excite modes fell within a lower range of nonlinearity which was not emphasized in subcomponent testing, leading to larger inaccuracies in modes such as the 9th mode, (bending of the tail). More rigor could be used to determine the correct model form for each mode that fits data over a large range of modal amplitudes. This would enable the modal model to be more accurate even when excited from a weaker or stronger drive point.

References

1. Mayes, R.L., Pacini, B.R., Roettgen, D.R.: A modal model to simulate typical structural dynamic nonlinearity. In: International Modal Analysis Conference XXXIV, Orlando, FL (2016)
2. Roettgen, D., Allen, M.: Nonlinear characterization of a bolted, industrial structure using a modal framework. Mech. Syst. Signal Process. **84**, 152–170 (2016)
3. Segalman, D.: A Modal Approach to Modeling Spatially Distributed Vibration Energy Dissipation (2010)
4. Segalman, D.J., Allen, M.S., Eritern, M., Hoppman, K.: Experimental assessment of joint-like modal models for structures. In: ASME Mechanical Vibration and Noise XXVII, Boston, MA (2015)
5. Allen, M.S., Mayes, R.L., Bergman, E.L.: Experimental modal substructuring to couple and uncouple substructures with flexible fixtures and multi-point connections. Sound Vib. **329**(23), 4891–4906 (2010)
6. Roettgen, D.R., Allen, M.S., Kammer, D.C., Mayes, R.L.: Substructuring of a nonlinear beam using a modal Iwan framework: Part I: Nonlinear modal model identification. In: International Modal Analysis Conference XXXV, Garden Grove, CA (2017)
7. Roettgen, D.R., Seeger, B., Tai, W.C., Baek, S., Dossogne, T., Allen, M.S., Kuether, R.J., Brake, M.R.W., Mayes, R.L.: A comparison of reduced order modeling techniques used in dynamic substructuring. In: International Modal Analysis Conference XXXIV, Orlando, FL (2016)
8. Roettgen, D.R., Allen, M.S., Kammer, D.C., Mayes, R.L.: Substurcturing of a nonlinear beam using a modal Iwan framework, Part II: Nonlinear modal substructuring. In: International Modal Analysis Conference XXXV, Garden Grove, CA (2017)
9. Hensley, D.P., Mayes, R.L.: Extending SMAC to multiple references. In: International Modal Analysis Conference XXIV (2006)
10. Pacini, B.R., Mayes, R.L., Owens, B.C., Schultz, R.: Nonlinear finite element model updating, Part I: Experimental techniques and nonlinear modal model parameter extraction. In: International Modal Analysis Conference XXXV, Garden Grove, CA (2017)
11. Gaëtan, K., Worden, K., Vakakis, F.A., Golinval, J.: Past, present, and future of nonlinear system identification in structural dynamics. Mech. Syst. Signal Process. **20**, 505–592 (2006)
12. Mayes, R.L., Johansen, D.D.: A modal parameter extraction algorithm using best-fit reciprocal vectors. In: International Modal Analysis Conference XVI (1998)

Chapter 9
Dynamic Substructuring with a Sliding Contact Interface

Jacopo Brunetti, Walter D'Ambrogio, and Annalisa Fregolent

Abstract In this paper, the general framework for dynamic substructuring is extended to time-variant interfaces among connecting substructures. Specifically, a time-variant interface due to a sliding contact is considered. The contact can be without or with friction. The problem can be tackled in time domain using primal assembly and numerical time integration, and in time-frequency domain using dual assembly, thus obtaining a Time Dependent Frequency Response Function (TD-FRF) of the assembled structure. The method is applied to lumped parameter models of substructures, and, under some assumptions, it can be extended to simple finite element models.

Keywords Experimental dynamic substructuring · Time-variant interface · Sliding contact · Friction · Time-frequency analysis

9.1 Introduction

Typical applications of dynamic substructuring deal with time-invariant systems [1]. An interesting extension of the results deriving by the huge effort that the scientific community has produced in this area over several decades can be to consider, among time-variant systems, those built from time-invariant component subsystems subjected to time-variant coupling conditions, such as those encountered in contact problems (e.g. when a sliding velocity exists between two contacting bodies).

The link between substructuring and contact problems can lead to interesting perspectives both for the substructuring problems and for the contact problems. For instance, the macroscopic analysis of the time-frequency behaviour of bodies in relative sliding with friction can benefit of the substructuring approach.

Since time-variant systems are considered, a time domain approach seems to be the most appropriate. Substructuring in time domain has been considered by Rixen and van der Valk [2, 3], by using impulse response functions. However, the coupling conditions (compatibility and equilibrium) are considered as time-invariant.

Therefore, in this paper, the general framework for dynamic substructuring is extended to time-variant interfaces among connecting substructures. Specifically, a time-variant interface due to a sliding contact is considered by introducing time dependent compatibility and equilibrium conditions. The sliding contact can be without or with friction. In presence of friction, the set of degrees of freedom to which equilibrium conditions apply includes tangential directions at the contact interface, which are not considered in the compatibility conditions because of sliding. This is in fact a non collocated interface as defined in decoupling problems [4, 5].

The problem is tackled in the time domain using primal assembly and numerical time integration, and in time-frequency domain using dual assembly, thus obtaining Time Dependent Frequency Response Functions (TD-FRFs) of the assembled structure. The approach is applied to a lumped parameter model in order to evaluate its feasibility and effectiveness. The approach can be extended to simple finite element models under the assumption that the relative sliding motion at the contact interfaces is known a-priori, at least approximately.

J. Brunetti · W. D'Ambrogio (✉)
Dipartimento di Ingegneria Industriale e dell'Informazione e di Economia, Università dell'Aquila, L'Aquila (AQ), Italy
e-mail: jacopo.brunetti@univaq.it; walter.dambrogio@univaq.it

A. Fregolent
Dipartimento di Ingegneria Meccanica e Aerospaziale, Università di Roma La Sapienza, Rome, Italy
e-mail: annalisa.fregolent@uniroma1.it

A. Linderholt et al. (eds.), *Dynamics of Coupled Structures, Volume 4*, Conference Proceedings of the Society for Experimental Mechanics Series, https://doi.org/10.1007/978-3-319-74654-8_9

9.2 Theoretical Background

9.2.1 Substructuring

Let us consider a structural system consisting of n coupled subsystems. The equation of motion of a linear time-invariant subsystem r may be written as:

$$[M]^{(r)}\{\ddot{u}(t)\}^{(r)} + [C]^{(r)}\{\dot{u}(t)\}^{(r)} + [K]^{(r)}\{u(t)\}^{(r)} = \{f(t)\}^{(r)} + \{g(t)\}^{(r)} \tag{9.1}$$

where:

$[M]^{(r)}$, $[C]^{(r)}$, $[K]^{(r)}$ are the mass, damping and stiffness matrices of subsystem r;
$\{u(t)\}^{(r)}$ is the vector of displacements of subsystem r;
$\{f(t)\}^{(r)}$ is the vector of external forces on subsystem r;
$\{g(t)\}^{(r)}$ is the vector of connecting forces with other subsystems (internal constraint forces).

The equation of motion of the n subsystems to be coupled can be written in a block diagonal format as:

$$[M]\{\ddot{u}(t)\} + [C]\{\dot{u}(t)\} + [K]\{u(t)\} = \{f(t)\} + \{g(t)\} \tag{9.2}$$

with

$$[M] = \begin{bmatrix} [M]^{(1)} & & \\ & \ddots & \\ & & [M]^{(n)} \end{bmatrix}, \quad [C] = \begin{bmatrix} [C]^{(1)} & & \\ & \ddots & \\ & & [C]^{(n)} \end{bmatrix}, \quad [K] = \begin{bmatrix} [K]^{(1)} & & \\ & \ddots & \\ & & [K]^{(n)} \end{bmatrix} \tag{9.3}$$

$$\{u(t)\} = \begin{Bmatrix} \{u(t)\}^{(1)} \\ \vdots \\ \{u(t)\}^{(n)} \end{Bmatrix}, \quad \{f(t)\} = \begin{Bmatrix} \{f(t)\}^{(1)} \\ \vdots \\ \{f(t)\}^{(n)} \end{Bmatrix}, \quad \{g(t)\} = \begin{Bmatrix} \{g(t)\}^{(1)} \\ \vdots \\ \{g(t)\}^{(n)} \end{Bmatrix} \tag{9.4}$$

The compatibility condition implies that any pair of matching DoFs at time t, i.e. DoF l on subsystem r and DoF m on subsystem s must have the same displacement, that is $u_l^{(r)}(t) - u_m^{(s)}(t) = 0$.

This condition can be generally expressed as:

$$[B_C(t)]\{u(t)\} = \{0\} \tag{9.5}$$

where each row of $[B_C(t)]$ corresponds to a pair of matching DoFs at time t. Note that $[B_C(t)]$ is, in most cases, a signed Boolean matrix and it can be written by distinguishing the contribution of the different subsystems:

$$[B_C(t)] = \begin{bmatrix} [B_C(t)]^{(1)} & \cdots & [B_C(t)]^{(n)} \end{bmatrix}$$

The equilibrium condition for internal constraint forces, arising from the compatibility conditions and eventually from sliding friction, implies that the sum of connecting forces at a pair of matching DoFs must be zero, i.e. $g_l^{(r)}(t) + g_m^{(s)}(t) = 0$: this holds for any pair of matching DoFs at time t. Furthermore, if DoF k on subsystem q is not a connecting DoF at time t, it must be $g_k^{(q)}(t) = 0$: this holds for any non-connecting DoF at time t. In presence of friction, the set of degrees of freedom to which the equilibrium condition applies includes tangential directions at the contact interface, which are not considered in the compatibility conditions due to sliding. This is in fact the definition of a non collocated interface.

Overall, the above conditions can be expressed as:

$$[L_E(t)]^T\{g(t)\} = \{0\} \tag{9.6}$$

where the matrix $[L_E(t)]$ is a Boolean localisation matrix. Note that the number of rows of $[L_E(t)]^T$ is equal to the number of non-interface DoFs plus the number of pairs of equilibrium interface DoFs.

Equations (9.2), (9.3), (9.4), (9.5), and (9.6) can be put together to get the three field formulation, i.e. the total system describing the coupling between any number of substructures:

$$\begin{cases} [M]\{\ddot{u}(t)\} + [C]\{\dot{u}(t)\} + [K]\{u(t)\} = \{f(t)\} + \{g(t)\} \\ [B_C(t)]\{u(t)\} = \{0\} \\ [L_E(t)]^T\{g(t)\} = \{0\} \end{cases} \tag{9.7}$$

Therefore, the three field formulation is very similar to the classical one for a stationary interface, except for $[B_C]$ and $[L_E]$ which now depend on time t.

9.2.2 Solution Approach in Time Domain

In principle, both primal assembly and dual assembly could be used [1, 6].

In the primal assembly, a unique set of interface DoFs is defined and the interface forces are eliminated by automatically satisfying the interface equilibrium. This is obtained by stating that:

$$\{u(t)\} = [L_C(t)]\{q(t)\} \tag{9.8}$$

where $\{q\}$ is the unique set of DoFs, including also non-interface DoFs, and $[L_C(t)]$ is different from the localisation matrix $[L_E(t)]$ introduced previously, when friction forces are exchanged at the interface. Since Eq. (9.8) states that the DoFs of all subsystems are obtained from the unique set $\{q\}$, the compatibility condition holds for any set $\{q\}$, i.e.

$$[B_C(t)]\{u(t)\} = [B_C(t)][L_C(t)]\{q(t)\} = \{0\} \qquad \forall\{q\} \tag{9.9}$$

Hence, $[L_C(t)]$ represents the nullspace of $[B_C(t)]$:

$$[L_C(t)] = \text{null}\left([B_C(t)]\right) \tag{9.10}$$

Since the compatibility condition in Eq. (9.7) is satisfied by the choice of the unique set $\{q\}$, the system of equations is:

$$\begin{cases} [M][L_C(t)]\{\ddot{q}(t)\} + [C][L_C(t)]\{\dot{q}(t)\} + [K][L_C(t)]\{q(t)\} = \{f(t)\} + \{g(t)\} \\ [L_E(t)]^T\{g(t)\} = \{0\} \end{cases} \tag{9.11}$$

Pre-multiplying the dynamic equilibrium equation by $[L_E(t)]^T$ and noting that $[L_E(t)]^T\{g(t)\} = \{0\}$, the assembled system reduces to:

$$[L_E(t)]^T[M][L_C(t)]\{\ddot{q}(t)\} + [L_E(t)]^T[C][L_C(t)]\{\dot{q}(t)\} + [L_E(t)]^T[K][L_C(t)]\{q(t)\} = [L_E(t)]^T\{f(t)\} \tag{9.12}$$

i.e.

$$\left[\tilde{M}(t)\right]\{\ddot{q}(t)\} + \left[\tilde{C}(t)\right]\{\dot{q}(t)\} + \left[\tilde{K}(t)\right]\{q(t)\} = \{\tilde{f}(t)\} \tag{9.13}$$

where

$$\begin{aligned} \left[\tilde{M}(t)\right] &= [L_E(t)]^T[M][L_C(t)] & \left[\tilde{C}(t)\right] &= [L_E(t)]^T[C][L_C(t)] \\ \left[\tilde{K}(t)\right] &= [L_E(t)]^T[K][L_C(t)] & \{\tilde{f}(t)\} &= [L_E(t)]^T\{f(t)\} \end{aligned} \tag{9.14}$$

Equation (9.13) can be recast in state space form:

$$\begin{Bmatrix} \{\dot{q}(t)\} \\ \{\ddot{q}(t)\} \end{Bmatrix} = \begin{bmatrix} [0] & [1] \\ -\left[\tilde{M}(t)\right]^{-1}\left[\tilde{K}(t)\right] & -\left[\tilde{M}(t)\right]^{-1}\left[\tilde{C}(t)\right] \end{bmatrix} \begin{Bmatrix} \{q(t)\} \\ \{\dot{q}(t)\} \end{Bmatrix} + \begin{Bmatrix} \{0\} \\ \left[\tilde{M}(t)\right]^{-1}\{\tilde{f}(t)\} \end{Bmatrix} \tag{9.15}$$

i.e.

$$\{\dot{y}(t)\} = [A(t)]\{y(t)\} + \{F(t)\} \tag{9.16}$$

where

$$\{y\} = \begin{Bmatrix} \{q\} \\ \{\dot{q}\} \end{Bmatrix} \qquad [A] = \begin{bmatrix} [0] & \lceil 1 \rfloor \\ -[\tilde{M}]^{-1}[\tilde{K}] & -[\tilde{M}]^{-1}[\tilde{C}] \end{bmatrix} \qquad \{F\} = \begin{Bmatrix} \{0\} \\ [\tilde{M}]^{-1}\{\tilde{f}\} \end{Bmatrix} \tag{9.17}$$

having omitted the explicit dependence on t in the last equation for the sake of simplicity.

Since $[A(t)]$ is time dependent, Eq. (9.16) can not be solved analytically but needs numerical time step integration.

In the dual assembly, the total set of DoFs is retained, i.e. each interface DoF is present as many times as there are substructures connected through that DoF. The equilibrium condition $g_l^{(r)} + g_m^{(s)} = 0$ at a pair of interface DoFs is ensured by choosing, for instance, $g_l^{(r)} = -\lambda$ and $g_m^{(s)} = \lambda$. Therefore, the overall interface equilibrium can be ensured by writing the connecting forces in the form:

$$\{g(t)\} = -[B_E(t)]^T\{\lambda(t)\} \tag{9.18}$$

where $\{\lambda\}$ are Lagrange multipliers corresponding to connecting force intensities and $[B_E(t)]$ is different from the signed Boolean matrix $[B_C(t)]$ used to enforce the compatibility condition, because $[B_E(t)]$ must also account for friction forces at the interface.

The interface equilibrium condition (9.6) is thus written:

$$[L_E(t)]^T\{g(t)\} = -[L_E(t)]^T[B_E(t)]^T\{\lambda(t)\} = \{0\} \tag{9.19}$$

Hence $[B_E]^T$ represents the nullspace of $[L_E]^T$:

$$[B_E(t)]^T = \mathrm{null}\left([L_E(t)]^T\right) \tag{9.20}$$

Since Eq. (9.19) is always satisfied, Eq. (9.7) becomes:

$$\begin{cases} [M]\{\ddot{u}(t)\} + [C]\{\dot{u}(t)\} + [K]\{u(t)\} + [B_E(t)]^T\{\lambda(t)\} = \{f(t)\} \\ [B_C(t)]\{u(t)\} = \{0\} \end{cases} \tag{9.21}$$

and in block matrix form:

$$\begin{bmatrix} [M] & [0] \\ [0] & [0] \end{bmatrix} \begin{Bmatrix} \{\ddot{u}(t)\} \\ \{\ddot{\lambda}(t)\} \end{Bmatrix} + \begin{bmatrix} [C] & [0] \\ [0] & [0] \end{bmatrix} \begin{Bmatrix} \{\dot{u}(t)\} \\ \{\dot{\lambda}(t)\} \end{Bmatrix} + \begin{bmatrix} [K] & [B_E]^T \\ [B_C] & [0] \end{bmatrix} \begin{Bmatrix} \{u(t)\} \\ \{\lambda(t)\} \end{Bmatrix} = \begin{Bmatrix} \{f(t)\} \\ \{0\} \end{Bmatrix} \tag{9.22}$$

Equation (9.22) is singular and cannot be explicitly solved in the time domain. Therefore, dual assembly is performed by computing time dependent frequency response functions (TD-FRFs).

9.2.3 Time Dependent Frequency Response Function

To have an idea of the influence of the time dependent interface on the dynamics of the system at different frequencies, Eq. (9.21) can be expressed as:

$$\begin{cases} [Z(j\omega)]\{u(\omega,t)\} + [B_E(t)]^T\{\lambda(\omega,t)\} = \{f(\omega)\} \\ [B_C(t)]\{u(\omega,t)\} = \{0\} \end{cases} \tag{9.23}$$

where $[Z(j\omega)] = -\omega^2[M] + j\omega[C] + [K]$ is a dynamic stiffness; $\{u\}$ and $\{\lambda\}$ are considered as functions of ω and t, while $[B_E]^T$ and $[B_C]$ remain functions only of time. These assumptions hold if the characteristic time of the motion of the sliding interface is different from the characteristic period of vibration of the system by several orders of magnitude.

To eliminate $\{\lambda\}$, the first of Eq. (9.23) can be written:

$$\{u(\omega,t)\} = -[Z(j\omega)]^{-1}[B_E(t)]^T\{\lambda(\omega,t)\} + [Z(j\omega)]^{-1}\{f(\omega)\} \tag{9.24}$$

which substituted in the second of Eq. (9.23) gives:

$$[B_C(t)][Z(j\omega)]^{-1}[B_E(t)]^T\{\lambda(\omega,t)\} = [B_C(t)][Z(j\omega)]^{-1}\{f(\omega)\} \tag{9.25}$$

from which

$$\{\lambda(\omega,t)\} = \left([B_C(t)][Z(j\omega)]^{-1}[B_E(t)]^T\right)^{-1}[B_C(t)][Z(j\omega)]^{-1}\{f(\omega)\} \tag{9.26}$$

Substituting $\{\lambda\}$ in the first of Eq. (9.23), it is obtained:

$$[Z(j\omega)]\{u(\omega,t)\} + [B_E(t)]^T\left([B_C(t)][Z(j\omega)]^{-1}[B_E(t)]^T\right)^{-1}[B_C(t)][Z(j\omega)]^{-1}\{f(\omega)\} = \{f(\omega)\} \tag{9.27}$$

Therefore, $\{u(\omega,t)\}$ is obtained as

$$\{u(\omega,t)\} = \left([H(j\omega)] - [H(j\omega)][B_E(t)]^T\left([B_C(t)][H(j\omega)][B_E(t)]^T\right)^{-1}[B_C(t)][H(j\omega)]\right)\{f(\omega)\} \tag{9.28}$$

where $[H(j\omega)] = [Z(j\omega)]^{-1}$ is a block diagonal matrix, too.

In Eq. (9.28), $\{u(\omega,t)\}$ represents the response of the assembled system to an external force $\{f(\omega)\}$. Therefore, the time dependent frequency response function $\left[H^C(j\omega,t)\right]$ of the assembled system with a sliding contact interface is:

$$\left[H^C(j\omega,t)\right] = [H(j\omega)] - [H(j\omega)][B_E(t)]^T\left([B_C(t)][H(j\omega)][B_E(t)]^T\right)^{-1}[B_C(t)][H(j\omega)] \tag{9.29}$$

9.3 Numerical Model

The considered system (Fig. 9.1) consists of two subsystems connected through a sliding interface with a constant friction coefficient μ. Subsystem 1 is a bar of length l and mass m_1, suspended by two vertical springs k_1 and k_2, and one horizontal spring k_3. The bar is modeled as a 2 node bar element with lumped mass matrix ($m_1/2$ at each end), and an axial stiffness k_6. Subsystem 2 is a lumped mass m_2 connected, by a vertical spring k_4 and a horizontal spring k_5, to a support moving with constant velocity v in the horizontal direction. A vertical preload F_2 acts on the lumped mass m_2 to push it on the bar (Table 9.1).

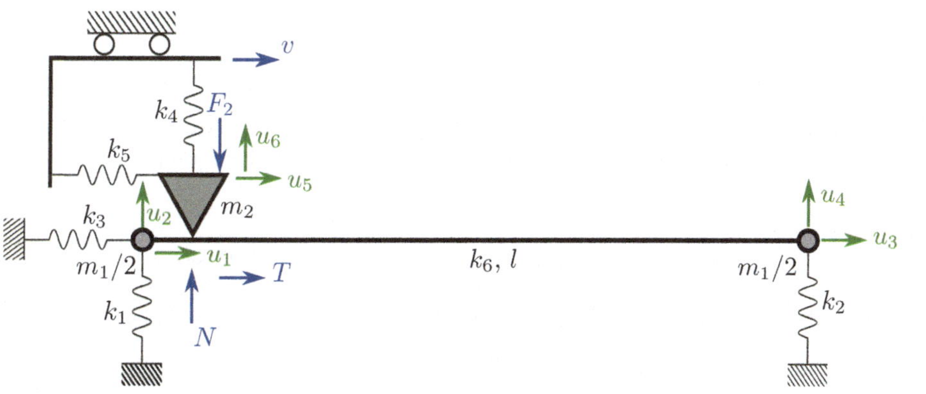

Fig. 9.1 Sketch of the system model

Table 9.1 Values of the physical parameters of the model

k_1	k_2	k_3	k_4	k_5	k_6		m_1	m_2		l	F_2	v
50	200	3000	200	1e5	6.41e6	N/m	1	0.3	kg	2 m	10 N	9.77 mm/s

Subsystem 1 has 4 DoFs $\{u\}^{(1)} = [u_1\ u_2\ u_3\ u_4]^T$, and the mass and stiffness matrices are:

$$[M]^{(1)} = \begin{bmatrix} m_1/2 & 0 & 0 & 0 \\ 0 & m_1/2 & 0 & 0 \\ 0 & 0 & m_1/2 & 0 \\ 0 & 0 & 0 & m_1/2 \end{bmatrix} \qquad [K]^{(1)} = \begin{bmatrix} k_3 + k_6 & 0 & -k_6 & 0 \\ 0 & k_1 & 0 & 0 \\ -k_6 & 0 & k_6 & 0 \\ 0 & 0 & 0 & k_2 \end{bmatrix} \tag{9.30}$$

The vector of external forces $\{f\}^{(1)}$ acting on subsystem 1 is zero.
Subsystem 2 has 2 DoFs $\{u\}^{(2)} = [u_5\ u_6]^T$, and the mass and stiffness matrices are:

$$[M]^{(2)} = \begin{bmatrix} m_2 & 0 \\ 0 & m_2 \end{bmatrix} \qquad [K]^{(2)} = \begin{bmatrix} k_5 & 0 \\ 0 & k_4 \end{bmatrix} \tag{9.31}$$

The vector of external forces acting on subsystem 2 is $\{f\}^{(2)} = [0\ F_2]^T$.

The block diagonal mass and stiffness matrices of the whole system are expressed according to Eq. (9.3), while the vectors of the full set of DoFs and of the external forces are expressed according to Eq. (9.4). A proportional viscous damping matrix is defined as $[C] = \beta[K]$ with $\beta = 1e - 4\,\text{s}$. .

When the two subsystems are connected together, it is assumed that mass m_2 is always in contact with the bar and that the tangential force is $|T| = \mu|N|$ where N is the normal contact force between the moving mass and the bar.

9.3.1 Compatibility and Equilibrium Matrices for Primal Assembly

To perform the primal assembly, the independent set of DoFs is selected as $\{q\} = [u_1\ u_2\ u_3\ u_4\ u_5]^T$. Therefore:

$$[L_C(t)] = \begin{bmatrix} 1 & 0 & 0 & 0 & 0 \\ 0 & 1 & 0 & 0 & 0 \\ 0 & 0 & 1 & 0 & 0 \\ 0 & 0 & 0 & 1 & 0 \\ 0 & 0 & 0 & 0 & 1 \\ 0 & n_1(t) & 0 & n_2(t) & 0 \end{bmatrix} \qquad [L_E(t)] = \begin{bmatrix} 1 & 0 & 0 & 0 & 0 \\ \mu & 1 & 0 & 0 & 0 \\ 0 & 0 & 1 & 0 & 0 \\ 0 & 0 & \mu & 1 & 0 \\ 0 & 0 & 0 & 0 & 1 \\ 0 & n_1(t) & 0 & n_2(t) & \mu \end{bmatrix} \tag{9.32}$$

where

$$n_1(t) = 1 - vt/l \qquad n_2(t) = vt/l \tag{9.33}$$

are time dependent coefficients used in $\{u\} = [L_C]\{q\}$ to obtain the vertical position u_6 of mass m_2 as functions of the vertical positions u_2 and u_4 of the end nodes of the bar, and in $[L_E]^T\{g\} = 0$ to satisfy the equilibrium of constraint forces in the vertical direction, e.g.

$$g_2 + n_1(t)g_6 = 0 \qquad g_4 + n_2(t)g_6 = 0 \tag{9.34}$$

Note that the relation between the friction force and the normal force ($|T| = \mu|N|$) is also accounted in $[L_E]^T\{g\} = 0$ as

$$g_1 + \mu g_2 = 0 \qquad g_3 + \mu g_4 = 0 \qquad g_5 + \mu g_6 = 0 \tag{9.35}$$

9.3.2 *Compatibility and Equilibrium Matrices for Dual Assembly*

To perform the dual assembly, the matrices $[B_C(t)]$ and $[B_E(t)]$ to be used in the compatibility equation $[B_C(t)]\{u\} = 0$ and in the equilibrium equation $\{g\} = -[B_E(t)]^T\{\lambda\}$ are

$$[B_C(t)] = \begin{bmatrix} 0 & n_1(t) & 0 & n_2(t) & 0 & -1 \end{bmatrix} \qquad [B_E(t)] = \begin{bmatrix} -\mu n_1(t) & n_1(t) & -\mu n_2(t) & n_2(t) & \mu & -1 \end{bmatrix} \tag{9.36}$$

where $n_1(t)$ and $n_2(t)$ are defined as above. Note that using the values in Eq. (9.36) and in Eq. (9.32), it can be verified that $[B_C][L_C] = 0$ and $[L_E]^T[B_E]^T = 0$ as stated by Eqs. (9.9) and (9.19).

9.4 Results

Two values of the friction coefficient μ are used: $\mu = 0$ and $\mu = 0.2$.

9.4.1 *Time Domain Results Using Primal Assembly*

The problem defined by Eq. (9.15) is solved numerically by using the Matlab function `ode45`, using fourth order Runge Kutta time step integration. The simulation is performed for the time employed by the mass m_2 to slide along the total length of the bar, i.e. 204.75 s.

The sampling time chosen to observe the results is $\Delta t = 0.05$ s. Time histories of displacements u_1, u_2, u_3, u_4 at the ends of the bar are shown in Fig. 9.2 for $\mu = 0$. Without friction, the horizontal displacements u_1, u_3 are equal to zero, because the horizontal motion is not excited by the moving mass m_2. Vertical displacements u_2, u_4 behave as expected.

Spectrograms of vertical velocities $\dot{u}_2, \dot{u}_4$ are shown in Fig. 9.3a, b. It can be noticed that the first natural frequency of the system goes from 2.81 to 1.59 Hz when the mass m_2 moves from left to right, while the second natural frequency increases from 3.18 to 3.55 Hz.

Time histories of displacements u_1, u_2, u_3, u_4 at the ends of the bar are shown in Fig. 9.4a for $\mu = 0.2$. With friction, the horizontal displacements u_1, u_3 in Fig. 9.4b are very low but not equal to zero, because the horizontal motion is excited by the friction force exerted by the moving mass m_2. Vertical displacements u_2, u_4 are very similar to the frictionless case.

Spectrograms of velocities $\dot{u}_3, \dot{u}_4$ are shown in Fig. 9.5a, b. The spectrogram of the horizontal velocity $\dot{u}_3$ shows a third natural frequency corresponding to horizontal motion of the bar. The spectrogram of the vertical velocity $\dot{u}_4$ is very similar to the frictionless case.

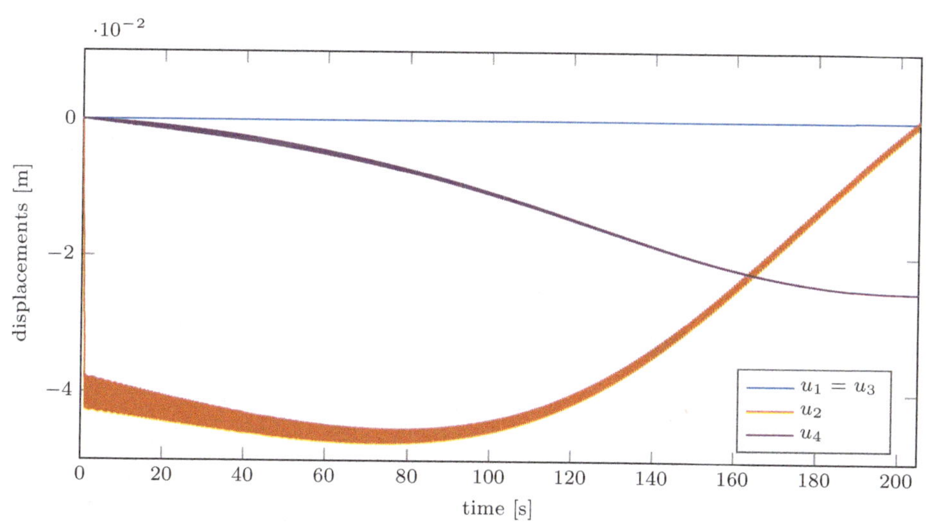

Fig. 9.2 Time histories of displacements at the ends of the bar for $\mu = 0$

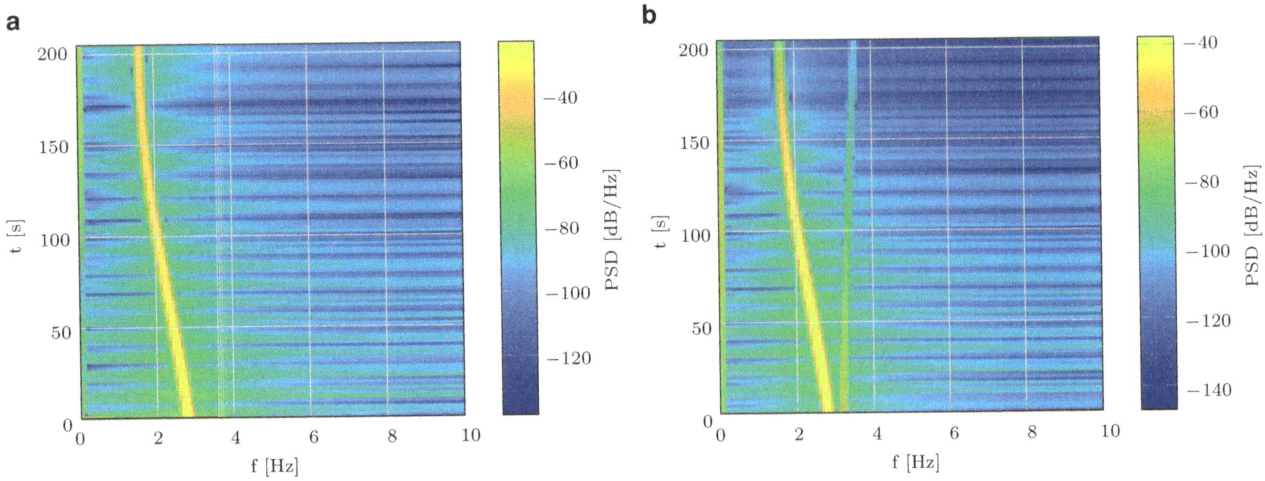

Fig. 9.3 Velocity spectrograms for $\mu = 0$. (**a**) $\dot{u}_2$. (**b**) $\dot{u}_4$

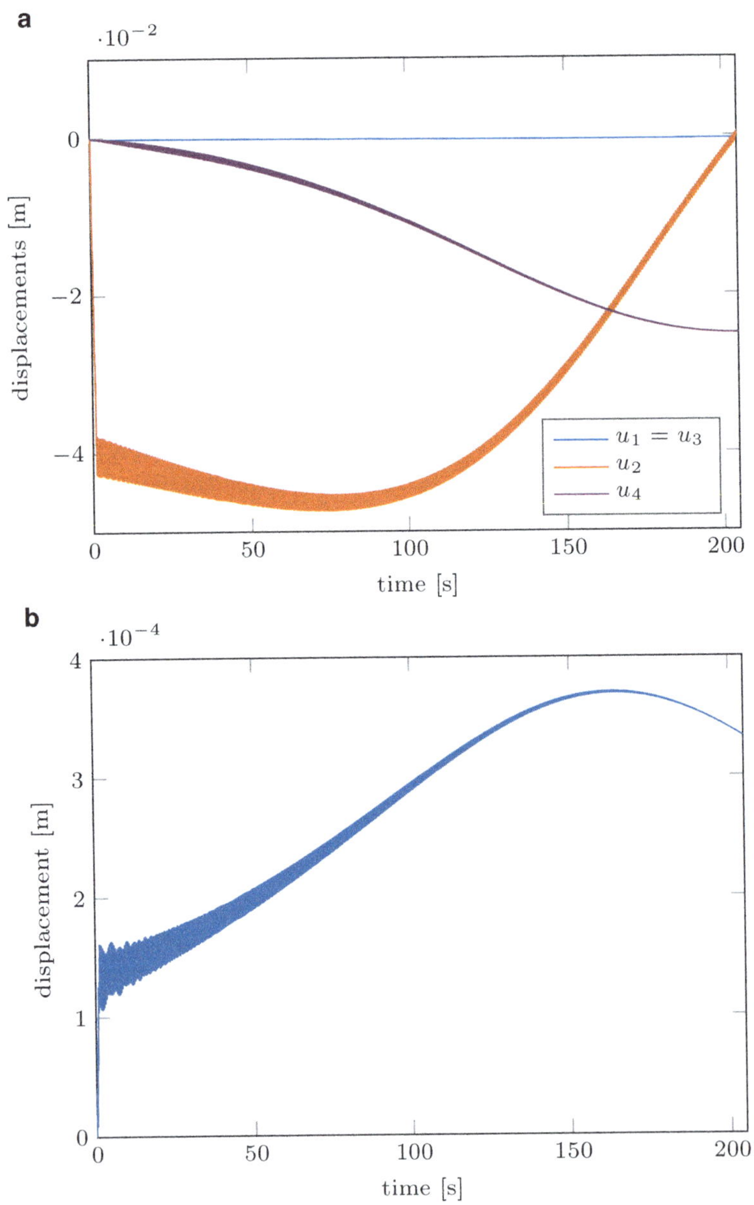

Fig. 9.4 Time histories of displacements at the ends of the bar for $\mu = 0.2$. (**a**) DoFs u_1, u_2, u_3 and u_4. (**b**) DoF $u_1 = u_3$

9.4.2 Time Dependent FRF Using Dual Assembly

Equation (9.29) is used to compute time dependent FRFs of the coupled system when mass m_2 is sliding along the bar with the same speed as in the previous case. The time and frequency intervals are [0, 204.75] s and [0, 10] Hz. Time and frequency increments are $\Delta t = 0.05$ s and $\Delta f = 0.1$ Hz.

Figure 9.6a, b show the time dependent FRFs for $\mu = 0$ by considering an input force acting at DoF 6 (vertical DoF of mass m_2) and output displacements at DoFs 2 and 4 (vertical DoFs at the ends of the bar). The peaks of the time dependent FRFs are quite similar to those of the corresponding velocity spectrograms, i.e. those related to the same output DoF.

Figure 9.7a, b, c, d show the time dependent FRFs for $\mu = 0.2$ by considering additional output displacements at DoFs 3 (horizontal DoFs at the right end of the bar) and 6.

The peaks of the time dependent FRFs of H_{26} and H_{46} are nearly identical to those of the frictionless case. Conversely, while in the frictionless case the time dependent FRF H_{36} (not shown) was identically zero, with $\mu = 0.2$ the FRF is not zero and the peaks are quite similar to those of the velocity spectrogram $\dot{u}_3$. Finally, the FRF H_{66} is in fact a linear combination of H_{26} and H_{46} with coefficients $n_1(t)$ and $n_2(t)$. (DoF u_6 could not be observed in primal assembly because it was not selected as an independent DoF.)

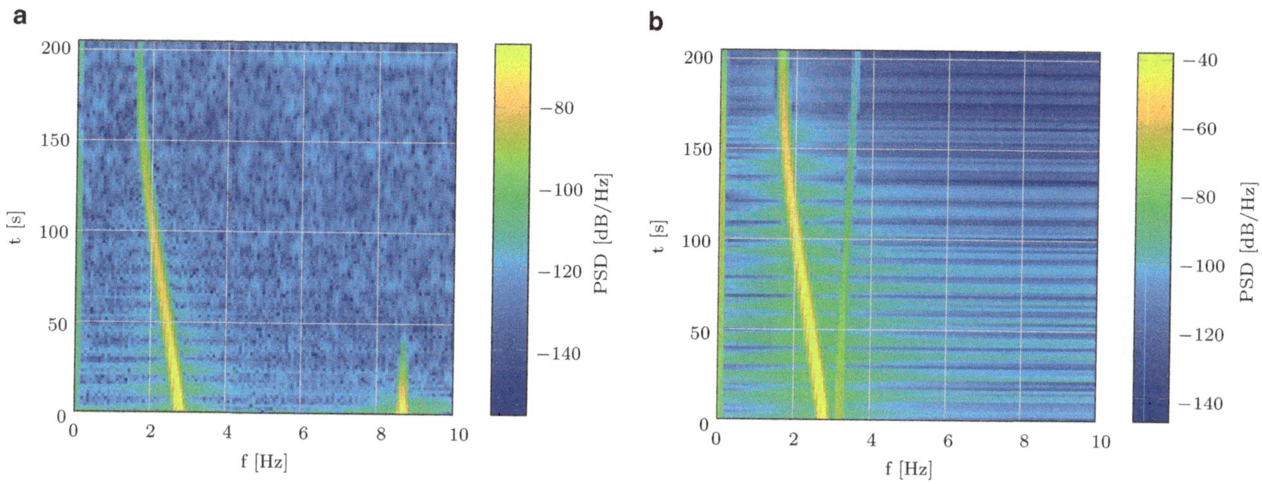

Fig. 9.5 Velocity spectrograms for $\mu = 0.2$. (**a**) $\dot{u}_3$. (**b**) $\dot{u}_4$

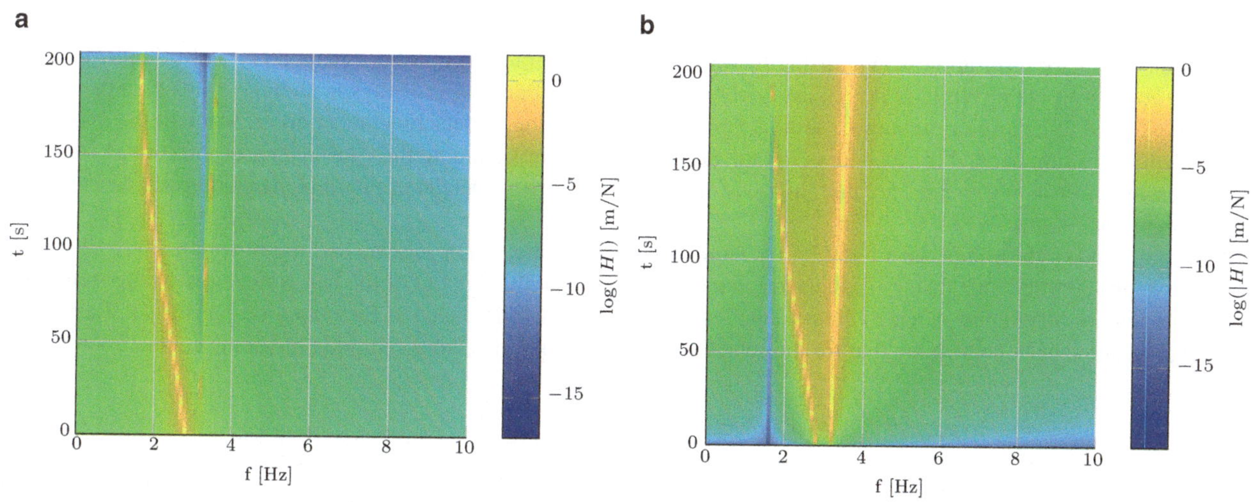

Fig. 9.6 Time dependent FRFs for $\mu = 0$. (**a**) $|H_{26}(\omega, t)|$. (**b**) $|H_{46}(\omega, t)|$

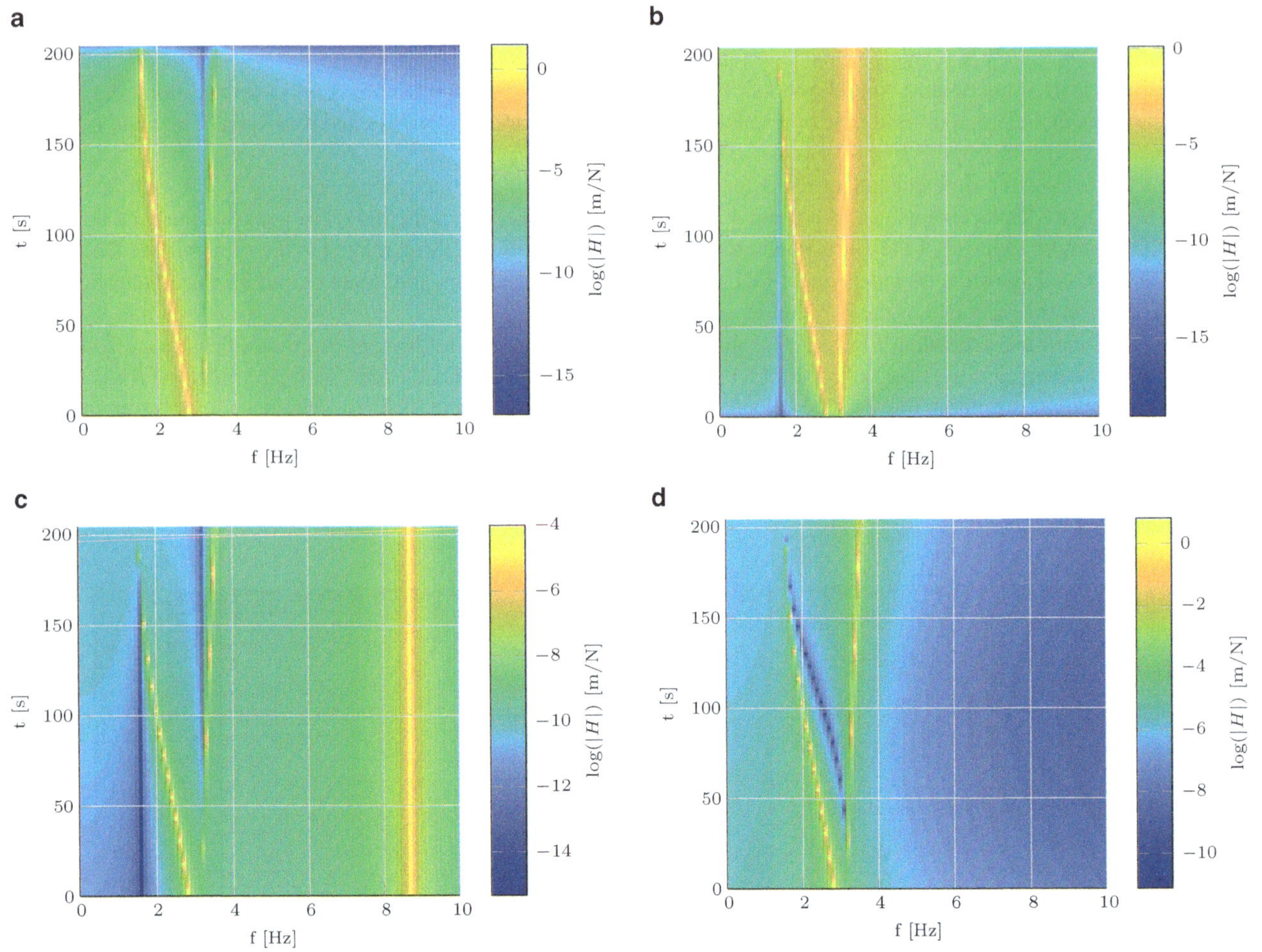

Fig. 9.7 Time dependent FRFs for $\mu = 0.2$. (**a**) $|H_{26}(\omega, t)|$. (**b**) $|H_{46}(\omega, t)|$. (**c**) $|H_{36}(\omega, t)|$. (**d**) $|H_{66}(\omega, t)|$

9.4.3 Time Domain Validation Accounting for Real Contact Status

An explicit Newmark integration scheme using forward increment Lagrange multipliers method [7] is also implemented to solve the contact problem. While in primal assembly it is assumed that contact is always in sliding, in this case the contact status is assumed to be a-priori unknown: sliding, reverse sliding, stiking or detachment can occur. Some of the main steps of the method are summarized in Appendix A.1: the iteration procedure leading to the convergence of the contact force and the discussion of the contact status are not reported for the sake of brevity. Details of the implementation will be the object of a future work.

Figure 9.8 shows the comparison between the explicit Newmark integration scheme and the time domain results using primal assembly, for the time history of the vertical displacement u_2 at the left end of the bar. Note that the time step of the explicit Newmark integration scheme is much smaller than the time step produced as the output of the `ode45`. Results are in very good agreement, confirming that for the selected boundary conditions the assumption of sliding contact is correct.

9.5 Conclusions

In this paper, the general framework for dynamic substructuring is extended to time-variant interfaces among connecting substructures. Specifically, a time-variant interface due to a sliding contact is considered by introducing time dependent compatibility and equilibrium conditions. The sliding contact can be without or with friction. In presence of friction, equilibrium conditions involve tangential directions at the contact interface, that do not appear in the compatibility conditions.

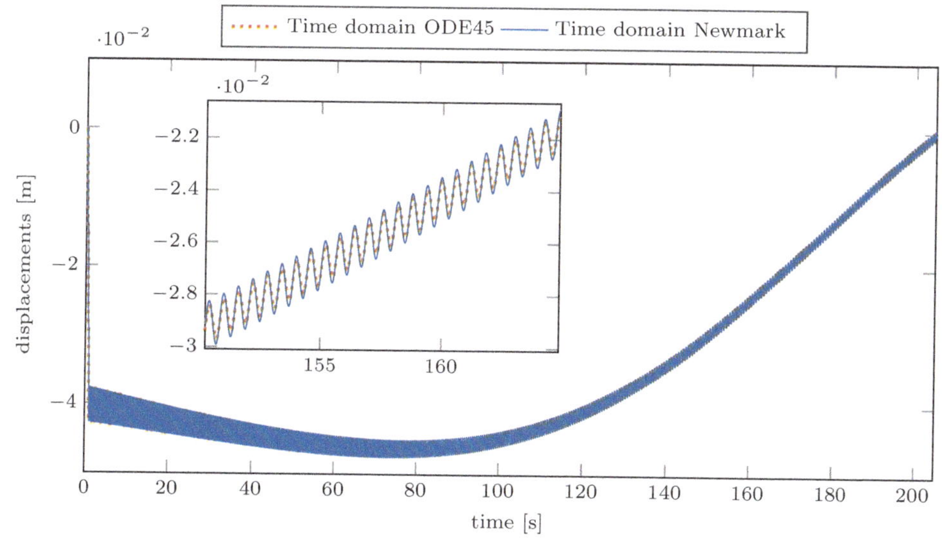

Fig. 9.8 Time history of displacement u_2 using primal assembly (ode45) and explicit time integration (Newmark)

The solution approach in time domain uses primal assembly and numerical time integration (ode45). The dynamic characteristics of the response can be highlighted in time-frequency domain using dual assembly, to obtain Time Dependent Frequency Response Functions (TD-FRFs) of the assembled structure.

Results on a lumped parameter model show the feasibility and effectiveness of the proposed approach. Furthermore the solution is validated using a different computational approach. Future developments involve the extension of the approach to simple finite element models under the assumption that the relative sliding motion at the contact interfaces is known a-priori, at least approximately.

Acknowledgements This research is supported by University of Rome La Sapienza and University of L'Aquila.

A.1 Newmark with Forward Lagrange Multipliers

For a mechanical system the equation of motion can be expressed as Eq. (9.2). The contact conditions can be considered by constraint equation such as:

$$[G]\left(\{u\}+\{X\}\right)=\{0\} \tag{9.37}$$

where $\{X\}$ is the vector of nodal coordinates at equilibrium, $(\{u\}+\{X\})$ is the current vector of nodal coordinates and $[G]$ is an a-priori unknown time-varying contact displacement constraint matrix. Hence, in Eq. (9.2), the interfaces forces can be expressed using Lagrange multipliers $\{g(t)\}=-[G]^T\{\lambda\}$.

With respect to Eq. (9.21) the matrix $[G]$, which replaces the matrices $[B_C]$ and $[B_E]$ is a-priori unknown. To overcome the singularity problem already mentioned for Eq. (9.22), Forward increment Lagrange multipliers method can be used. This formulation relates constraints at time $t_{n+1}=t_n+h$ with Lagrange multipliers at time t_n, and the equation of motion becomes:

$$\begin{cases} [M]\{\ddot{u}\}_n+[C]\{\dot{u}\}_n+[K]\{u\}_n+[G]_{n+1}^T\{\lambda\}_n=\{f\}_n \\ [G]_{n+1}\left(\{u\}_{n+1}+\{X\}\right)=\{0\} \end{cases} \tag{9.38}$$

For explicit integration of the equation, the Newmark β_2 scheme can be used, that for $\beta_2=0.5$ reduces to the central difference scheme:

$$\{\dot{u}\}_n=\frac{1}{2h}\left(\{u\}_{n+1}-\{u\}_{n-1}\right) \tag{9.39}$$

$$\{\ddot{u}\}_n = \frac{1}{h^2}\left(\{u\}_{n+1} - 2\{u\}_n + \{u\}_{n-1}\right) \tag{9.40}$$

The equation of motion can be rewritten as:

$$\{\ddot{u}\}_n = [M]^{-1}\left[\{f\}_n - [C]\{\dot{u}\}_n - [K]\{u\}_n - [G]_{n+1}^T\{\lambda\}_n\right] \tag{9.41}$$

By substituting the central difference expression of $\{\ddot{u}\}_n$ in the equation of motion, and by using $\{\dot{u}\}_n = \left(\{u\}_n - \{u\}_{n-1}\right)/h$ one obtains:

$$\{u\}_{n+1} = h^2[M]^{-1}\left[\{f\}_n - \frac{1}{h}[C]\left(\{u\}_n - \{u\}_{n-1}\right) - K\{u\}_n\right] + 2\{u\}_n - \{u\}_{n-1} - h^2[M]^{-1}[G]_{n+1}^T\{\lambda\}_n \tag{9.42}$$

By defining the predictor of displacement $\{u\}_{n+1}^{\star}$ as:

$$\{u\}_{n+1}^{\star} = h^2[M]^{-1}\left(\{R\}_n - [C]\{\dot{u}\}_n - [K]\{u\}_n\right) + 2\{u\}_n - \{u\}_{n-1} \tag{9.43}$$

the displacement at time t_{n+1} is:

$$\{u\}_{n+1} = \{u\}_{n+1}^{\star} - h^2[M]^{-1}[G]_{n+1}^T\{\lambda\}_n \tag{9.44}$$

Therefore, by substituting Eq. (9.44) in the second line of Eq. (9.38) one obtains:

$$[G]_{n+1}\left(\{u\}_{n+1}^{\star} + \{X\}\right) = h^2[G]_{n+1}[M]^{-1}[G]_{n+1}^T\{\lambda\}_n \tag{9.45}$$

Hence, an estimate of the contact forces $\{\lambda\}_n$ can be computed from the predictor $\{u\}_{n+1}^{\star}$:

$$\{\lambda\}_n = \left(h^2[G]_{n+1}[M]^{-1}[G]_{n+1}^T\right)^{-1}[G]_{n+1}\left(\{u\}_{n+1}^{\star} + \{X\}\right) \tag{9.46}$$

Finally, the corrector of the displacement $\{u\}_{n+1}^{c}$ is defined as:

$$\{u\}_{n+1}^{c} = -h^2[M]^{-1}[G]_{n+1}^T\{\lambda\}_n \tag{9.47}$$

and the displacement at time t_{n+1} is expressed as:

$$\{u\}_{n+1} = \{u\}_{n+1}^{\star} + \{u\}_{n+1}^{c} \tag{9.48}$$

References

1. de Klerk, D., Rixen, D.J., Voormeeren, S.: General framework for dynamic substructuring: history, review, and classification of techniques. AIAA J. **46**(5), 1169–1181 (2008)
2. Rixen, D.J., van der Valk, P.L.: An impulse based substructuring approach for impact analysis and load case simulations. J. Sound Vib. **332**(26), 7174–7190 (2013)
3. van der Valk, P.L., Rixen, D.J.: An impulse based substructuring method for coupling impulse response functions and finite element models. Comput. Methods Appl. Mech. Eng. **275**(Supplement C), 113–137 (2014)
4. Voormeeren, S.N., Rixen, D.J.: A family of substructure decoupling techniques based on a dual assembly approach. Mech. Syst. Signal Process. **27**, 379–396 (2012)
5. D'Ambrogio, W., Fregolent, A.: Inverse dynamic substructuring using direct hybrid assembly in the frequency domain. Mech. Syst. Signal Process. **45**(2), 360–377 (2014)
6. D'Ambrogio, W., Fregolent, A.: Direct decoupling of substructures using primal and dual formulation. In: Conference Proceedings of the Society of Experimental Mechanics Series, Volume 4: Linking Models and Experiments, 31 Jan–3 Feb 2011, vol. 2, pp. 47–76. Springer, Jacksonville (2011)
7. Carpenter, N.J., Taylor, R.L., Katona, M.G.: Lagrange constraints for transient finite element surface contact. Int. J. Numer. Methods Eng. **32**(1), 103–128 (1991)

Chapter 10
Introducing SEMM: A Novel Method for Hybrid Modelling

S. W. B. Klaassen and M. V. van der Seijs

Abstract A method named System Equivalent Model Mixing (SEMM) is presented. SEMM allows for a mixing of two equivalent frequency based models which can be created by either numerical simulation or direct measurements. The resulting hybrid FRF model is full-rank, consists of the DoF of both input models, and contains a physically relevant weighted mix of input dynamics. It is demonstrated that with SEMM a numerical DoF-set can be used to extend an experimental model with limited measurement points; specifically, it is shown how complete interface dynamics can be obtained with just a handful of sensors. The purpose of SEMM is similar to the well-known SEREP and VIKING concepts, yet instead applies Frequency Based Substructuring (FBS) techniques to form a hybrid dynamic model.

Keywords SEMM SEREP · System equivalent model mixing · Hybrid model · Dynamic substructuring · Frequency based substructuring · Model expansion

Nomenclature

DoF	degree of freedom
FRF	frequency response function
$\mathbf{u}$	dynamic displacements/rotations
$\mathbf{f}$	applied forces/moments
$\mathbf{g}$	interface forces/moments
$\mathbf{Y}$	admittance FRF matrix
$\mathbf{Z}$	impedance FRF matrix
$\mathbf{T}$	transformation matrix
$\mathbf{B}$	signed Boolean coupling matrix
$\star^{\text{par}}$	pertaining to the parent model
$\star^{\text{ov}}$	pertaining to the overlay model
$\star^{\text{rem}}$	pertaining to the removed model
$\star^{\text{SEMM}}$	pertaining to the SEMM hybrid model
$\star^{+}$	pseudo inverse
$\star_b$	boundary or interface DoF
$\star_i$	internal DoF
$\star_g$	global DoF

10.1 Introduction

High-tech systems consisting of increasingly complex components advance the need for high quality dynamic models. These models are created either numerically, or now also more and more experimentally. This is a result of increased measurement accuracy and validity, such that standalone experimental models are now considered legitimate. However, in order for a dynamic model to be deemed valid it must comply with some basic requirements: it must adequately represent the dynamic properties of the component (e.g. Resonance frequencies, damping values) and the model must contain enough DoF or control points in order for it to be used in future operations, e.g. modeshape observation or interface construction in Dynamic Substructuring (DS).

S. W. B. Klaassen (✉)
Faculty of Mechanical Engineering, Institute of Applied Mechanics, Technische Universität München, Garching, Germany

Department of Precision and Microsystems Engineering, Delft University of Technology, Delft, The Netherlands
e-mail: s.w.b.klaassen@outlook.com

M. V. van der Seijs
VIBES.technology, Delft, The Netherlands

Department of Precision and Microsystems Engineering, Delft University of Technology, Delft, The Netherlands

A. Linderholt et al. (eds.), *Dynamics of Coupled Structures, Volume 4*, Conference Proceedings of the Society for Experimental Mechanics Series, https://doi.org/10.1007/978-3-319-74654-8_10

The latter is particularly troublesome for experimental models which are, in practice, limited to the number of sensors available. Yet, cases exist where experimental models are nevertheless preferable because they represent the actual measured physics, rather than predict user-assumed physics. Consequently, much effort is put in the development of methods to expand the information provided by measurements. Iterative updating of numerical models based on experimental data is probably the most commonplace: The basic physics used to construct a numerical Finite Element (FE)-model is updated such that it converges to measurement results. A more involved method identifies the modeshapes present in the experimental model and fits a numerical model by means of pole/residue parametrisation [1]. Entirely different approach are DoF-set expanding methods that use experimental models as their foundation. These methods use numerical manifolds to expand the lacking DoF-span of the measurements. Existing DoF-expansion methods include physical expansion such as Guyan-, Dynamic- or IRS-expansion [2, 3]; or modal expansion methods such as the SEREP, or VIKING[1] [4, 5].

The approach taken with System Equivalent Model Mixing (SEMM) is different. With SEMM equivalent models (i.e. models of the same component yet founded on varying assumptions) are dynamically coupled following DS theory [6]. The result is a model that is comprised of dynamic properties from either one of the input models and which contains the DoF-space spanned by both input models. Theoretically, a numerical and experimental model can be combined to form a new hybrid model which consists of experimental dynamic properties and contains the DoF-set provided by a numerical model. Accordingly, although it is not a expansion method per se, SEMM can be used to expand the measured DoF-set to a larger numerical DoF-set, much like the aforementioned techniques. This justifies the comparison between SEMM and said methods.

10.1.1 Paper Outline

The paper does not intend to fully explain the complete working of SEMM, for this we refer to [6]. Instead, a main application of the method is tested and compared to a similar well-known techniques used throughout the community to offer a viable test-case. The method is therefore explained only in a basic sense. After a small recap of SEREP, it is compared to SEMM by means of explanation of the physical differences and similarities, as well as through a small numerical test-case.

10.2 Theory

In short, with SEMM, two FRF models of the same component are coupled on all their shared DoF while at the same time unwanted dynamics are removed by decoupling a reduced form of the unwanted model. This is illustrated in Fig. 10.1. The result is a hybrid dynamic model which consists of the DoF-space spanned by both DoF-sets and which has the dynamic properties of one of the (chosen) input models.

For SEMM two FRF models, named parent models, are required. Both can be acquired either by numerical or experimental means. One of the models acts as the provider of the DoF-set; therefore it is generally of a numerical nature. The second, often experimentally based, provides the dynamic properties. Generally both models have different DoF-sets and can contain DoF which are unique to their own ($\mathbf{u}_i$). However, their DoF-sets must overlap such that they contain shared DoF. The shared DoF are the interface or boundary DoF ($\mathbf{u}_b$); with these, dynamic coupling can be performed.

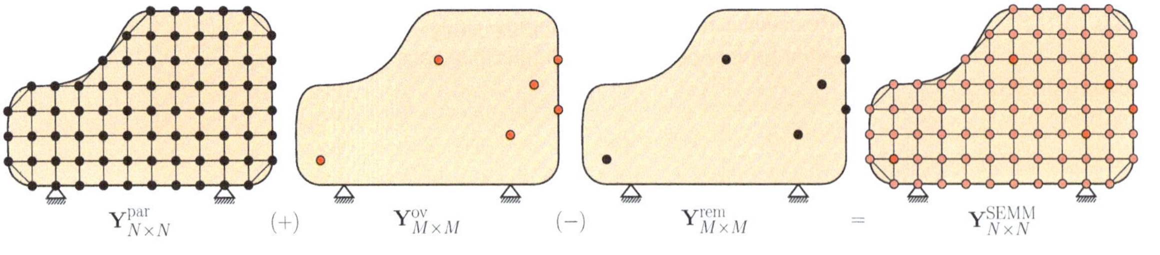

Fig. 10.1 A large DoF-set (numerical) model (far-left) is coupled to a small DoF-set (experimental) model (middle left) at all shared DoF. From this a reduced small DoF-set (numerical) model (middle-right) is decoupled to obtain a large DoF-set SEMM model (far-right)

[1]Improved Reduction System (IRS), System Equivalent Reduction Expansion Process (SEREP), Variability Improvement of Key Inaccurate Node Groups (VIKING).

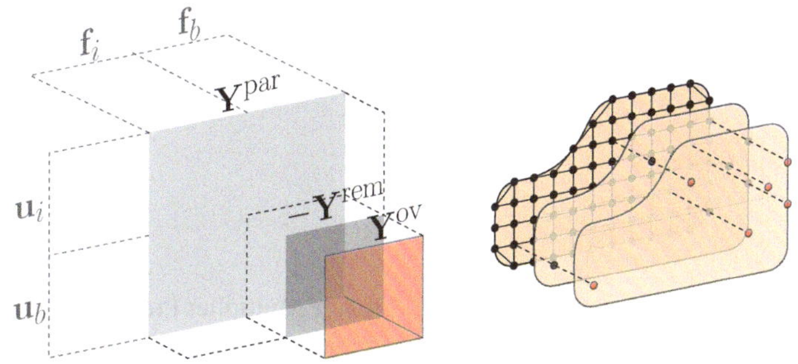

Fig. 10.2 The relation of the different sized models and how they are linked to one another to form a SEMM model. Left is a graphical representation of the models and right is graphical representation of what is physically happening with SEMM

The interface of the second (experimental) model is called the overlay; it will determine the dynamic properties of the SEMM expanded model. Finally, a tertiary part is needed to remove the unwanted dynamic properties of the first (numerical) parent model. This is the removed interface, and is a sub-set (based on the same interface) of the first parent model. Figure 10.2 provides a graphical representation of the different models.

To summarize:

1. Parent models: The models used in the SEMM process. One (often numerical) provides a DoF set, the other (often experimental) provides the dynamic properties.

$$\mathbf{Y}^{\text{par}} = \begin{bmatrix} \mathbf{Y}^{\text{par}}_{ii} & \mathbf{Y}^{\text{par}}_{ib} \\ \mathbf{Y}^{\text{par}}_{bi} & \mathbf{Y}^{\text{par}}_{bb} \end{bmatrix} \tag{10.1}$$

2. Overlay: The interface of the (often experimental) parent model that provides the dynamic properties.

$$\mathbf{Y}^{\text{ov}} = \begin{bmatrix} \mathbf{Y}^{\text{ov}}_{bb} \end{bmatrix} \tag{10.2}$$

3. Removed interface: The interface of the parent model of which the dynamic properties are removed.

$$\mathbf{Y}^{\text{rem}} = \begin{bmatrix} \mathbf{Y}^{\text{rem}}_{bb} \end{bmatrix} \tag{10.3}$$

These form the building blocks in SEMM modelling. A graphical representation is provided in Fig. 10.2. Note that there are experimental (overlay) and numerical components (removed interface, numerical parent model), thus it is considered a hybrid modelling form. The different models will be coupled using Frequency Based Substructuring (FBS). An equation of motion can be formulated:

$$\mathbf{u} = \mathbf{Y}\,(\mathbf{f} - \mathbf{g}) \quad \text{where} \quad \mathbf{Y} = \begin{bmatrix} \mathbf{Y}^{\text{par}} & & \\ & -\mathbf{Y}^{\text{rem}} & \\ & & \mathbf{Y}^{\text{ov}} \end{bmatrix} \tag{10.4}$$

Here the minus sign exists to indicate the removal of the removed model $\mathbf{Y}^{\text{rem}}$. The system is subject to constraints on the chosen boundary which enforce the coupling of the system: these are the compatibility and equilibrium conditions, which can both be reformulated to a matrix notation with the use of a signed boolean coupling matrix $\mathbf{B}$.

In the compatibility the displacements are set equal on the boundary:

$$\mathbf{u}^{\text{par}}_b = \mathbf{u}^{\text{rem}} = \mathbf{u}^{\text{ov}} \Longrightarrow \begin{cases} \mathbf{u}^{\text{par}}_b - \mathbf{u}^{\text{rem}} = \mathbf{0} & (10.5\text{a}) \\ \mathbf{u}^{\text{rem}} - \mathbf{u}^{\text{ov}} = \mathbf{0} & (10.5\text{b}) \end{cases}$$

Which in matrix notation becomes:

$$\mathbf{B}\mathbf{u} = \mathbf{0} \quad \text{where} \quad \mathbf{B} = \left[\begin{array}{c|c|c} \mathbf{I} & -\mathbf{I} & \mathbf{0} \\ \mathbf{0} & \mathbf{I} & -\mathbf{I} \end{array}\right] \tag{10.6}$$

In the equilibrium the forces are equal but opposite on the boundary:

$$\mathbf{g}_b^{\text{par}} + \mathbf{g}^{\text{rem}} + \mathbf{g}^{\text{ov}} = \mathbf{0} \tag{10.7}$$

Likewise, in matrix notation this becomes:

$$\mathbf{g} = -\mathbf{B}^T \boldsymbol{\lambda} \tag{10.8}$$

These can be solved using standard FBS, more precisely the Lagrange Multiplier Frequency Based Substructuring (LM-FBS) method [7]. It is shown in [6] that solving the LM-FBS problem and reformulating it to a primal notation results in a single line formulation for the SEMM model described by Eq. (10.9)

$$\mathbf{Y}^{\text{SEMM}} = \mathbf{Y}_{gg} - \mathbf{Y}_{gb}^{\text{par}} \left(\mathbf{Y}_{bb}^{\text{rem}}\right)^{-1} \left(\mathbf{Y}_{bb}^{\text{rem}} - \mathbf{Y}_{bb}^{\text{ov}}\right) \left(\mathbf{Y}_{bb}^{\text{rem}}\right)^{-1} \mathbf{Y}_{bg}^{\text{par}} \tag{10.9}$$

Where g is regarded as the global DoF combining the boundary b and internal i sets.

When solving Eq. (10.9) one finds that spurious peaks occur. These peaks are a consequence of conflicting dynamics between the experimental modes in $\mathbf{Y}^{\text{ov}}$ and the residual numerical modes in $\mathbf{Y}^{\text{par}}$. The residual modes are the modes not removed by the condensed removed interface $\mathbf{Y}^{\text{rem}}$. This incomplete removal is due to the fact that, by mathematical definition, the smaller interface cannot observe nor control the full space of the parent model.

This problem is thus circumvented by expanding the size of the removed interface to include certain internal DoF while still remaining coupled to the interface prescribed by the overlay. The new removed interface can, for example, include all internal DoF as defined in Eq. (10.10). This ultimately results in a single-line notation provided by (10.11). It is shown in [6] that the extended interface removes spurious peaks caused by conflicting numerical dynamics.

$$\mathbf{Y}^{\text{rem}} = \left[\mathbf{Y}_{gg}^{\text{rem}}\right] = \begin{bmatrix} \mathbf{Y}_{ii}^{\text{rem}} & \mathbf{Y}_{ib}^{\text{rem}} \\ \mathbf{Y}_{bi}^{\text{rem}} & \mathbf{Y}_{bb}^{\text{rem}} \end{bmatrix} \tag{10.10}$$

$$\mathbf{Y}^{\text{SEMM}} = \mathbf{Y}_{gg} - \mathbf{Y}_{gg}^{\text{par}} \left(\mathbf{Y}_{bg}^{\text{rem}}\right)^{+} \left(\mathbf{Y}_{bb}^{\text{rem}} - \mathbf{Y}_{bb}^{\text{ov}}\right) \left(\mathbf{Y}_{gb}^{\text{rem}}\right)^{+} \mathbf{Y}_{gg}^{\text{par}} \tag{10.11}$$

10.3 Comparing SEMM to SEREP

A currently well-known and often used method to extend a DoF set is the SEREP method [5]. SEREP extends an experiment's DoF-set by projecting it on a predefined set of modes computed from an equivalent numerical model. This modal projection can be expanded to a numerical DoF set.

In short, the displacements $\mathbf{u}$ are a weighted summation of the modeshapes present in the natural mode-matrix $\boldsymbol{\Phi}$. The weight is determined by the modal contributions $\boldsymbol{\eta}$. The mode-matrix can be decomposed in retained r and truncated t mode columns and internal i and boundary b DoF rows. It is assumed that the influence of the truncated modes is negligible; thus the modal decomposition (10.12) can be approximated by (10.13).

$$\begin{bmatrix} \mathbf{u}_i \\ \mathbf{u}_b \end{bmatrix} = \begin{bmatrix} \boldsymbol{\Phi}_{ir} & \boldsymbol{\Phi}_{it} \\ \boldsymbol{\Phi}_{br} & \boldsymbol{\Phi}_{bt} \end{bmatrix} \begin{bmatrix} \boldsymbol{\eta}_r \\ \boldsymbol{\eta}_t \end{bmatrix} \tag{10.12}$$

$$\begin{bmatrix} \mathbf{u}_i \\ \mathbf{u}_b \end{bmatrix} \approx \begin{bmatrix} \boldsymbol{\Phi}_{ir} \\ \boldsymbol{\Phi}_{br} \end{bmatrix} \boldsymbol{\eta}_r \Longrightarrow \begin{cases} \mathbf{u}_i = \boldsymbol{\Phi}_{ir}\boldsymbol{\eta}_r & \text{(10.13a)} \\ \mathbf{u}_b = \boldsymbol{\Phi}_{br}\boldsymbol{\eta}_r & \text{(10.13b)} \end{cases}$$

The modal contributions are then obtained by inverting Eq. (10.13b).

$$\boldsymbol{\eta}_r = \left(\boldsymbol{\Phi}_{br}\right)^{-1} \mathbf{u}_b \tag{10.14}$$

Such that the expansion matrix **T** can be defined:

$$\begin{bmatrix} \mathbf{u}_i \\ \mathbf{u}_b \end{bmatrix} = \begin{bmatrix} \mathbf{\Phi}_{ir} \\ \mathbf{\Phi}_{br} \end{bmatrix} (\mathbf{\Phi}_{br})^{-1} \mathbf{u}_b = \mathbf{T}\mathbf{u}_b \tag{10.15}$$

We see that if size(r) = size(b) then the lower part of the expansion matrix **T** is identity and consequently the boundary DoF remain unchanged; this is the case for a specific form of SEREP called SEREPa. In standard SEREP size(r) < size(b) is true; therefore over-projection of the boundary DoF b over the modal space r filters the boundary unto the retained numerical modes. In short:

SEREPa: Non-filtering modal projection (size(r) = size(b))

$$\begin{bmatrix} \mathbf{u}_i \\ \mathbf{u}_b \end{bmatrix} = \begin{bmatrix} \mathbf{\Phi}_{ir} (\mathbf{\Phi}_{br})^{-1} \\ \mathbf{I} \end{bmatrix} \mathbf{u}_b = \mathbf{T}^{\mathrm{a}}\mathbf{u}_b \tag{10.16}$$

SEREP: Filtering modal projection (size(r) < size(b))

$$\begin{bmatrix} \mathbf{u}_i \\ \mathbf{u}_b \end{bmatrix} = \begin{bmatrix} \mathbf{\Phi}_{ir} (\mathbf{\Phi}_{br})^{+} \\ \mathbf{\Phi}_{br} (\mathbf{\Phi}_{br})^{+} \end{bmatrix} \mathbf{u}_b = \mathbf{T}\mathbf{u}_b \tag{10.17}$$

Note that the reciprocal of this same expansion matrix **T** is used to expand the input DoF, i.e. $(\mathbf{T})^T$ for SEREP or $(\mathbf{T}^a)^T$ for SEREPa.

One could determine that a distinction between SEMM and SEREP is in the use of different domains; however, the use of other domains does not result in a difference in outcomes per se. In fact, even though the methodology for acquiring the expanded model is different and admittedly based on other physical principles (Dynamic substructuring for SEMM and mode-projection for SEREP), both for SEMM and SEREP(a) the expansion is based on numerical modeshapes.

The actual distinction is twofold: For SEREP(a) the modes used in the expansion are explicitly chosen; they are the natural modes retained in the r mode set. For SEMM, the modes used are implicit in the numerical model. The modes are no longer evenly distributed, rather a distribution is decided by the natural prevalence of modes in the numerical FRF-model.

The second difference is that SEREP(a) expands the experimental vectors themselves whereas SEMM expands the subtraction of experimental from the numerical vectors. This expanded difference is removed from the full numerical model. Therefore physically, the method includes internal dynamics, i.e. the internal-to-internal dynamics retained from the numerical model. Mathematically, this ensures that the SEMM model's rank is determined by the full-numerical model and will generally be full-rank. This is not the case for the SEREP(a) expanded models; here the rank is generally determined by the number of modes r used in the projection. This is of little importance if the application of the expanded DoF-set is limited to visualization or fitting of the included modes. However, the rank is important if the expanded DoF are expected to be included in further DS or other applications that require, for example, matrix-inversions or transformations.

10.4 SEMM and SEREP in Practice

For the test-cases presented, a structure previously used in experimental substructuring method validation [8]. This component is modelled with FEM to obtain a full-model stiffness and mass matrix (15630 DoF). A perturbed model is subsequently created by adjusting the stiffness matrix by a pre-defined factor which is the same as a perturbation in Young's Modulus of the material. Figure 10.3 along with the same mass-matrix, it can be used to construct the true 'experimental' model which will be used for the expansion techniques, as well as for the reference model. Both the original and perturbed models are redefined as FRF-models consisting of 57 DoF and different modal-damping values are assigned to each model. From the perturbed model a subset of DoF is extracted, the size of which is varied for each sub-case. The DoF used, as well as the DoF contained in the to-be-expanded boundary in each sub-case are shown in Fig. 10.4.

Property	Num.	Exp.
Rel. E-modulus	1	1.1
Modal Damping	2%	4%

(a) (b)

Fig. 10.3 (**a**): The modelled component. (**b**): The different properties for the numerical and perturbed experimental model

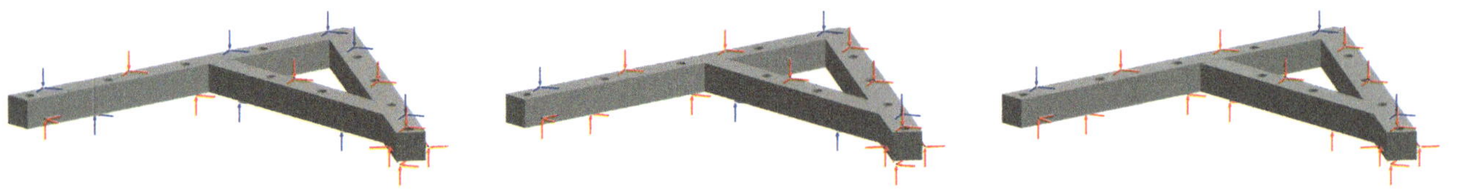

(a) (b) (c)

Fig. 10.4 The locations of the DoF used in the normal and perturbed FRF-model are displayed. In (**a**) the locations of the large boundary DoF-set used in case 1 is displayed in blue. This is reduced to the medium DoF-set in (**b**), and even further to the small DoF-set in (**c**)

These subsets are expanded by the SEMM and SEREP techniques and later compared. The comparison is done by means of testing the average coherence of all FRF provided by the different expansion techniques when compared to the reference FRF; this ensures valid full-model testing. Along with the average coherence test a typical non-driving point FRF is also provided for extra physical insight. Finally a full-system Frequency Response Assurance Criterion (FRAC)-matrix [9] provides an insight into which FRF fail by means of comparing the FRF vectors between the expanded and the reference model (Performed for the large DoF-set case).

The number of modes used in the SEREPa expansion is determined by the number of boundary DoF, i.e. size(r) = size(b). For standard SEREP, the number of modes is best determined by a constraint placed on the condition of $\mathbf{\Phi}_{br}$ thus ensuring proper inversion. In these particular cases, we require that $\text{cond}(\mathbf{\Phi}_{br}) < 5$. N_{modes} is the number of modes to include in SEREP and is found as follows:

$$N_{\text{modes}} = \max_{N \in \mathbb{N}+} \{\mathbf{\Phi}_{br} : r = 1, \ldots, N)\} \quad \text{s.t.} \quad \text{cond}(\mathbf{\Phi}_{br}) < 5 \tag{10.18}$$

This has resulted in the following number of modes for each case:

For case 1 the largest set of boundary DoF is used, namely 9 nodes or 27 DoF. The FRF displayed in Fig. 10.5a show that the SEMM FRF follows the reference throughout the relevant frequency band. Both SEREP FRF follow the reference for the majority of the band. However, at 1.7 kHz both diverge from the reference and at 2.2 kHz are deemed invalid. The average coherence of the all the FRF displayed in Fig. 10.5b supports these findings: SEMM shows stronger coherence throughout the band and SEREP outperforms SEREPa for the entire band.

When the boundary DoF-set is reduced to the medium set of 15 DoF for Case 2, the expansion's performance is worsened for all three cases, as is seen in Fig. 10.6a. Only slightly for SEMM, but more so for SEREP and SEREPa. SEMM is less affected by the reduce in boundary DoF. It should be noted that although SEREP outperforms SEREPa in the lower frequency band, SEREPa in fact shows better performance than SEREP at higher frequencies. This could provide an insight into the cause to the problems SEREP and SEREPa face with decreasing boundary size (Table 10.1).

10.4.1 Case 1: Expansion of a 27 DoF Perturbed Set to a 57 DoF Numerical Set

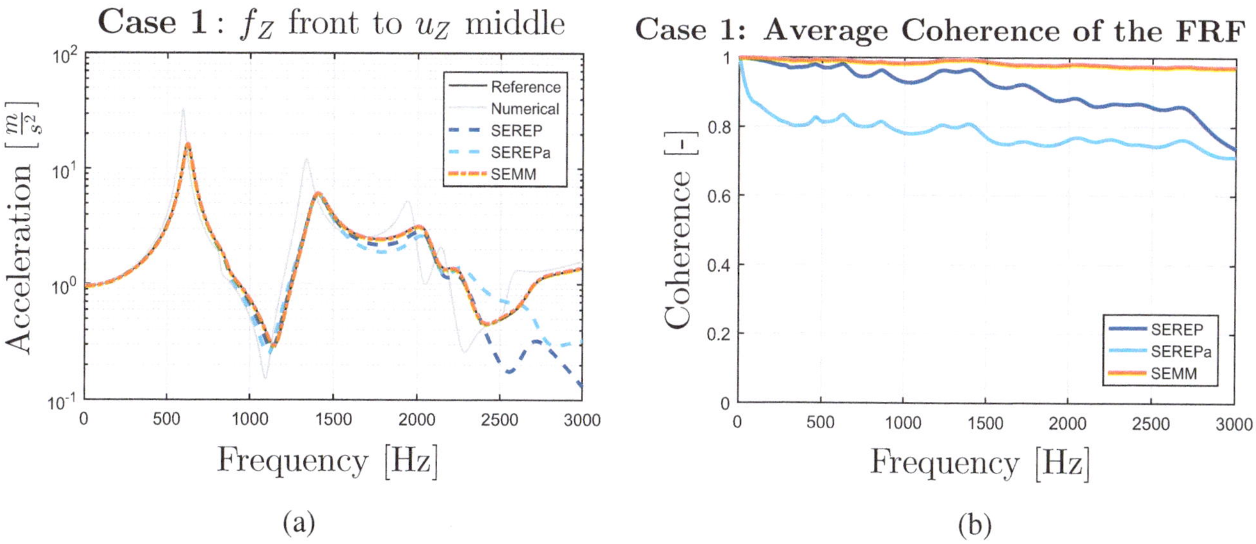

Fig. 10.5 The FRF (**a**) provided by all three methods follow the reference quite closely. At around 1.7 kHz the FRF provided by the SEREP methods veer off. The FRF provided by SEMM follows the reference up to at least 3 kHz. This is confirmed by the average coherence (**b**) of the FRF compared to the 'experimental' reference

10.4.2 Case 2: Expansion of a 15 DoF Perturbed Set to a 57 DoF Numerical Set

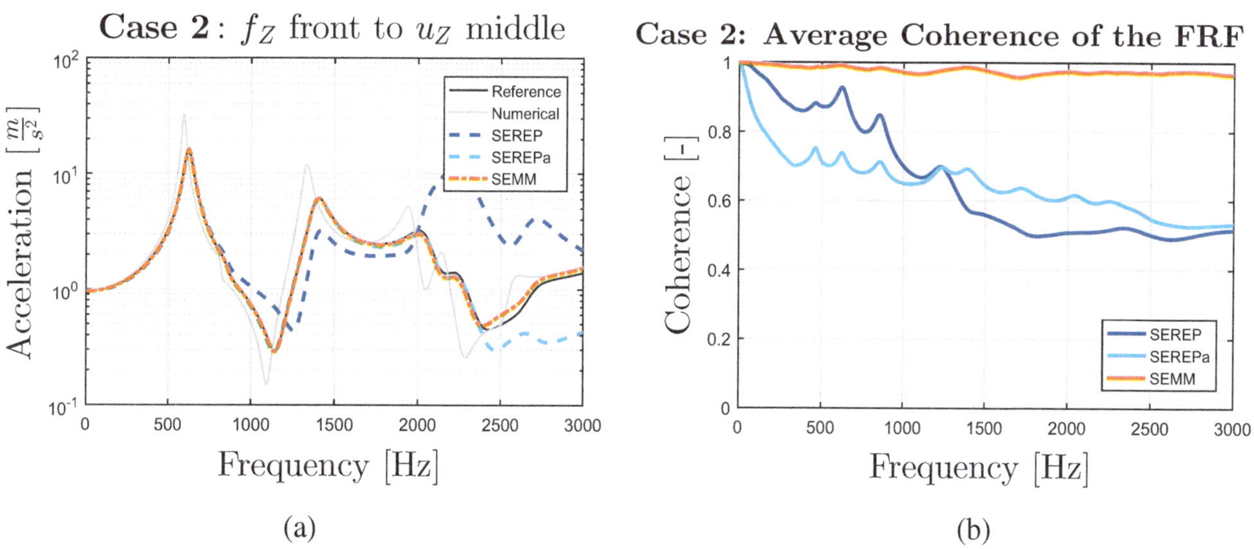

Fig. 10.6 The FRF (**a**) show strong coherence for all methods, at 2.2 kHz SEREP and SEREPa start to diverge strongly from the reference. The average coherence (**b**) of the FRF show that SEMM is more coherent on average than SEREP and SEREPa

Table 10.1 Modes used in each case for SEREP

Case	DoF (b)	Modes (r)	Cond. number
1	27	17	3.9
2	15	10	4.0
3	9	7	2.3

10.4.3 *Case 3: Expansion of a 9 DoF Perturbed Set to a 57 DoF Numerical Set*

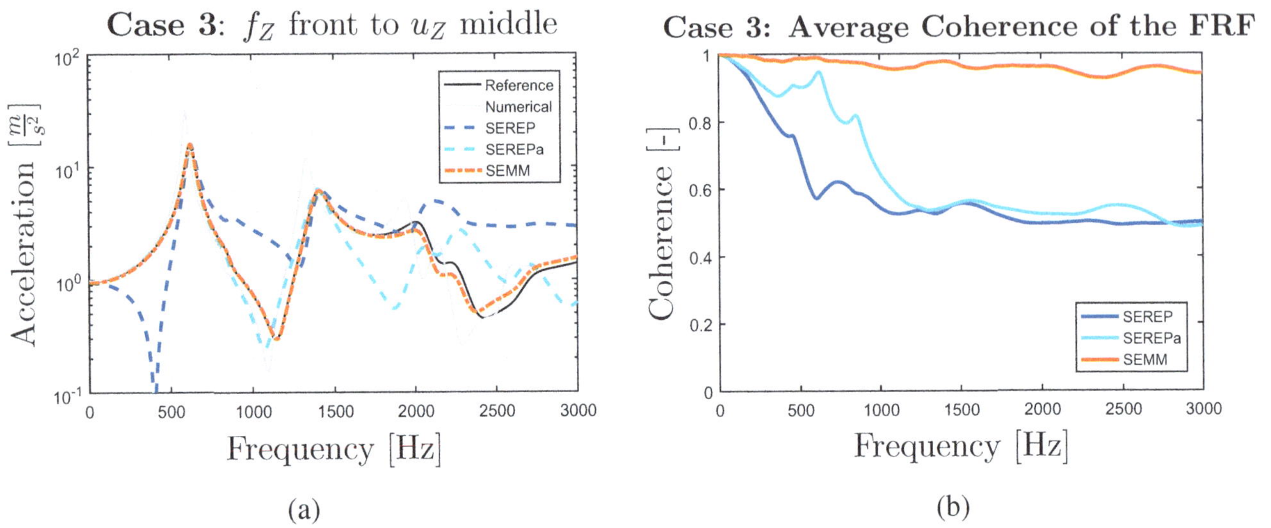

Fig. 10.7 The SEREP methods both fail with such a small boundary. SEMM is still strongly coherent on average (**b**), as is also shown in the typical FRF in (**a**)

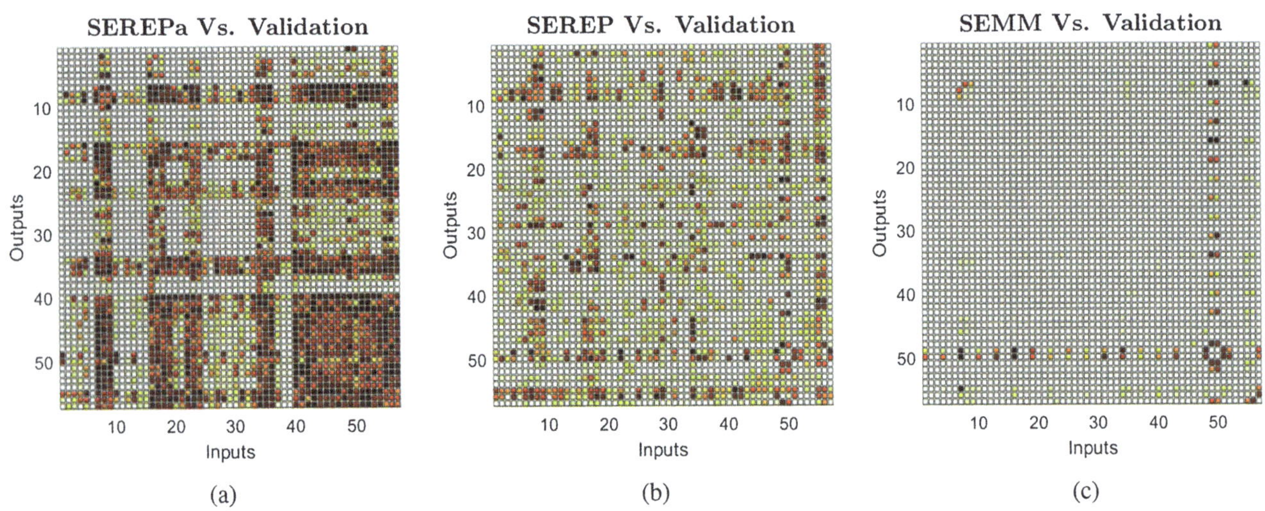

Fig. 10.8 The FRAC results for the different expansion methods. The comparison is made between the reference perturbed FRF against the SEREP(**a**) method in (**a**), the SEREP method in (**b**), and the SEMM method in (**c**)

Finally in case 3, SEREP and SEREPa are no longer valid throughout almost the entire frequency band. SEMM, which is largely unaffected by the decrease in boundary size still shows valid FRF. Interestingly, SEREPa now outperforms SEREP throughout the frequency band (Fig. 10.7).

Figure 10.8 shows the FRAC between the different expanded models and the reference model for every FRF. It is performed for the setup of case 1: Making use of a 27 DoF boundary. In Fig. 10.8a it is readily shown that SEREPa shows great results for the rows and columns of the boundary DoF: boundary-to-internal and internal-to-boundary are well represented. Naturally, the boundary DoF show perfect one-to-one coherence since, by definition of Eq. (10.16), these DoF remain unchanged. The expanded internal FRF however, are very poor on average. In Fig. 10.8b the results for the SEREP model are shown. The boundary DoF are no longer perfectly coherent as they are filtered, however are still strongly coherent. Additionally, the expanded FRF strongly represent the reference FRF, although weaker FRF exist. Finally in Fig. 10.8c it is shown that the SEMM expanded DoF are coherent with respect to the reference FRF, both at the boundary (which just like SEREPa remains unchanged) as well as internally.

10.5 Conclusion and Discussion

SEMM can create valid FRF, as demonstrated by the test-case shown. The FRF of the models expanded by SEREP and SEREPa are less coherent for all the cases presented above. Furthermore, it is shown that these methods are more sensitive to the number of boundary DoF as compared to SEMM.

The fact that SEMM is less sensitive to the boundary DoF size is in part explained by the fact that SEMM, unlike SEREP covers non-observed dynamics with its numerical counterpart. SEREP would fare better when more control points are used in the boundary DoF-set. It can be observed that the coherence of SEREP FRF converge to unity as more boundary DoF are included; more control points and therefore more modes are included in the projection manifold increasing the accuracy of the method. This is further supported by the fact that the SEREPa method, which by definition uses more modes, fares better than SEREP for small boundary sets.

It can be concluded that SEMM is the preferable method for applications where fewer measurement points can be made. For larger measurement DoF sizes the results are inconclusive and more research into the methods is warranted. Seeing as the sensitivity to the number of DoF in the control set for SEMM is low due to the inclusion of numerical internal dynamic information; it may be valid to test SEMM against methods that include some notion of internal dynamics, such as the VIKING method [4].

References

1. Balmès, E.: Frequency domain identification of structural dynamics using the pole/residue parametrization. In: IMAC XIV – 14th International Modal Analysis Conference – Noise and Vibration Harshness (NVH), Proceedings of the 14th IMAC, A Conference and Exposition on Structural Dynamics (1996)
2. Guyan, R.J.: Reduction of stiffness and mass matrices. AIAA J. **3**(2), 380 (1964). ISSN:0001–1452. https://doi.org/10.2514/3.2874
3. Avitabile, P.: Model reduction and model expansion and their applications part 1 – theory. In: Proceedings of the XXIII International Modal Analysis Conference (IMAC). Society for Experimental Mechanics, Orlando/Bethel (2005)
4. Thibault, L., Butland, A., Avitabile, P.: Variability improvement of key inaccurate node groups – viking. In: Topics in Modal Analysis II, Volume 6. Conference Proceedings of the Society for Experimental Mechanics Series, pp. 603–624. Springer, New York (2012). https://doi.org/10.1007/978-1-4614-2419-2_61
5. O'Callahan, J.C., Avitabile, P., Riemer, R.: System equivalent reduction expansion process (SEREP). In: Proceedings of the VII International Modal Analysis Conference (IMAC), vol. 1, pp. 29–37. Society for Experimental Mechanics, Boston/Bethel (1989)
6. Klaassen, S.W.B., van der Seijs, M.V., de Klerk, D.: System equivalent model mixing. Mech. Syst. Sig. Process. **105**, 90–112 (2018). ISSN 0888–3270. https://doi.org/10.1016/j.ymssp.2017.12.003. (http://www.sciencedirect.com/science/article/pii/S0888327017306301)
7. de Klerk, D., Rixen, D.J., de Jong, J.: The frequency based substructuring method reformulated according to the dual domain decomposition method. In: Proceedings of the XXIV International Modal Analysis Conference (IMAC). Society for Experimental Mechanics, St. Louis/Bethel (2006)
8. van der Seijs, M.V., Pasma, E.A., van den Bosch, D.D., Wernsen, M.W.F.: A benchmark structure for validation of experimental substructuring, transfer path analysis and source characterisation techniques. In: Dynamics of Coupled Structures, Volume 4. Conference Proceedings of the Society for Experimental Mechanics Series, pp. 295–305, chapter 26. Springer, New York (2017). https://doi.org/10.1007/978-3-319-54930-9_26.
9. Allemang, R.J.: The modal assurance criterion: twenty years of use and abuse. Sound Vib. Mag. 14–21 (2003). ISSN:1541–0161. http://www.sandv.com/downloads/0308alle.pdf

Chapter 11
Transmission Simulator Mass Loading Effects in Experimental Substructuring – A Study of the Ampair 600 Benchmark System

Andreas Linderholt

Abstract During the last years, a lot of research focusing on appropriate interfaces between substructures has been made; the transmission simulator method has become a tool in that strive. In this paper, the end effects on assembled structures consisting of finite element substructure models representing experimental setups with different levels of transmission simulator mass loadings at their interfaces are studied. Here, components of the Society of Experimental Mechanics, SEM, substructuring focus group's benchmark; the Ampair A600 wind turbine, constitute the structure studied. Models of an A600 blade and bracket system attached to dummy masses of different sizes are coupled to models representing an A600 hub together with two blades attached to dummy masses of different sizes after numerical subtraction of the dummy masses on each of the substructures. The results are compared to data stemming from a model of the assembled system.

Keywords Experimental substructuring · A600 wind turbine · Transmission simulator · Mass loading · Decoupling

11.1 Introduction

At IMAC XXX, a modified version of the Ampair 600 W wind turbine was presented [1] and became the benchmark structure for the Society of Experimental Mechanics', SEM's, substructuring focus group. One of the existing benchmark structures, located at Linnaeus University in Sweden, is shown in Fig. 11.1. As the SEM Substructuring Focus Group's benchmark test bed for substructuring, it has been used in many studies made within both the USA and Europe [2–4].

Dynamic substructuring in the context of numerical modelling is a mature tool. However, the research area has been subjected to renewed interest in recent years; now in the context of experimental substructuring [5]. The concept of dynamic substructuring is to divide a complex structure into a number of components, which are referred to as substructures. To form an assembled system, the substructures interact with each other. These interactions are described by the compatibility of motion and the equilibrium of forces, which have to be fulfilled at the interfaces between the substructures. Comparisons of FRF and modal techniques have been presented by de Klerk et al. [5] and Allen and Mayes [6] in which also the history of dynamic substructuring is presented.

One shortcomings of experimental dynamic substructuring is that models identified from measurements in conditions intended to be as close to free-free as possible are poor representations of the motion of the substructures when assembled with other substructures [6–8]. A technique to overcome this problem is referred to as the transmission simulator (TS) method. The idea is to add a well-known structure, the transmission simulator, to the substructure, aimed to be experimentally modelled, to enable the substructure to move more similar to its movement when assembled. The TS's effects are then removed, after the measurements, by adding numerical data representing the negative of the TS to the test data.

In this paper, the frequency response function (FRF) based form of the transmission simulator technique is applied to the hub and blades of the dynamic substructuring benchmark structure. Studies of the transmission simulator technique, using the benchmark structure, have been made by studied of Rohe [9] and Macknelly [3]. The substructure focus group has also a wiki page [10]. This study aims at investigating the effects of changing the mass of the transmission simulators by adding TSs of different weights to one of the blades in combination with adding TSs of different weights to the hub with two blades and then couple the blade models, after numerically removing the TSs, with the models representing the two bladed hub. Here, all coupling dofs are used. When the true FRFs are used, the coupling/de-coupling exercises should not introduce any errors. All FRFs stem from the modal based frequency response analysis in MSC Nastran (SOL 111). All modes upto 20 kHz are used to calculate the receptances. Synthetic test data (receptances) upto 1 kHz are used in the analyses.

A. Linderholt (✉)
Department of Mechanical Engineering, Linnaeus University, Växjö, Sweden
e-mail: andreas.linderholt@lnu.se

A. Linderholt et al. (eds.), *Dynamics of Coupled Structures, Volume 4*, Conference Proceedings of the Society for Experimental Mechanics Series, https://doi.org/10.1007/978-3-319-74654-8_11

Fig. 11.1 The modified Ampair 600 W turbine at Linnaeus University, Sweden

11.2 Theory

The motion of a deformable body, here looked upon as substructure s, is described by a set of partial differential equations which, when discretized within the spatial domain, gives rise to ordinary differential equations of the form:

$$M^{(s)} \cdot \ddot{u}^{(s)}(t) + C^{(s)} \cdot \dot{u}^{(s)}(t) + K^{(s)} \cdot u^{(s)}(t) = f^{(s)}(t) + g^{(s)}(t)$$

where $M^{(s)}$, $C^{(s)}$ and $K^{(s)}$, are the mass, damping and stiffness matrices and $u^{(s)}, f^{(s)}$ and $g^{(s)}$ denote displacements, external and connection forces, respectively. The equations of motion for all substructures can be written in a block diagonal form as

$$M \cdot \ddot{u}(t) + C \cdot \dot{u}(t) + K \cdot u(t) = f(t) + g(t)$$

Two conditions are required to be fulfilled by a substructure within an assembly, compatibility and equilibrium. The compatibility condition implies that the displacements must be equal for degrees-of-freedom (dofs) that are rigidly connected. A matrix B is formed to contain the information about compatibility between paired dofs on substructures' interfaces according to

$$B \cdot u(t) = 0.$$

Another matrix, L, is designed to contain the information that the sum of forces and moments have to be equal to zero at the interacting interfaces of the substructures according to

$$L^T \cdot g(t) = 0$$

in which g represents the interface forces.

The above equations form the well-known three-field formulation:

$$\begin{cases} M \cdot \ddot{u}(t) + C \cdot \dot{u}(t) + K \cdot u(t) = f(t) + g(t) \\ B \cdot u(t) = 0 \\ L^T \cdot g(t) = 0 \end{cases}$$

In the dual formulation, each interface dof appears in the displacement vector, u, as many times as the number of substructures connected through that dof. The interface forces, g, are given by

$$g = -B^T \cdot \lambda$$

where λ is a Lagrange multiplier. The equations can then be assembled as

$$\begin{cases} M \cdot \ddot{u}(t) + C \cdot \dot{u}(t) + K \cdot u(t) + B^T \cdot \lambda(t) = f(t) \\ B \cdot u(t) = 0 \end{cases}$$

In matrix form, the set of equations become:

$$\begin{bmatrix} M & 0 \\ 0 & 0 \end{bmatrix} \begin{Bmatrix} \ddot{u}(t) \\ \ddot{\lambda}(t) \end{Bmatrix} + \begin{bmatrix} C & 0 \\ 0 & 0 \end{bmatrix} \begin{Bmatrix} \dot{u}(t) \\ \dot{\lambda}(t) \end{Bmatrix} + \begin{bmatrix} K & B^T \\ B & 0 \end{bmatrix} \begin{Bmatrix} u(t) \\ \lambda(t) \end{Bmatrix} = \begin{Bmatrix} f(t) \\ 0 \end{Bmatrix}$$

Within the frequency domain and with Z denoting the dynamic stiffness, $Z(\omega) = -\omega^2 M + j\omega C + K$ the set of equations becomes

$$\begin{bmatrix} Z(\omega) & B^T \\ B & 0 \end{bmatrix} \begin{Bmatrix} u(\omega) \\ \lambda(\omega) \end{Bmatrix} = \begin{Bmatrix} f(\omega) \\ 0 \end{Bmatrix}$$

It should be clear which quantities depend on the driving frequency and from now on, (ω) is not written out. Let Y denote the reacceptance of the substructures, i.e.

$$Y = \begin{bmatrix} Y^{(1)} & 0 & \dots & 0 \\ 0 & Y^{(2)} & \dots & 0 \\ \vdots & \vdots & \ddots & 0 \\ 0 & 0 & 0 & Y^{(h)} \end{bmatrix} = \begin{bmatrix} Z^{(1)^{-1}} & 0 & \dots & 0 \\ 0 & Z^{(2)^{-2}} & \dots & 0 \\ \vdots & \vdots & \ddots & 0 \\ 0 & 0 & 0 & Z^{(h)-1} \end{bmatrix} = Z^{-1}$$

Then, the displacement amplitude vector of the coupled system becomes

$$u = \left[Y - Y \cdot B^T \left(B \cdot Y \cdot B^T \right)^{-1} \cdot B \cdot Y \right] \cdot f$$

The effect of subtracting a substructure can be achieved by simply adding the negative of that structure. That is, adding a substructure that has negative mass, damping and stiffness terms, to the coupled system in the generalized domain as described above.

11.3 Test Setup

The idea is to study the TS mass loading effects, related to sensitivities due to ill-conditioned matrices, on the components of the SEM focus group's benchmark structure. Here, it was decided to cut the shaft between the hub and one blade to form the two substructures used in this study. Hence, one substructure consists of one blade together with its holder and a part of the shaft whereas the other substructure is composed by the hub and the remaining two blades. The main idea for the study is schematically depicted in Fig. 11.2.

Decoupling of the transmission simulators are carried out through adding negative versions of them. The resulting models are then numerically coupled to achieve representations of the dynamics of the three bladed hub. Harvie and Avitabile defined 20 measurement positions for the A600 blades [11]. For each blade, eleven of these are assumed to be instrumented with sensors in this study, see Fig. 11.3. Among the eleven positions is the measurement position at the tip of the blade. Receptances from one blade tip to another blade tip are used in the evaluation later.

Vibrational testing has been prepared; some tests have been made but a throughout evaluation remains. A part of the preparations has been made as master's theses at Linnaeus University [12, 13]. A previous study, limited to TSs on the blade and with a different test setup, was also made [14].

The manufactured TSs intended to be added to the two bladed hub are shown in Fig. 11.4. The middle mass is made of aluminum and the other two are made of steel.

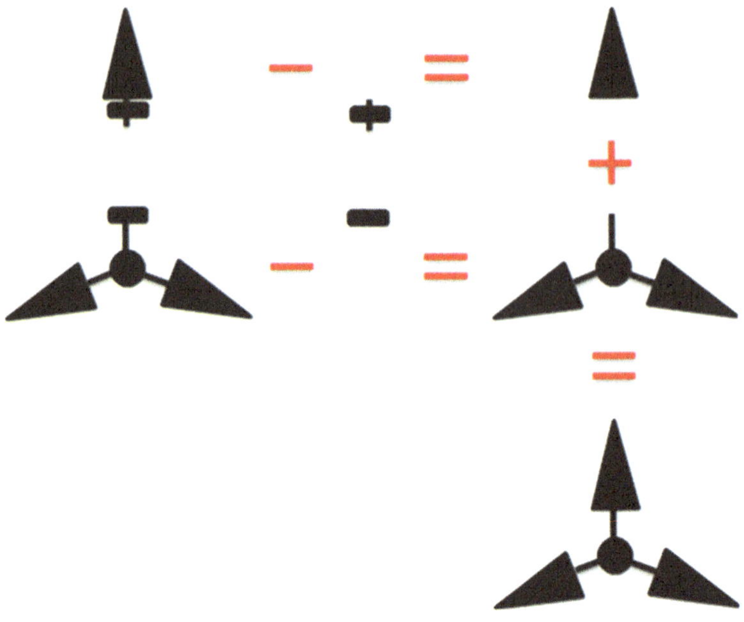

Fig. 11.2 The substructure decoupling- coupling process

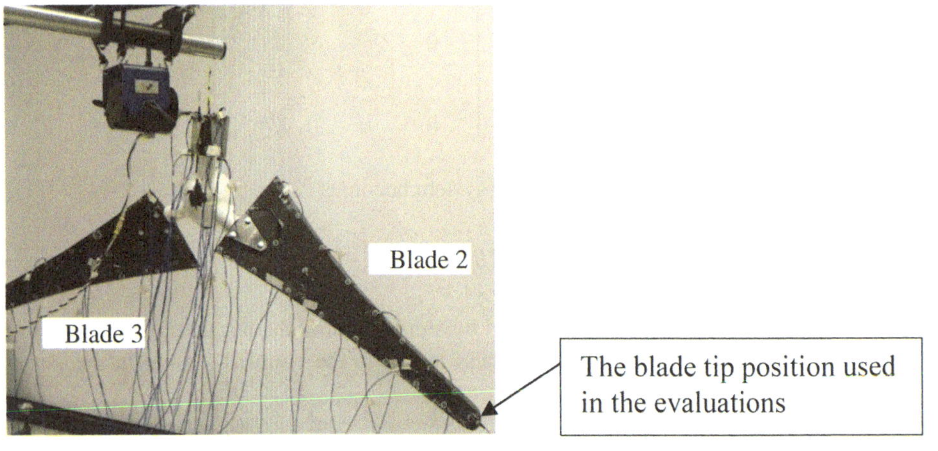

Fig. 11.3 The two-bladed hub with a transmission simulator mass

The manufactured TS aimed to be added to the blade substructure are shown in Fig. 11.5. The hole in the middle of each mass fits the shaft on the bracket. A screw fixates the mass to the shaft.

11.4 Preparatory Numerical Studies

The model used here sets out from an MSC NASTRAN finite element model provided by the substructuring focus group's wiki [10]. It consists of a laminate skin on an isotropic core. The material in the skin is glass fibers weave reinforced polypropylene, while the core consists of polypropylene only. The blades used in this study together with 17 other blades were examined in a previous study [15]. The properties for the blades, used in the finite element models, are taken from this study. Four equal layers in a 0°-90°-0°-90° stack where each layer has the stiffness properties, shown in Table 11.1, stemming from model calibrations made in the previous study, model the blade skins, which are 2.8 mm thick. The out-of-plane shear modulus $G_{1z} = G_{2z} = 0.72$GPa are used. The blade core modulus is $E = 1.67$GPa and the corresponding Poisson's ratio is $\nu = 0.3$. The density of the skin in the models is 272 kg/m^3 and the core material's density is 818 kg/m^3. The skin is modeled using six-noded triangular shell elements and the core is discretized by ten-noded tetrahedral elements. The model

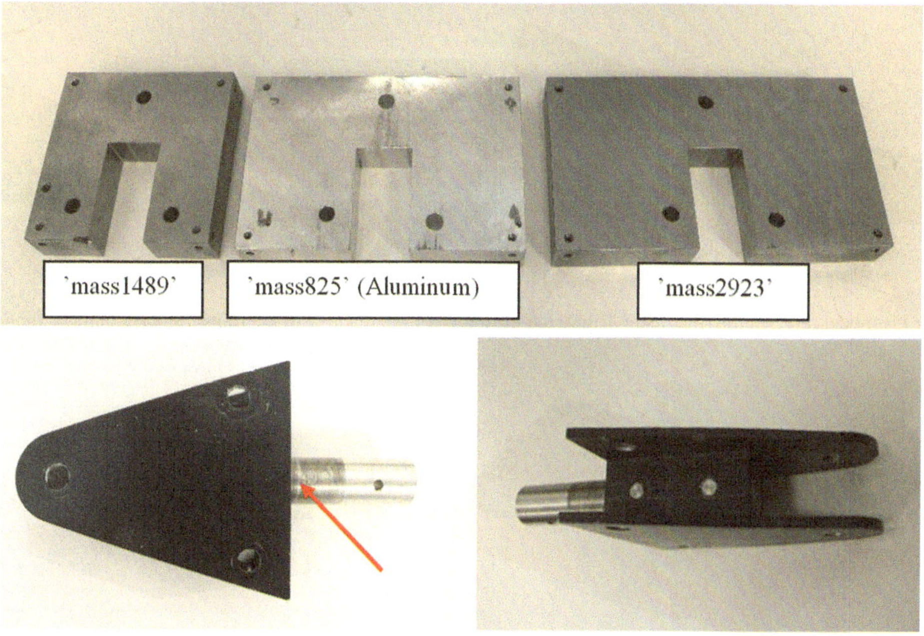

Fig. 11.4 Manufactured transmission simulator masses aimed to be added to the two-bladed hub. The red arrow at the bottom left sub-figure points at the interface between the substructure (the two bladed hub) and the TS [12]

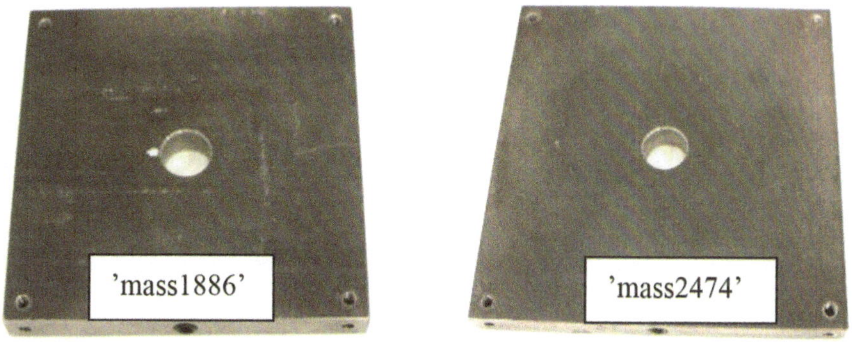

Fig. 11.5 Manufactured transmission simulator masses aimed to be added to the single blade substructure [13]

Table 11.1 Properties used to represent the three blades within the FE-model

	Blade 1 (single blade)	Blade 2	Blade 3
E_1	7.90 GPa	7.31 GPa	10.38 GPa
E_2	5.20 GPa	4.94 GPa	1.97 GPa
G_{12}	0.40 GPa	0.40 GPa	0.40 GPa
E	1.74 GPa	1.72 GPa	1.83 GPa

representing the hub with three blades is built up by 77,453 elements (MSC Nastran CTRIA6 shell elements, CTETRA10 and CHEXA solid elements together with CONM2 point masses and RBE2 rigid elements. This model has 117,869 nodes. The model of the TS, which is connected to the blade, consists of 722 elements and 1144 nodes. The model of the TS which is connected to the two bladed hub is modelled using 2198 elements and 4082 nodes.

A model of the hub with three attached blades is used for comparison. The blade models derived through coupling/decoupling of TSs are coupled to models, also derived through coupling/decoupling of TSs, representing the two bladed hub. The intersections have been augmented with rigid elements (MSC Nastran RBE2 elements) to create a single coupling node at each part. For the aluminum parts, the Young's modulus, Poisson's ratio and density equal to 69 GPa, 0.33 and 2700 kg/m^3, were used. For the steel parts, 210 GPa, 0.3 and 7800 m^3 were used. In this study, the mass equal to 1.886 kg is used as the nominal mass of the TS attached to the single blade. The 2.993 kg mass is chosen as the nominal TS to be attached to the two bladed hub. Changes of TS masses are then made numerically by changing the densities of the

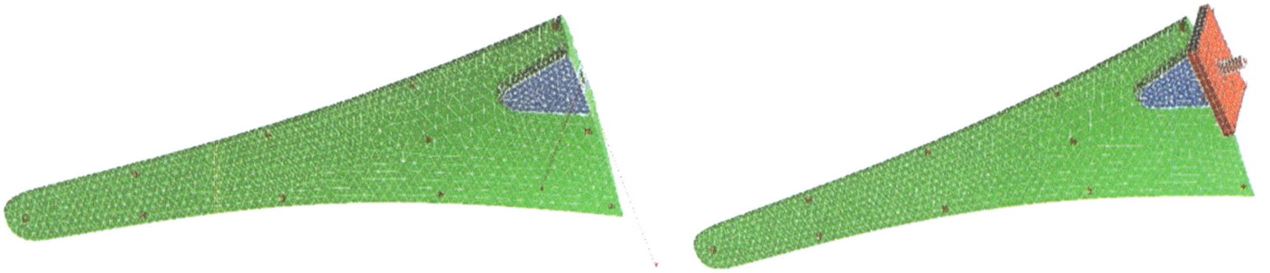

Fig. 11.6 Left, the FE-model representing the blade. Right, the FE model of the blade with the transmission simulator (nominally with the mass 1.886 kg) attached

Fig. 11.7 FE-models of the transmission simulators that are attached to the two bladed hub (the left model) and to the single blade (the right model)

Fig. 11.8 Left, the FE-model representing the two bladed hub. Right, the FE model of the two bladed hub with the transmission simulator attached

TS masses. Studies using the nominal masses times 1/16, 1, 4, 16 and 100 are reported here. In addition, the substructure representations found without any coupling-decoupling,i.e. the models alone, are also studied. All models used, which are shown in Figs. 11.6, 11.7, 11.8 and 11.9, are undamped.

11.5 Results

First, the models representing the single blade are compared. In Fig. 11.10, the direct point receptances at the interface point of the single blade models are shown. The receptance for the single blade model, which has not been involved in any coupling-decoupling, is compared to the resulting models for which 1/16, 1, 4, 16 and 100 times the nominal mass (1.886 kg) have been used in the coupling-decoupling. In a) the dof, translation in the x-direction (along the blade) is shown. In b) the dof, translation in the y-direction (edgewise, perpendicular to the normal of the nearly flat surface of the blade) is shown and in c) the dof, translation in the z-direction (flap wise, roughly in the direction of the normal to the flat surface of the blade). In d)-f), the rotational dofs are shown. As can be seen, although some spurious spikes can be found, the direct point receptances become almost the same no matter which of the studied TS that is added and then de-coupled. That should be the case according to theory but it shows that the process of decoupling the TSs, from the blade together with the TSs, is not sensitive. In Fig. 11.11, the associated analyses for TSs attached to the two bladed hub are made. The presentation follows

Fig. 11.9 FE-model of hub with three blades attached. This model serves as the true solution to which dynamics the assembled, coupled-decoupled, substructures are compared

the same logic as for the single blade. That de-coupling process is much more dependent on which TS that is de-coupled. For some, heavier, TSs the process renders in quite poor representations of the two bladed hub. The cause is that the matrix to be inverted; $(B \cdot Y \cdot B^T)$becomes ill-conditioned. It seems that for large masses, in this case, the dynamics is governed by the TS so that the difference between the receptances for the structure with a TS and the TS alone becomes numerically close to each other. In Fig. 11.12, the condition numbers, as functions of the excitation frequency, for the matrices to be inverted, i.e. the differences in coupling receptances, are shown. It is clear that the condition numbers are larger for the two bladed hub de-couplings. The condition numbers increase when heavier TSs are used.

The different representations of the single blade are finally coupled to the different representations of the two bladed hub. Since six representations of each of the two substructures have been found from the analyses, 36 single blade-two bladed hub combinations can be made. In Fig. 11.13, some transfer receptances; excitation on the tip of the single blade (blade one) and response at the tip of blade two (both along the z-axes which are almost perpendicular to the flat surfaces of the blades). The receptances for the FE-model of the full system, denoted as "full structure" in Fig. 11.13, are compared to the receptances found from the substructuring exercises. No de-coupling means that the substructure was not subjected to the coupling-decoupling whereas the other substructure was.

11.6 Conclusions and Future Work

Here, the sensitivity within the decoupling-coupling procedure using transmission simulators is studied. It is seen that the choice of transmission simulator can make the matrix which is to be inverted, more or less ill-conditioned. For test data, noise and deviations from perfect data are omni-present. Hence, de-coupling transmissions simulators that renders in ill-conditioned matrices may jeopardize the process. It is shown that the large masses for the hub, here, make the assembly representation less useful.

The already achieved test data will be evaluated and then more measurements will be made in the vibration laboratory at the Linnaeus University. Other decoupling methods will also be investigated using the models in this paper.

Acknowledgements The author would like to acknowledge the former master students at Linnaeus University; Ms. Cwenarkiewicz, Mr. Johansson, Ms. AL – Mahdi and Mr. Tadeu do Santos. Further, Ms. Scheele and Mr. Nord (former Johansson) are acknowledged for the nice collaboration in a previous study, related to the present study.

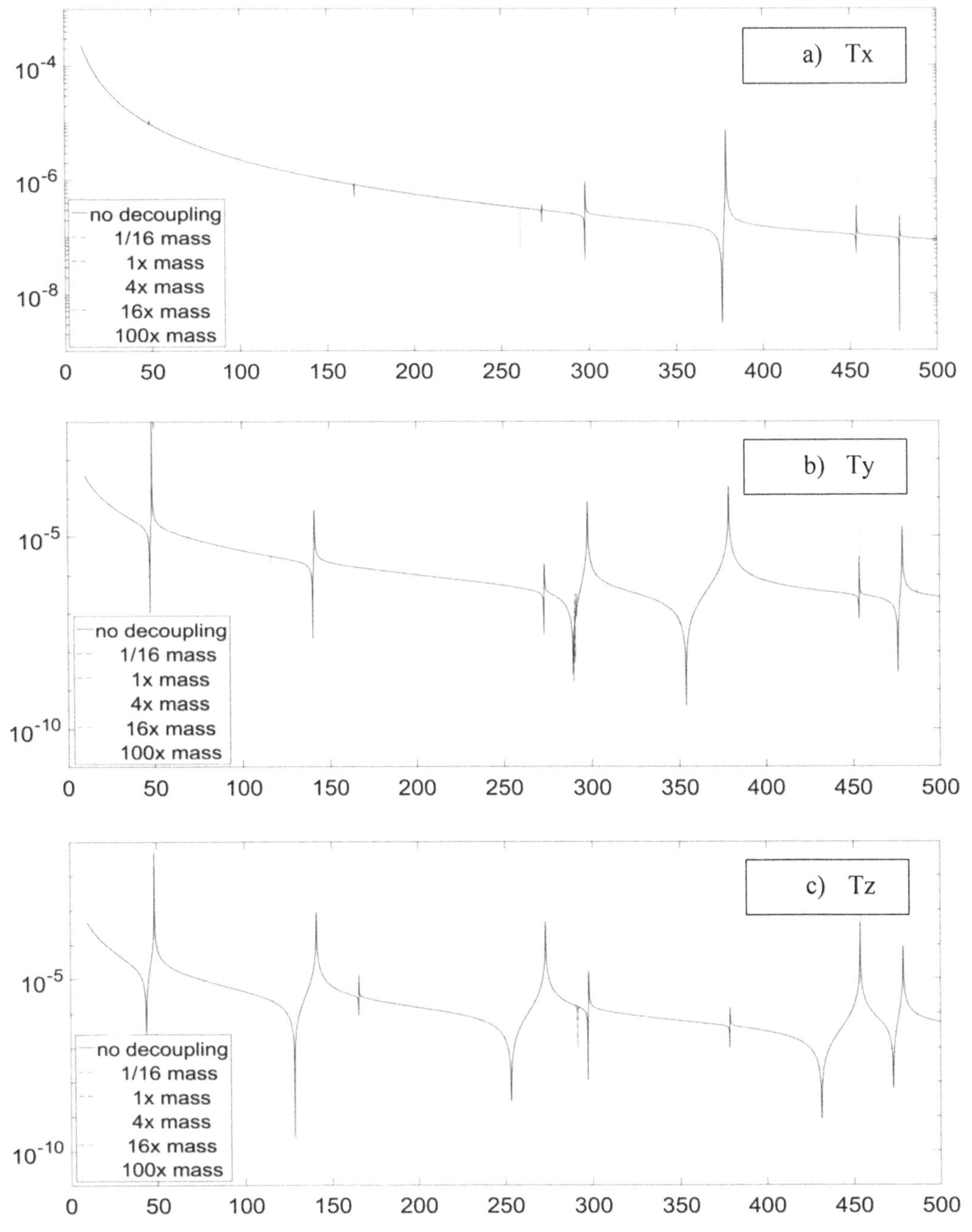

Fig. 11.10 Direct point receptances at the coupling node on the blade after de-coupling of the transmission simulator: (**a**) Translation in the x-direction, (**b**) Translation in the y-direction, (**c**) Translation in the z-direction, (**d**) Rotation around the x-axis, (**e**) Rotation around the y-axis, (**f**) Rotation around the z-axis

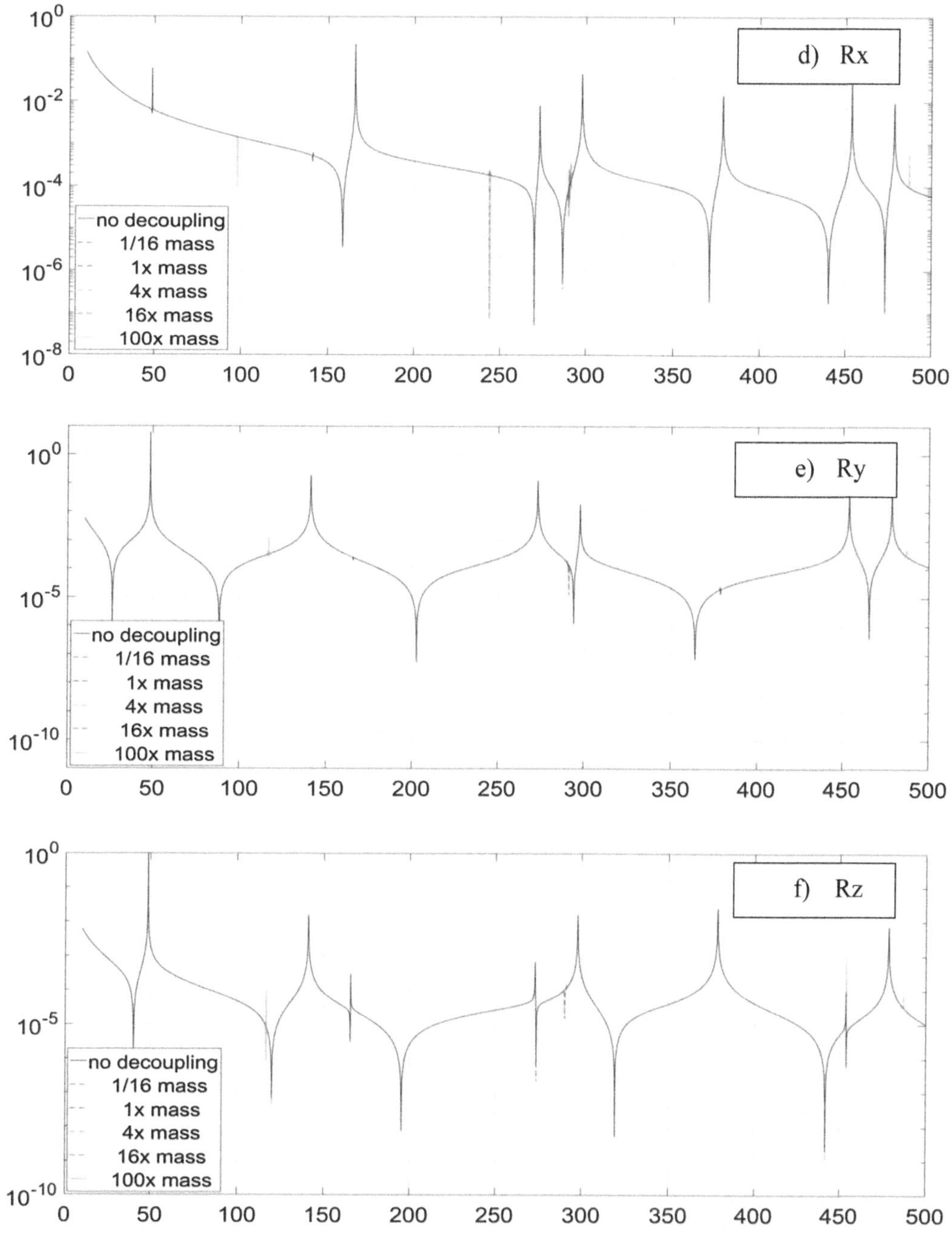

Fig. 11.10 (continued)

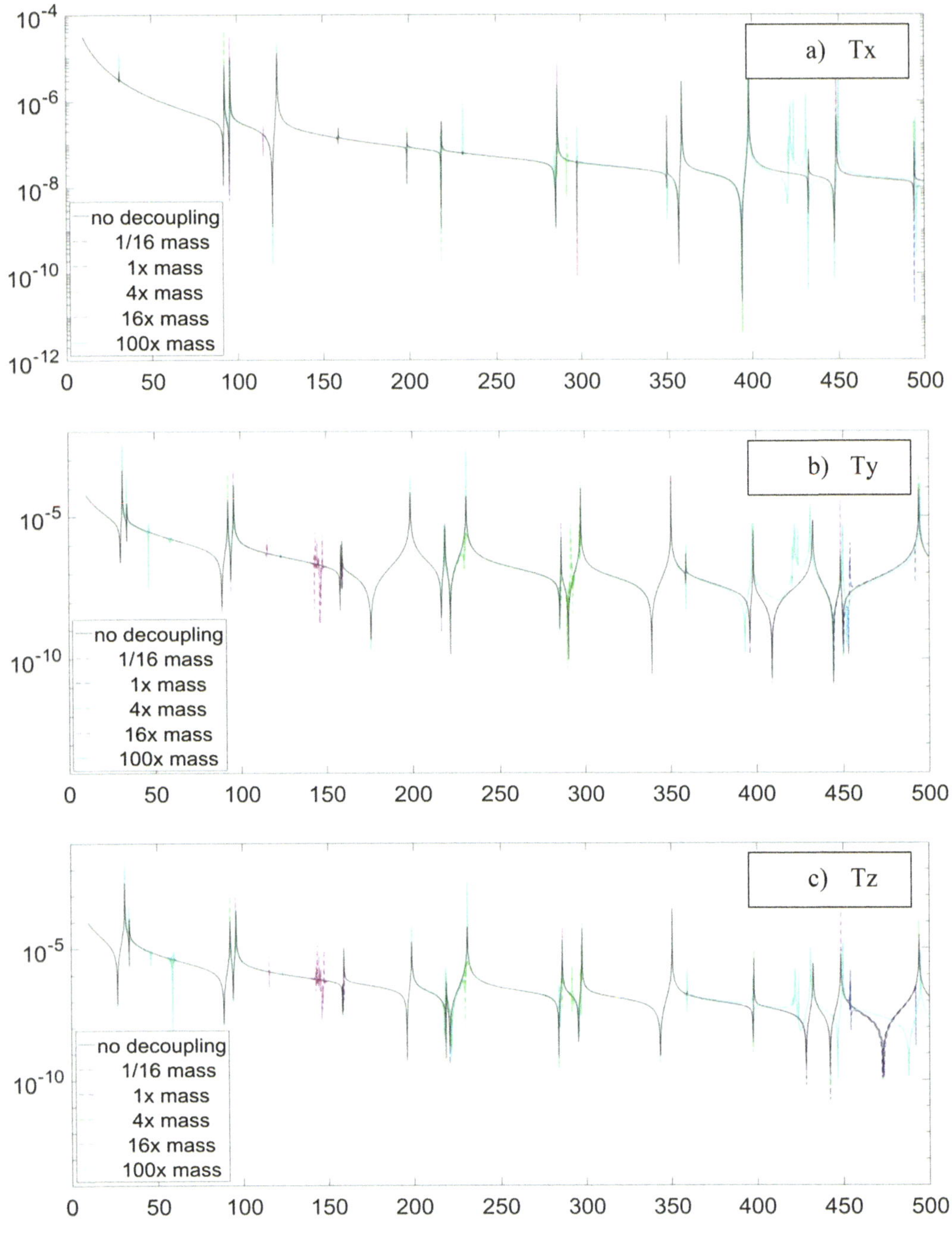

Fig. 11.11 Direct point receptances at the coupling node on the hub: (**a**) Translation in the x-direction, (**b**) Translation in the y-direction, (**c**) Translation in the z-direction, (**d**) Rotation around the x-axis, (**e**) Rotation around the y-axis, (**f**) Rotation around the z-axis

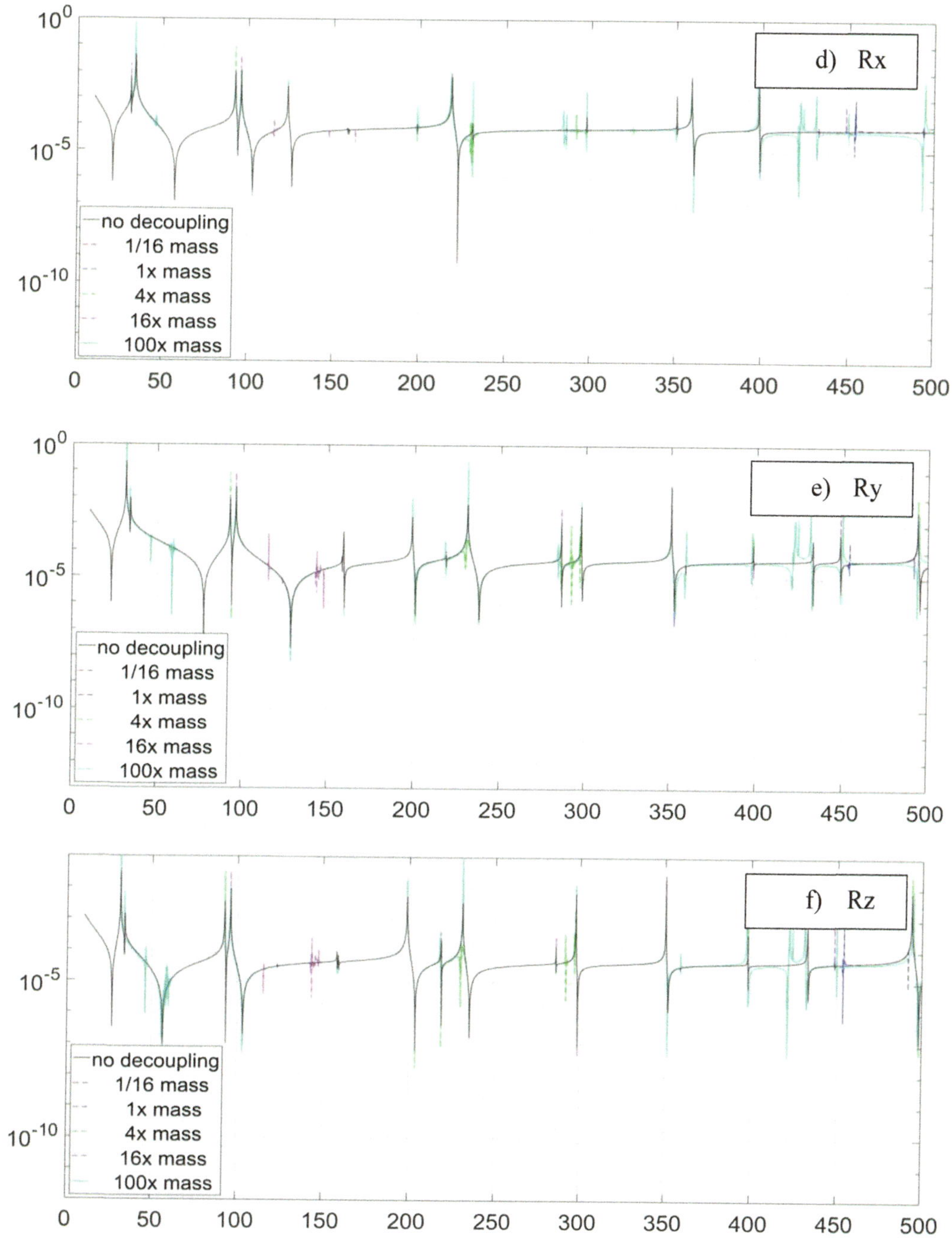

Fig. 11.11 (continued)

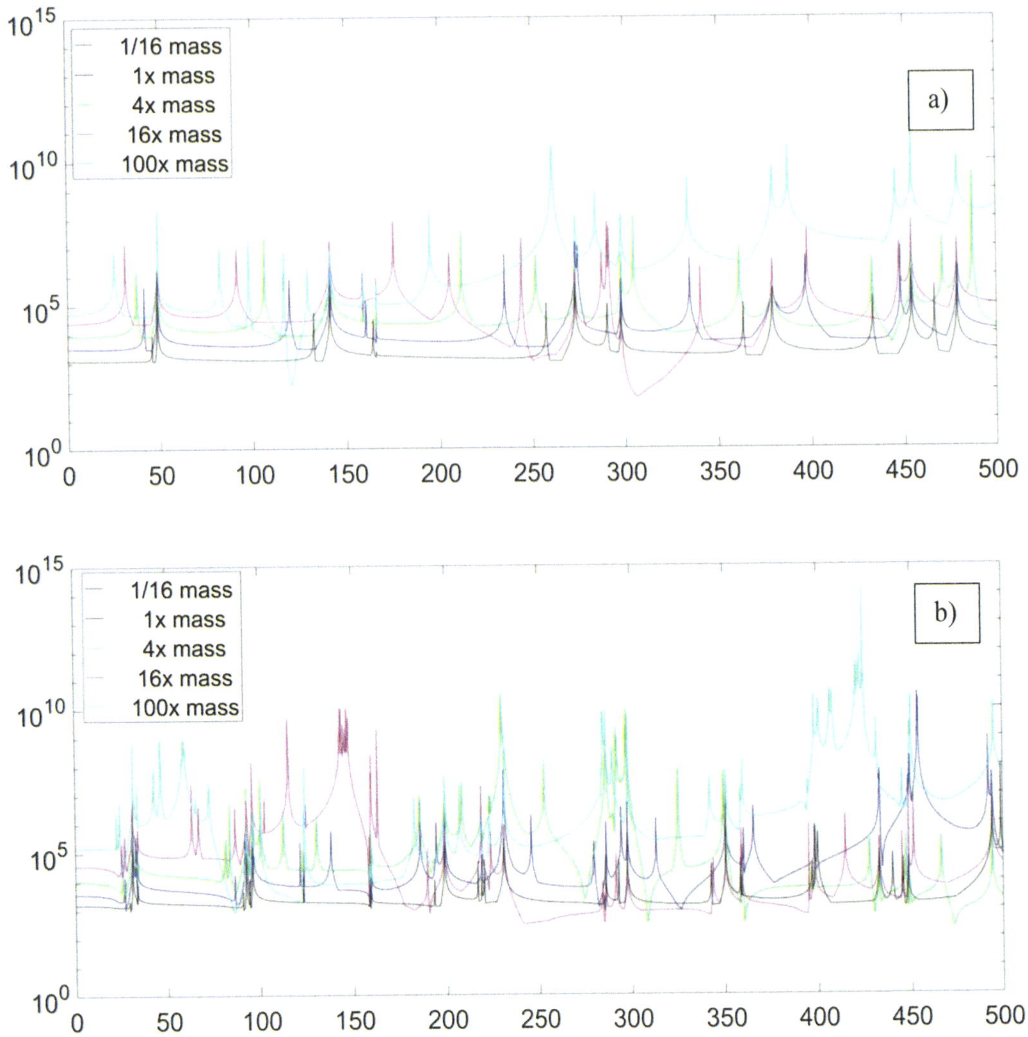

Fig. 11.12 Condition numer of the coupling matrix, size 6x6, which is inverted in the decoupling operation.: (**a**) decoupling of different masses from the blade, (**b**) decoupling of different masses from the two bladed hub

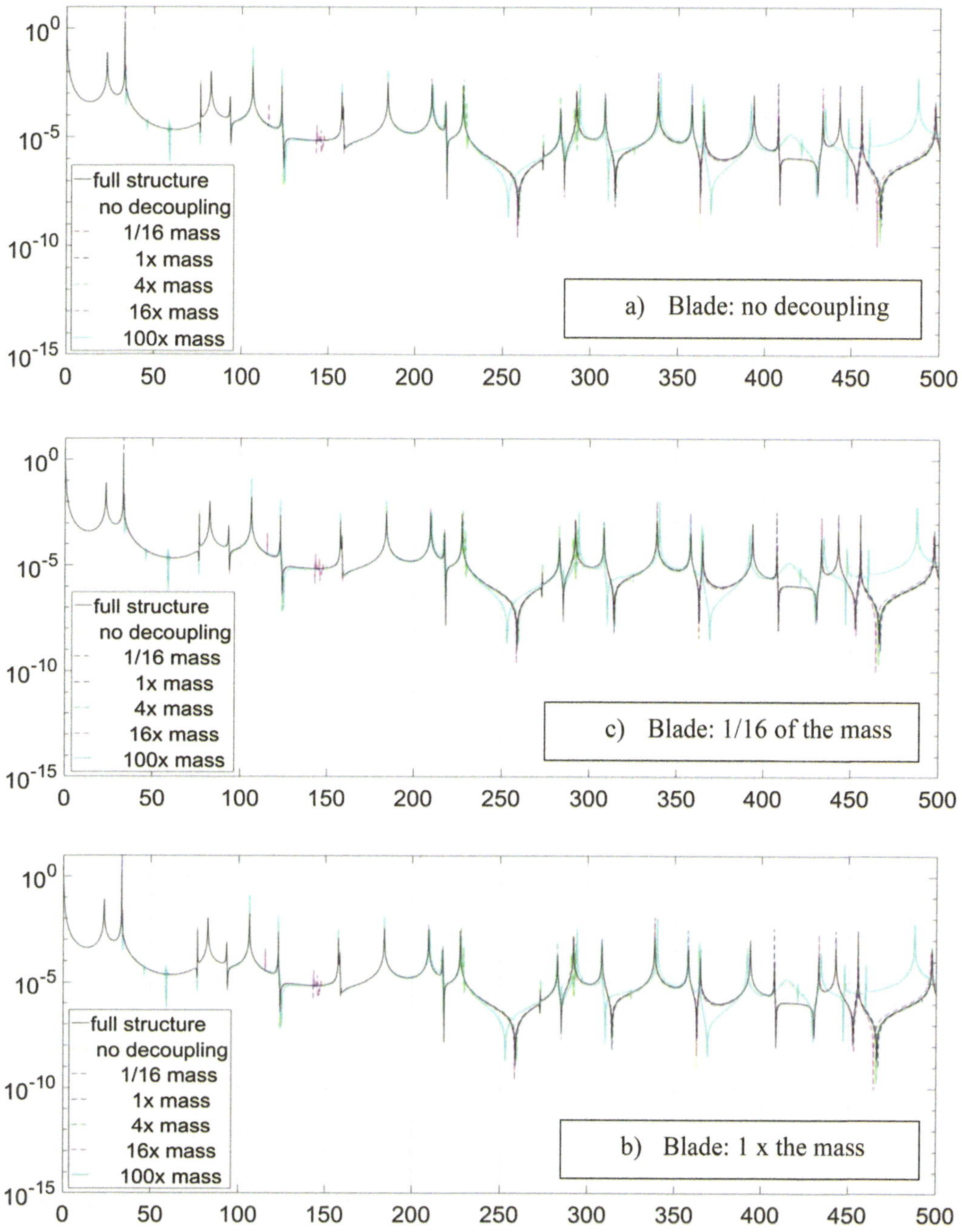

Fig. 11.13 Transfer receptances; excitation on the tip of blade no. two and response at the tip of blade no. one (both along the z-axes which are almost perpendicular to the blades). The FRFs from the fully assembled (full structure) are compared to the ones found from substructuring excersises. For each plot, the mass used for the blade is different. Within each plot the result for different de-coupled hub masses are shown. (**a**) The blade has not been coupled/decoupled, (**b**) 1/16 of the mass was used for the blade, (**c**) 1 mass (1884 g) was used for the blade, (**d**) 4 x the mass was used for the blade, (**e**) 16 x the mass was used for the blade, (**f**) 100 x the mass was used for the blade

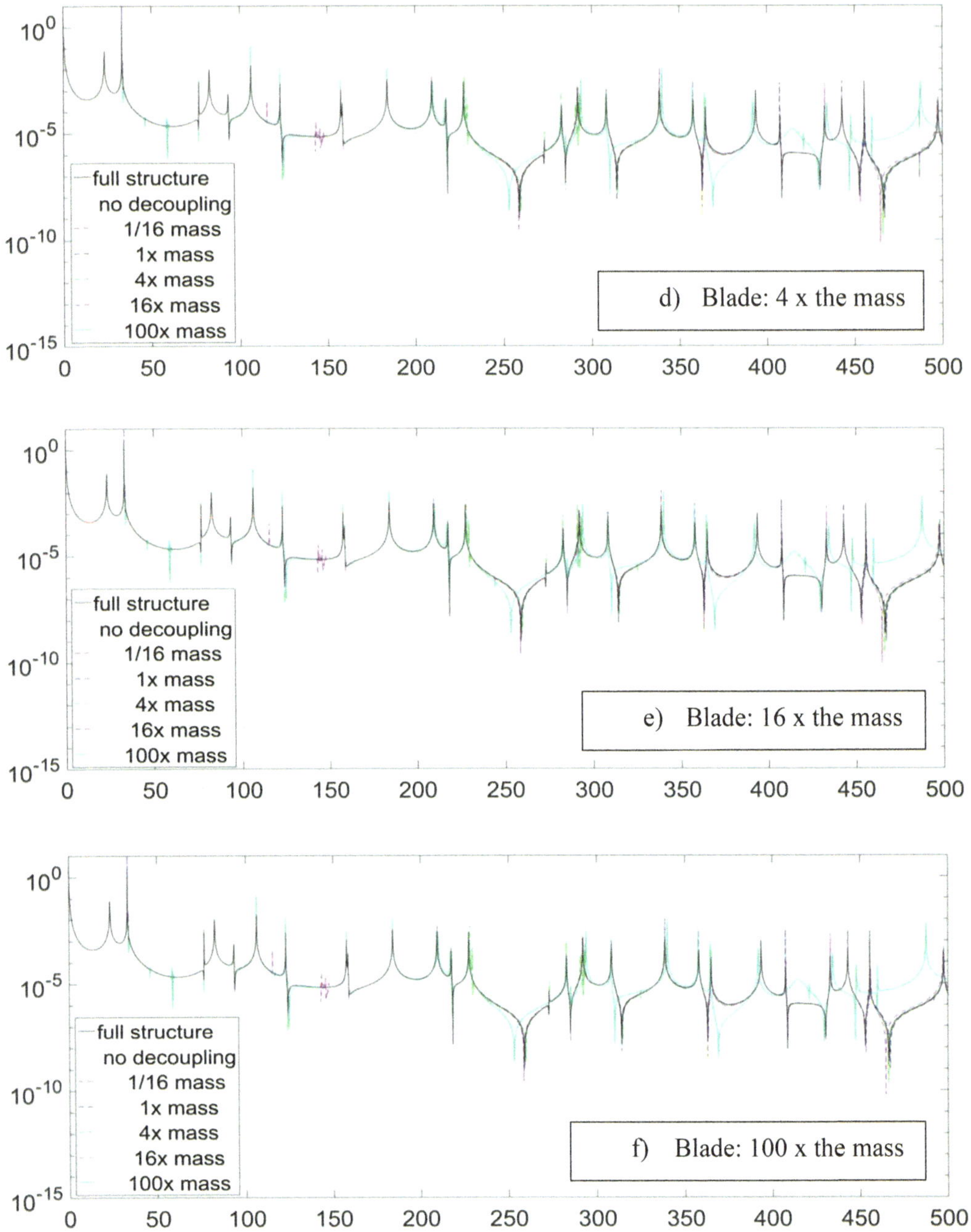

Fig. 11.13 (continued)

References

1. Mayes, R. L.: An introduction to the sem substructures focus group test bed - the ampair 600 wind turbine. Conference Proceedings of the Society for Experimental Mechanics Series. 30th IMAC, A Conference on Structural Dynamics (2012)
2. Rahimi, S., de Klerk, D., Rixen, D. J.: The ampair 600 wind turbine benchmark: Results from the frequency based substructuring applied to the rotor assembly. Conference Proceedings of the Society for Experimental Mechanics Series. 31th IMAC, A Conference on Structural Dynamics (2013)
3. Macknelly, D., Nurbhai, M., Monk, N.: Additional modal testing of turbine blades and the application of transmission simulator substructuring methodology for coupling. Conference Proceedings of the Society for ExperimentalMechanics Series. 31th IMAC, A Conference on Structural Dynamics (2013)
4. Rohe, D.P.: Validation of experimental–analytical substructuring using the transmission simulator method, MSc Thesis, Engineering Physics, University of Wisconsin-Madison, Madison, Wisconsin (2012)
5. de Klerk, D., Rixen, D.J., Voormeeren, S.N.: General framework for dy- namic substructuring: history, review and classification of techniques. AIAA J. **46**(5), 1169–1181 (2008)
6. Allen, M. S., Mayes, R. L.: Comparison of frf and modal methods for com- bining experimental and analytical substructures. Conference Proceedings of the Society for Experimental Mechanics Series. 25th IMAC, A Conference on Structural Dynamics (2007)
7. Allen, M.S., Mayes, R.L., Bergman, E.J.: Experimental modal sub- structuring to couple and uncouple substructures with flexible fixtures and multi-point connections. J. Sound Vib. **329**, 4891–4906 (2010)
8. Mayes, R. L.: Tutorial on experimental dynamic substructuring using the transmission simulator method. Conference Proceedings of the Society for Experimental Mechanics Series. 30th IMAC, A Conference on Structural Dynamics, pages 1–9 (2012)
9. Rohe, D. P.: Validation of experimental-analytical substructuring using the transmission simulator method, Master's thesis (2012)
10. SEM substructuring's wiki.: http://substructure.engr.wisc.edu/substwiki/ Oct 2017
11. Harvie, J., Avitabile, P.: Comparison of some wind turbine blade tests in various configurations. Conference Proceedings of the Society for Experimental Mechanics Series. 30th IMAC, A Conference on Structural Dynamics, pages 73–79 (2012)
12. Cwenarkiewicz, M., Johansson, T.: Experimental Dynamic Substructuring of an Ampair 600 WindTurbine Hub together with two Blades - A Study of the Transmission Simulator Method, Master's thesis, Linnaeus University (2016)
13. Al-Mahdi, N., Tadeu do Santos, J.: Experimental substructuring of an A600 wind turbine blade – A Study of the influence of interface loading, Master's thesis, Linnaeus University (2016)
14. Linderholt, A., Scheele, M., Johansson, A.: Effects of interface loading in dynamic substructuring, The International Conference of Engineering Vibration (ICoEV), Ljubljana, Slovenia (2015)
15. Johansson, A., Linderholt, A., Abrahamsson, T.: Model calibration and uncertainty of A600 wind turbine blades. Conference Proceedings of the Society for Experimental Mechanics Series. 32nd IMAC, A Conference and Exposition on Structural Dynamics (2014)

Chapter 12
Modeling Transverse Vibration in Spider Webs Using Frequency-Based Dynamic Substructuring

Andrew W. Otto, Damian O. Elias, and Ross L. Hatton

Abstract Orb weaving spiders and their webs have co-evolved into a highly efficient prey capture and retention system. Spiders also use their web as a sensory extension by "listening" to vibrations transmitted across the web such as struggling prey trapped in the web, courtship from potential mates, or the advance of a predator. What information is available to orb weaving spiders from web vibrations, and how might this information be used for localization and identification of cues? To better understand the information available to the spider in its web, we created both physical and computational models of spider webs. Enlarged, artificial webs suitable for physical measurements (1.2 m in diameter) were constructed from two types of parachute cord to mimic the different silks used by spiders in web construction. Accelerometers placed around the center of the web measured the vibration response of the artificial web. We formed a model for large networks of transversely vibrating strings, such as a web, using Frequency Based dynamic Substructuring (FBS). From the FBS model, we generated frequency response functions at locations corresponding to typical foot placements of orb weaving spiders for vibration sources at various points in the web. We explored the influence of web architecture on web frequency response by altering web composition, pre-tension, and geometry. Frequency response features sensitive to stimulus location were also identified from the FBS model. Our research indicates that localization of both bearing and range for a vibration stimulus is possible in the modeled webs.

Keywords Frequency based substructuring · Spider webs · Vibrations · Localization · Transmission

12.1 Introduction

Orb weaving spiders create webs to act as snares for catching prey. Once built, these webs also serve as a substrate to transmit vibrational information to the spider. How do vibrations propagate through the web? How does the spider use the information contained in vibrations to localize signals in its web? To answer these questions, we seek a physical understanding of the web as a dynamic structure.

The spider web is challenging to study due to its small size. If we seek to understand how the architecture of the web – its geometry and composition – influences web vibrations, then we must exert some form of control over its construction. Controlling variations in web architecture; however, is difficult because we cannot synthesize silk and must rely on the measurements taken from webs founds in nature. To overcome this limitation, we have constructed our own enlarged, artificial webs from parachute cord and elastic shock cord. By scaling up the web, we greatly ease the process of taking vibration measurements. Additionally, the web architecture can be freely specified to study the influence of factors such as web geometry, composition, and tension on web vibrations.

We have also developed a dynamic model of the orb web to study transverse web vibrations and to further explore the question of stimulus localization. We focus on transverse vibrations for two reasons: (i) prey impact is most frequently oriented perpendicular to the plane of the web, and (ii) prey impact occurs in the capture spiral portion of the web, and the spiral strands are better impedance-matched to the radial strands in the out-of-plane direction, and therefore transmit more energy to them in these modes. To model the web, we first identify web configurations (geometry and tension distribution) for which the web is in static equilibrium. We then build up a coupled oscillator model of the web as a network of interconnected vibrating strings. From the model, we generate pointwise Frequency Response Functions (FRFs) to describe motion of the

A. W. Otto · R. L. Hatton (✉)
Department of Mechanical, Industrial, and Manufacturing Engineering, Oregon State University, Corvallis, OR, USA
e-mail: ottoa@oregonstate.edu; ross.hatton@oregonstate.edu

D. O. Elias
Department of Environmental Science, Policy, and Management, University of California Berkeley, Berkeley, CA, USA

A. Linderholt et al. (eds.), *Dynamics of Coupled Structures, Volume 4*, Conference Proceedings of the Society for Experimental Mechanics Series, https://doi.org/10.1007/978-3-319-74654-8_12

web mesh in the frequency domain. We use this information to identify changes in the vibration response of the web due to variations in web architecture. Finally, we propose a pair of vibration cues that can be used to localize the position of an impulse in the web.

12.2 Related Work

This paper draws from research in biology of orb web vibrations as well as the field of dynamic substructuring for modeling structural vibrations.

12.2.1 Web Vibrations

Many investigations have focused on understanding the vibration properties of orb webs. Masters [1, 2], and later Landolfa and Barth [3], measured the transmission of different vibrational directions in the orb web (longitudinal, transverse, and lateral), and found that longitudinal vibrations were least attenuated, suggesting that the spider may use longitudinal vibrations for prey localization. Landolfa and Barth [3] also discovered a gradient in vibration amplitudes (in all vibration directions) arriving at the spider's tarsi that could be used to indicate stimulus direction. Due to the high measured wave speeds, they conclude that the spider would require equally high temporal fidelity to detect time of arrival differences to localize prey (approximately 1.5 ms). This finding suggests the spider may be exploiting the spatial distribution of the vibrations rather than the temporal to solve the localization problem. More recent works have closely studied the shock response and energy absorption of webs [4–6], but there remains a gap in understanding as to how these vibratory signals can be processed to determine prey location.

In addition to experimentally measuring the vibration response of orb webs, researchers have developed computational models for web vibrations. One of the first models was presented by Frohlich and Buskirk [7], which adopted a simplified representation of the web consisting of interconnected strings and lumped masses. Since then, many works have utilized finite element methods to model web vibrations in greater detail and investigate phenomena such as damage tolerance [8], the role of secondary structures [9], energy absorption [6, 10], and aerodynamic forces [11, 12]. Mortimer et al. [5] utilized computer models and experiments with orb webs to demonstrate the extent to which the spider can control various vibration properties of its web. Most recently, Morassi et al. [13] developed a continuous membrane model for the orb web with the goal of generalizing the web's vibration properties to analytical relationships. While numerical simulations provide detailed web displacement time-histories and analytical models bring generality to the vibration characteristics of orb webs, neither body of work establishes an input-output mapping between a stimulus (e.g. the impact of an insect) and the signals received by the spider.

A primary issue in comparing model results with web measurements lies in determining web pretension. Measuring web tension directly requires highly specialized tools that are not commercially available and are difficult to fabricate [1, 14]. Alternatively, estimating web tension through other experimental means (e.g. wave speed or natural frequencies) becomes a model fitting procedure using physical principles for taut strings. Secondly, validating model results on live webs raises issues of repeatability – spider silk production is highly influenced by environmental conditions, and no two webs will be alike in terms of composition, pre-tension, or geometry. To address these shortcomings, some researchers have turned toward making physical analogs of the orb web for vibration research [15–17]. We follow a similar approach, using an enlarged artificial web to validate results from our model.

12.2.2 Dynamic Substructuring

The field of dynamic substructuring is concerned with assembling the dynamic response of a complete system from individual subcomponents via a coupling procedure derived from the connectivity of the system [18]. One of the major advantages of dynamic substructuring is its ability to handle systems that are too large or complex to model as a whole, and instead build up a reduced model from detailed subcomponent representations. The origins of dynamic substructuring can be traced back to the method of component mode synthesis (CMS) [19, 20].

Substructuring can be performed in many different domains (e.g. physical, modal, time, and frequency). Substructuring in the frequency domain is most commonly employed in the experimental analysis of a structure's dynamics because it provides a straightforward method to generate a model of a complete system through testing of individual subcomponents at their interfaces. In this work, we use Frequency Based Substructuring (FBS) for its convenient algebraic representation of the substructures that comprise a web – strings and masses. Furthermore, modeling in the frequency domain lends itself to the same types of analysis used in the literature for web vibrations.

12.3 Materials and Methods

We carried out our investigation into spider web vibrations by creating both physical and computational models of the spider web. Enlarged, artificial spider webs were created to enable control over web architecture for experimentation. Our computational model was formed using dynamic substructuring methods.

12.3.1 Artificial Webs

We constructed webs using two different materials to mimic the different types of silk used by spiders during web construction. We selected a stiff, nylon parachute cord for the radial lines of the artificial web and an elastic shock cord for the capture spiral, both having a diameter of 3.2 mm and a mass per unit length of 7 g/m. The stress-strain data for these materials is shown in Fig. 12.2a along with representative data for dragline (radial) and viscid (capture spiral) silk.

The web geometry consisted of eight symmetric radial members and an Archimedean capture spiral. Figure 12.1 and Table 12.1 describe the parameters used to define web geometry. At the radial-spiral junctions, cable ties were used to clamp the strings together, and these joints were reinforced with cyanoacrylate glue. Webs were built slack (un-tensioned) in a conical shape and then tensioned to desired levels. The angle of the un-tensioned cone (β) was used to control the ratio of spiral to radial member tension. Different levels of radial tension were obtained by further stretching each web from its flattened state.

Webs were housed in an octagonal aluminum frame accommodating webs with a final stretched diameter D of 1.22 m (48 in), shown in Fig. 12.2b. Radial lines were terminated into a ratchet and load cell mechanism to measure and adjust the tension of peripheral radial members. A voice-coil shaker excited the web to measure transmissibility. Input signals to the shaker were in the form of a swept sine wave of increasing frequency. Three-axis Analog Devices ADXL326 MEMs-type accelerometers (1000 Hz X- and Y-axis bandwidth, 500 Hz Z-axis bandwidth) measured the vibration input at the

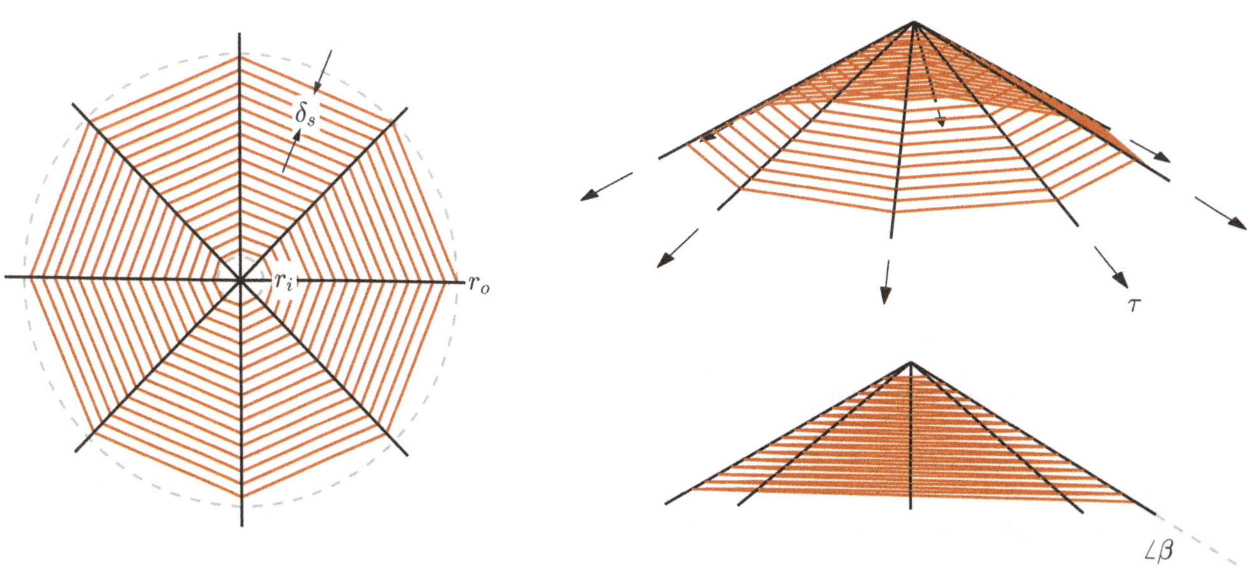

Fig. 12.1 Illustration of the un-tensioned web geometry and parameters used in this study

Table 12.1 Web geometry nomenclature

Parameter	Description
r_o	Outer radius of un-tensioned spiral
r_i	Inner radius of un-tensioned spiral
β	Angle of un-tensioned web cone
n_s	Number of spiral turns
δ_s	Inter-spiral spacing, $\delta_s = (r_o - r_i)/n_s$
τ	Peripheral radial pre-tension
D	Final diameter of tensioned web

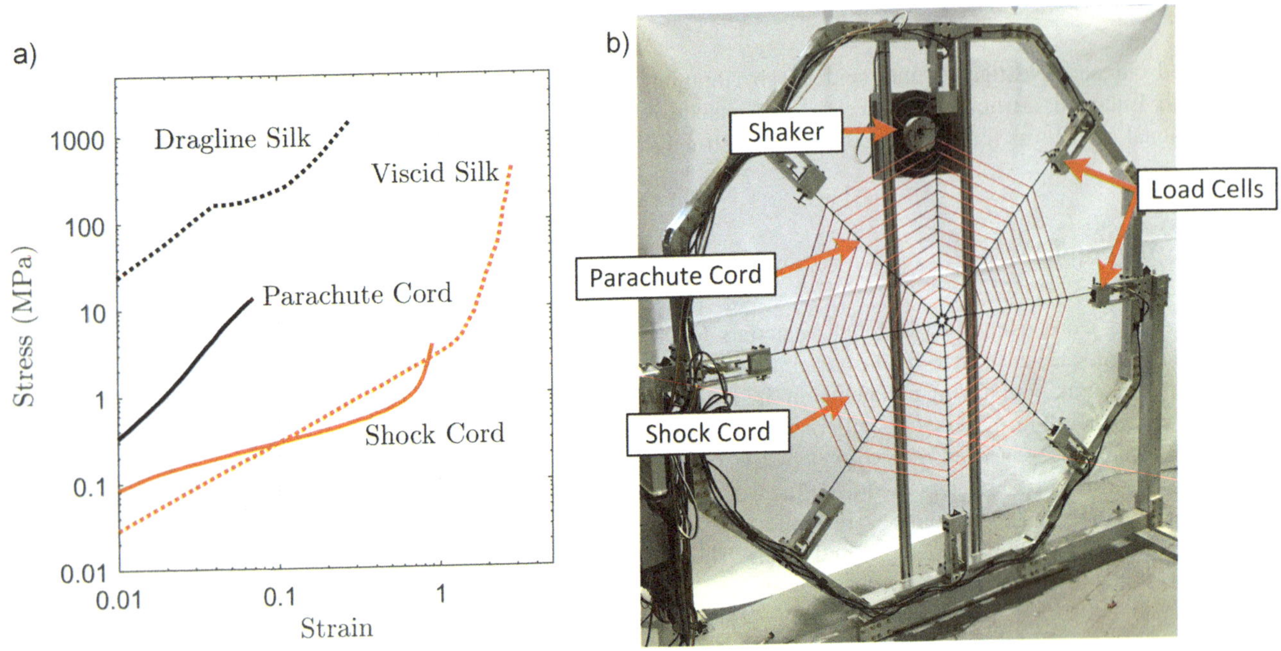

Fig. 12.2 (**a**) Stress-strain data for the parachute cord and shock cord used in the study as well as dragline and viscid spider silk adapted from [21], (**b**) the test stand and artificial web used in this study

shaker as well as the vibration response at various locations in the web. A personal computer and a National Instruments CompactDAQ collected, filtered, and processed signals arriving from the accelerometers. Data acquisition was triggered at the first output sample to the shaker and was sampled at 1000 Hz. The accelerometer waveforms were then filtered using a 3rd order Butterworth bandpass filter with lower and upper cutoff frequencies of 5 Hz and 250 Hz, respectively.

12.3.2 Vibration Model

Our model of the spider web has two components. The first component computes web pretension based on web boundary conditions and the mechanical properties of the strings in the web. The second component uses Frequency Based dynamic Substructuring (FBS) to assemble the web from individual string and mass elements to generate frequency response functions for points of interest in the web.

12.3.2.1 Preprocessing and Tensioning

The vibration of strings, such as those in a spider's web, is governed by the amount of tension in each string. Prior to modeling the vibrations in the web we must determine a pre-tensioned web state that satisfies static equilibrium. The tensioned web configuration under equilibrium is obtained using a minimum strain energy approach commonly employed in the analysis of cable roof structures [22]. To begin, a web geometry under zero tension is specified. Then, the tensioned equilibrium geometry is found from the initial un-tensioned geometry following a minimum strain energy approach as the set of nodal displacements x that satisfies

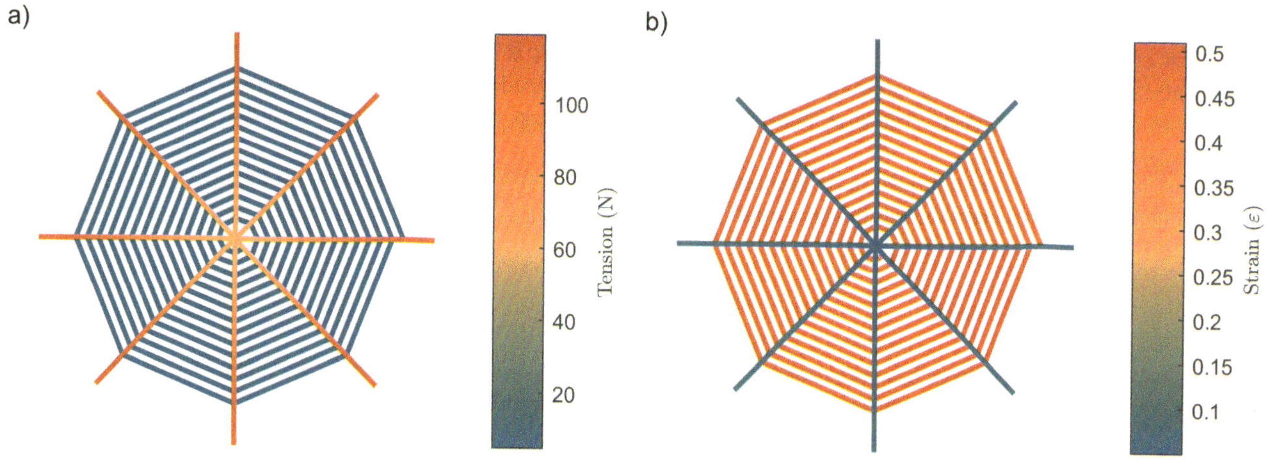

Fig. 12.3 (**a**) Tension distribution and (**b**) strain distribution output from the web pre-processing step for a sample web

$$\min_{\varepsilon} \sum_{j=1}^{M} \ell_j \int f_j(\varepsilon)\, d\varepsilon$$

$$\text{subject to } x_i = b_i, i = 1, \ldots, N, \tag{12.1}$$

where ε is the vector containing element strains ε_j, M is the total number of string elements in the web mesh, ℓ_j is the current length of element j, x_i is the nodal coordinate corresponding to constraint i, b_i is the constrained value for nodal coordinate x_i, N is the total number of nodal constraints, and $f_j(\varepsilon)$ is the force in element j which is obtained by interpolating the strain in the element into stress-strain data, such as that given in Fig. 12.2.

Strain (ε) is expressed in terms of the nodal coordinates using

$$\varepsilon = \frac{\ell_j - L_j}{L_j} = \frac{|x_{mn}| - L_j}{L_j}, \tag{12.2}$$

where L_j is the initial length of element j and x_{mn} is the vector spanning nodes m and n that defines element j. The advantage of computing web equilibrium using minimum strain energy is that it handles large displacements as well as nonlinear material properties in a compact manner, both of which are present in the process of web pre-tensioning. Additionally, this method provides a mapping from a pre-specified, zero-tension geometry (something that can be easily constructed in a physical system) to a tensioned structure. Solutions of the minimum strain energy problem can be found using standard numerical optimization packages, such as MATLAB's `fmincon`. A sample tension and strain distribution is displayed in Fig. 12.3.

12.3.2.2 Frequency Based Substructuring

Once the equilibrium geometry and tension are obtained, we use Frequency Based Substructuring (FBS) to model the transverse vibrations of the orb web. Frequency based substructuring computes the dynamics of a structure in the frequency domain by interconnecting smaller "substructures" through equilibrium and compatibility conditions. The method is covered in detail in [18]. We reproduce relevant portions here for convenience.

In general, a dynamic system can be expressed in the frequency domain using the equations

$$\boldsymbol{Z}(\omega)\, \boldsymbol{u}(\omega) = \boldsymbol{f}(\omega) + \boldsymbol{g}(\omega) \tag{12.3}$$

$$\boldsymbol{B}\boldsymbol{u}(\omega) = 0 \tag{12.4}$$

$$\boldsymbol{L}^T \boldsymbol{g}(\omega) = 0, \tag{12.5}$$

where m is the mass of the object. Assuming the point undergoes harmonic motion, the displacement of the point can be written as

$$x(t) = Xe^{i\omega t}, \tag{12.26}$$

where X represents the amplitude of motion. Differentiation of Eq. (12.26) and substitution into Eq. (12.25) results in

$$Fe^{i\omega t} = -m\omega^2 Xe^{i\omega t}. \tag{12.27}$$

From Eq. (12.27) the receptance matrix of the point mass can be identified as

$$\boldsymbol{Y}^s = \left[-\frac{1}{m\omega^2}\right], \tag{12.28}$$

which gives the dynamic stiffness matrix by the inverse operation

$$\boldsymbol{Z}^s = (\boldsymbol{Y}^s)^{-1} = \left[-m\omega^2\right]. \tag{12.29}$$

String and point mass substructures may be combined using mesh connectivity relationships between individual substructure degrees of freedom. From this connectivity information, the Boolean matrices $\boldsymbol{B}$ and $\boldsymbol{L}$ can be formed for calculation of the assembled system dynamic stiffness matrix $\tilde{\mathbf{Z}}$. Frequency response functions are obtained for the assembled system by applying unit external forces to a single degree of freedom and solving for the displacement vector $\boldsymbol{q}(\omega)$.

12.4 Results and Discussion

To evaluate the Frequency Based Substructuring (FBS) model we compared its output with measurements taken from the artificial web. Then, we use the FBS model to explore influences of web architecture on web frequency response at points around the center of the spider web corresponding to foot placements of a spider. Finally, we identified the presence of cues in the frequency response functions that can be used to localize a vibration stimulus in the web.

12.4.1 Model Validation

To validate our FBS model, we constructed an artificial web and measured its vibration response. The artificial web was created using a 45 degree cone angle and 14 spiral turns. The peripheral radial tension (τ) was held at two different levels, 115 N and 160 N, and a point mass was attached to the center of the web with two values of 10 g and 70 g. Excitation was applied at a single node at the outermost portion of the capture spiral. An accelerometer was located 12 cm away from the web center on the same radial as the excitation to measure web vibration. Mass loading of the web by the accelerometer was included in the FBS model by adding a 5 g mass at the point of measurement.

Experimental and modeled frequency responses are displayed in Fig. 12.4. For comparison, we focused on frequencies up to 50 Hz, which was slightly above the second web resonance. Damping for the FBS model was determined by matching the average Q factor (the ratio of the resonant frequency to the −3 dB width of the resonant peak) at the first resonant frequency in the experimental data across the four web treatments. As shown in Fig. 12.4, the model matches the first and second resonant peaks with fair accuracy, and also captures the anti-resonant trough occurring after the first resonance. In general, the model predicts the location of the first resonance and anti-resonance well, but the shape of the frequency response following the second resonance is less well-predicted. The small differences between the model and experimental web are likely due to imperfections in the web-building process, resulting in discrepancies in the final tension distribution as well as overall shape of the capture spiral portion of the web. The model clearly shows the same shifts in frequency response with increasing mass and increasing web tension as the experimental results (i.e. increasing resonant frequency with increasing tension, decreasing resonant frequency with increasing mass).

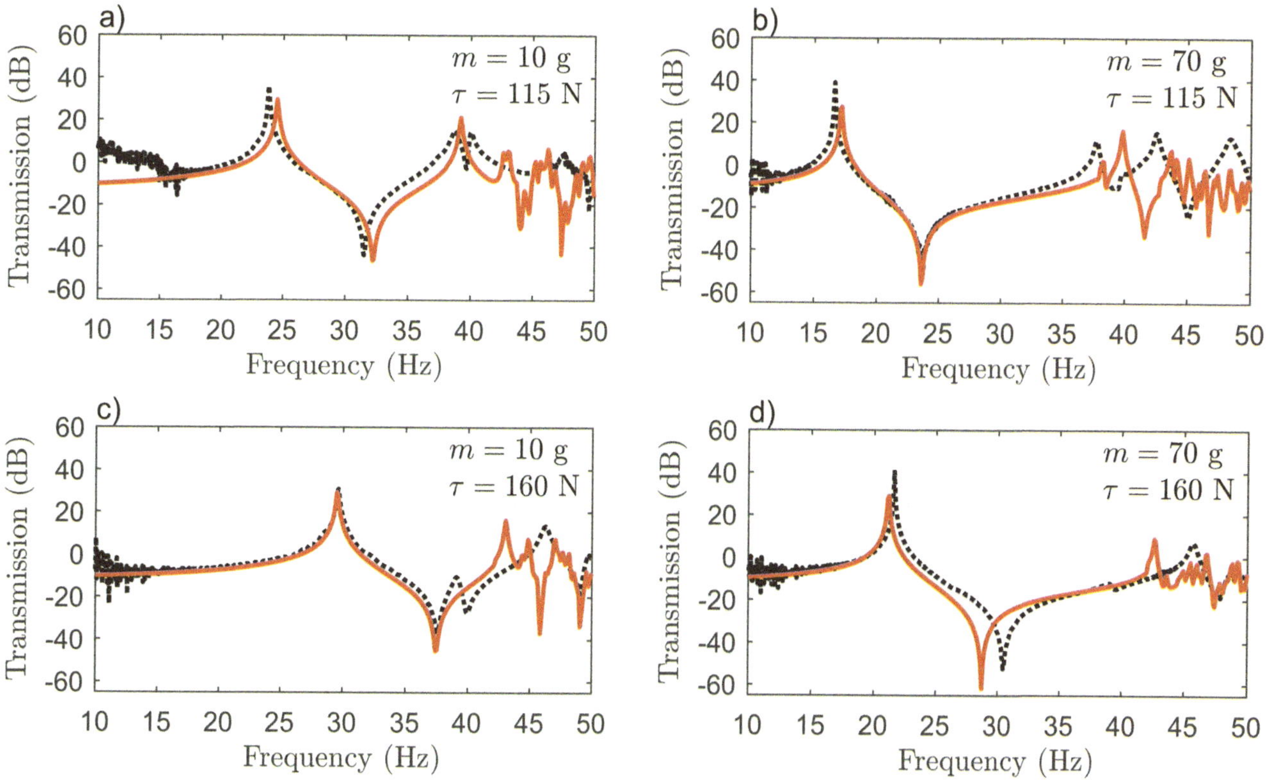

Fig. 12.4 Model (solid, orange) and experimental (dotted, black) frequency response (taken as the output acceleration magnitude divided by the input acceleration magnitude) for different values of point mass (m) and radial pretension (τ)

12.4.2 Web Architecture Effects

To characterize the influence of web architecture on web vibration, we varied the design parameters of the enlarged web in the FBS model around a "nominal" web configuration comprised of 16 spiral turns, 20 degree cone angle, 100 N peripheral radial tension, and 100 g of mass at the center of the web. One parameter was varied at a time from the nominal condition by selecting parameter levels above and below the nominal condition. Spiral turns were varied by ± 8 turns, cone angle was varied by $-20°$ and $+25°$, peripheral radial tension was varied by ± 40 N, and point mass was varied from 10 g at the low level to 1 kg at the high level. Figure 12.5 displays frequency responses for each of these parameter variations (a–d, e–f) as well as the nominal web (i). Each plot contains the frequency response function for eight spider leg placements located in a circle 10 cm from the web center on each radial, shown in Fig. 12.5j.

Several trends can be identified from Fig. 12.5: Increasing the point mass at the center of the web reduced the frequency of the first resonant peak and caused anti-resonances (seen as valleys or troughs) to occur in the frequency responses corresponding to several foot placement locations (Fig. 12.5d, h). These foot placement locations can be seen as the valleys located just following the first resonant peak, starting with the foot placement closest to the vibration source (orange). The frequency response following the second resonance was marginally affected by increasing the point mass. Increasing peripheral radial pretension (Fig. 12.5c, g) had the opposite effect of increasing mass, as it increased the frequency of the first resonant peak and eliminated several anti-resonant troughs that were present at decreased tension. Increasing the cone angle (Fig. 12.5b, f) of the web (i.e. decreasing the radial-spiral tension ratio) had a more complex effect on the frequency response structure, primarily increasing the number of anti-resonant troughs present following the first resonant peak as well as the location and magnitude of the second resonant peak. Finally, increasing the number of spirals in the web (Fig. 12.5a, e) reduced the frequency of the first resonance, caused additional anti-resonances between the first two resonant peaks, and increased the magnitude of the second web resonance.

While the parameters were only varied one-at-time from a nominal condition, they can also interact with one another, potentially allowing for a great degree of "tuning" of the web to achieve certain frequency response characteristics. For example, if it was desirable to have low amplitude motion for certain leg placements at specific frequencies (i.e. anti-resonances), the web could be constructed with a larger cone angle, which is analogous to decreasing the radial-spiral

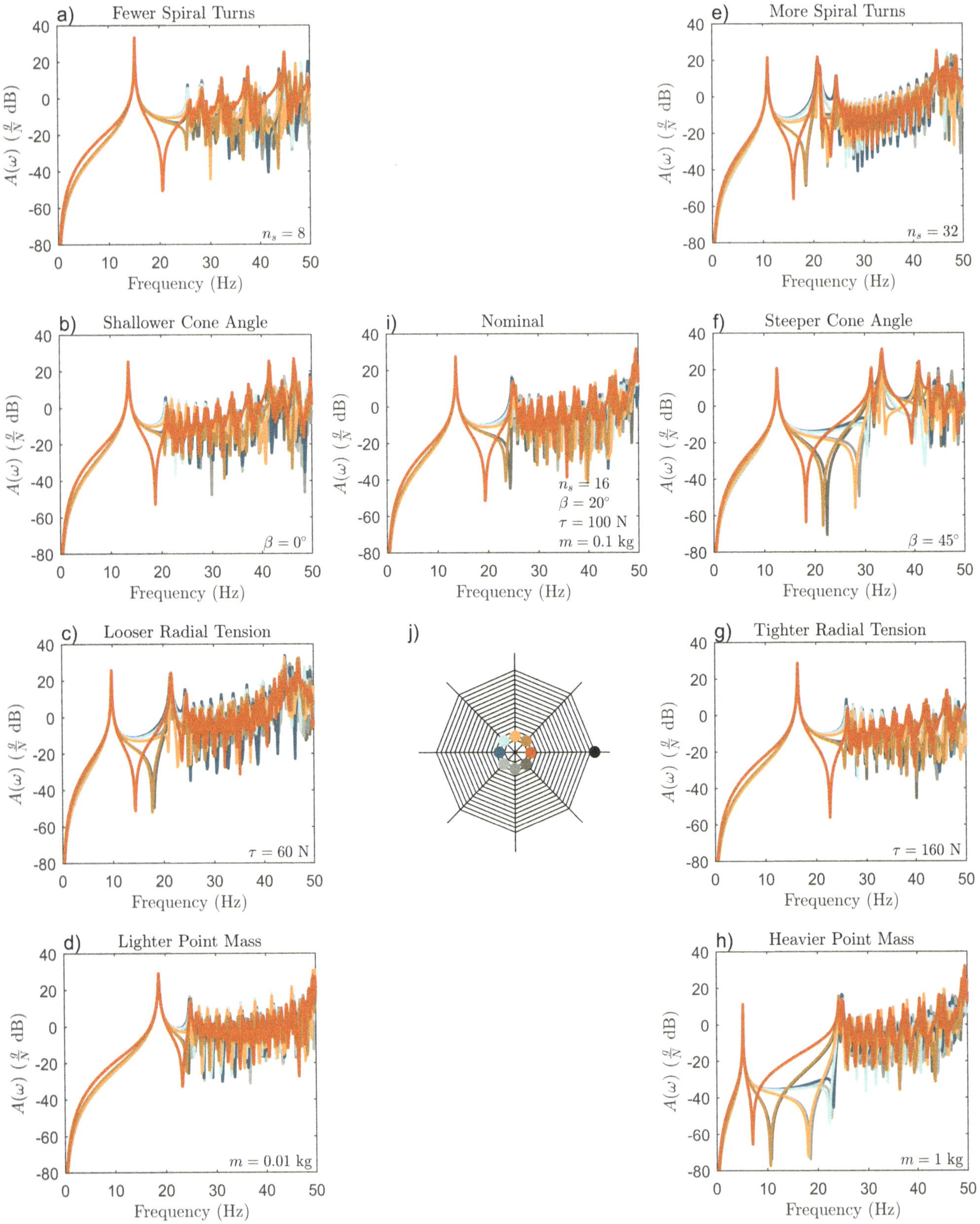

Fig. 12.5 Web architecture effects on web frequency response (accelerance $A(\omega)$), (**a**-**d**) web frequency response at low levels, (**e**-**h**) web frequency response at high levels, (**i**) web frequency response at nominal level, (**j**) input (black) and output (colors) locations for web frequency response

tension ratio, or the radial tension of the web could be decreased. In all, Fig. 12.5 demonstrates the complex way in which web architecture can influence web vibration, and how the vibration response of the web can be altered through changes in web geometry and composition given a particular end-goal.

12.4.3 Vibrational Cues for Stimulus Localization

Our primary interest in modeling spider web vibrations is to better understand the vibrational information presented to the spider in its web due to a stimulus (e.g. prey impact), and how that information can be used to determine the location of the stimulus. To investigate the presence of cues available to the spider for stimulus localization purposes, we generated frequency response functions for varying stimulus bearings and ranges relative to the center of the web using the FBS model. For this investigation, the measurement points of interest on the web were located at a radius of 10 cm from the center of the web on each of the web's eight radials, referred to here as leg placement locations. A single web architecture was used, composed of the paracord and shock cord in Fig. 12.2a for a 1.22 m diameter web and defined by 16 spiral turns, 45 degree cone angle, 120 N peripheral radial tension, and a point mass at the web center of 100 g. Stimuli were applied in varying locations within the capture spiral, corresponding to locations where prey is typically trapped. Leg placement locations, stimulus locations, and leg frequency responses are displayed in Fig. 12.6.

In terms of stimulus bearing to the spider (i.e. the angle between the spider heading and the stimulus location), prior work has shown that there is a vibration amplitude drop-off with increasing angular distance from a stimulus present in the webs of orb weaving spiders that can be used as a localization cue [3]. We interpret this cue as the total energy arriving at each leg in the frequency domain (obtained via integrating the frequency response), with legs that are closer to the vibration source receiving more energy. This energy is given as.3

$$E_i = \int_{\omega_0}^{\omega_n} A_i(\omega)\, d\omega \tag{12.30}$$

where E_i is the energy arriving at leg placement i and $A_i(\omega)$ is the frequency response function (accelerance) of the web at leg i. These leg energies are combined with the leg placement angles of the spider (taken counterclockwise from the positive X-axis in 45° increments for this case, with the web centered at the origin) to obtain a set of polar coordinates representing the energy arriving at each leg, shown at right in Fig. 12.6. To obtain a stimulus direction from the leg energy-angle pairs, we form a polygon and calculate its centroid. Stimulus direction is given by the angle (θ in Fig. 12.6) of the vector pointing from the origin to the polygon centroid. The length of this vector (r in Fig. 12.6) can be interpreted as the confidence in the detected bearing value.

The result of applying the bearing cue to the frequency response data in Fig. 12.6 demonstrates a clear ability to obtain accurate heading information from leg frequency responses. The radius values for the polygon centroid (r) show that confidence in stimulus heading decreases with increasing range for this specific web architecture, which intuitively agrees with the argument that distant disturbances are more difficult to localize than those that are close due to attenuation by the web.

Additionally, the frequency responses in Fig. 12.6 display a shift in frequency that accompanies changes in range of the vibration source – increasing frequency with decreasing range from the center of the web to the stimulus. We used the spectral centroid (ω_c) to quantify this frequency shift, defined as

$$\omega_c = \frac{\sum \omega_n X_n}{\sum X_n} \tag{12.31}$$

where X_n is the amplitude of the frequency response at bin n of the frequency spectrum and ω_n is the frequency value associated with bin n. Here, we use the averaged frequency response of all the leg placements in our calculation of the spectral centroid. For the three stimulus locations with decreasing range to the center of the web in Fig. 12.6, the respective spectral centroid values were 52.6 Hz, 70.3 Hz, and 80.6 Hz. The shift to higher frequency spectral centroids can be linked to the shorter length of the spiral member on which the vibration source is placed, i.e. at a given tension, shorter strings have higher natural frequencies. These findings demonstrate that our web model contains sufficient information to localize not only stimulus bearing as suggested by prior work, but also stimulus range.

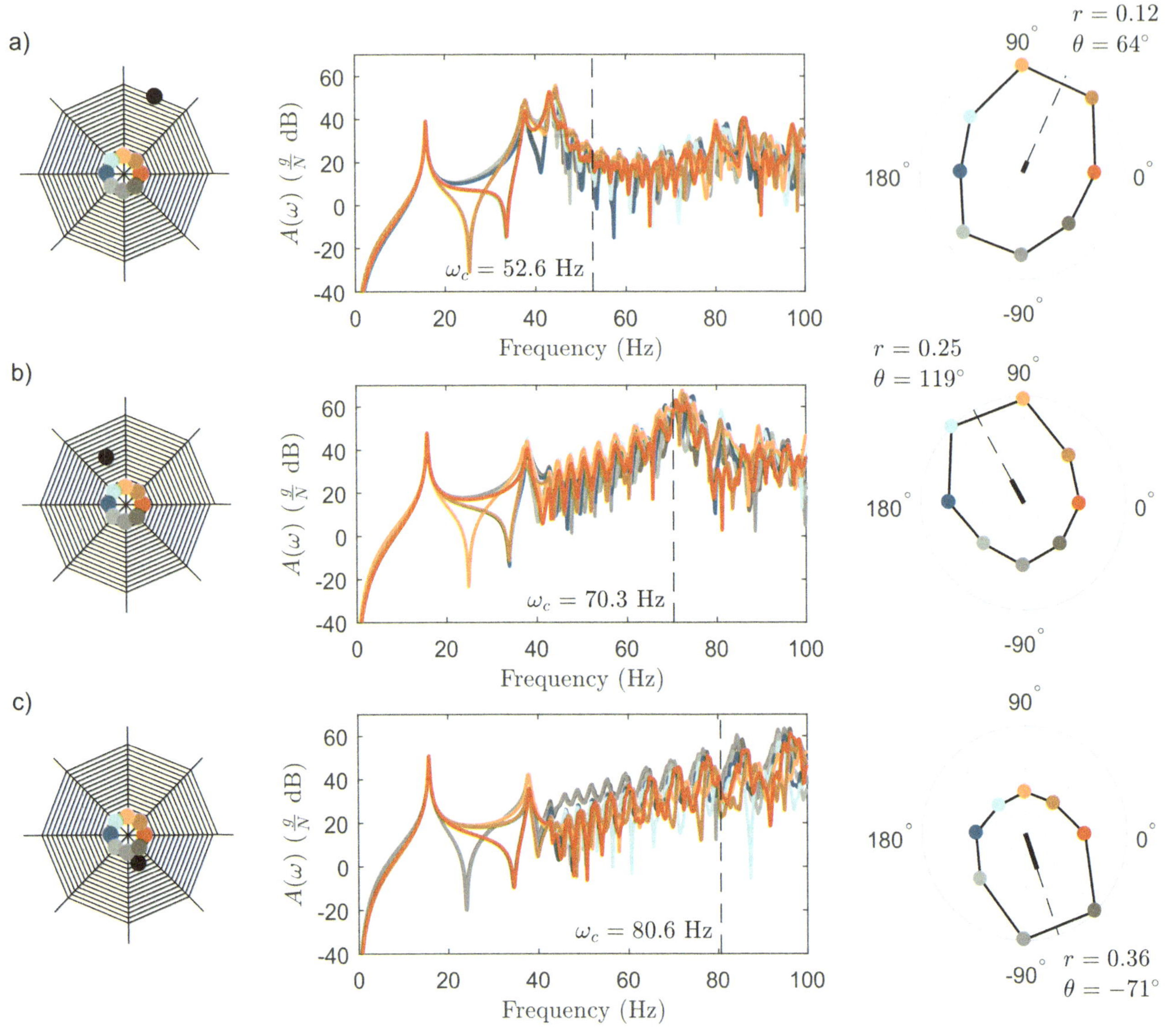

Fig. 12.6 Influence of vibration stimulus location (at left, black dot) on leg frequency responses (given here as accelerance) and spectral centroid ω_c (at center) as well as normalized leg energies (at right) for multiple vibration stimulus placements within the capture spiral for a single web architecture

12.5 Conclusion

This work modeled transverse vibrations in structures resembling the webs of orb-weaving spiders through the use of Frequency Based Substructuring (FBS). We validated the FBS model using an enlarged, artificial web instrumented with off-the-shelf MEMs-type accelerometers. Using the FBS model, we showed how different aspects of web architecture (e.g. capture spiral density, radial-spiral tension ratio, web pre-tension, and center mass) influence web frequency response. We also proposed a method for obtaining stimulus bearing and range cues from web frequency response by calculating the spectral energy at each leg placement location as well as the spectral centroid of the averaged leg spectrums.

Our results highlight the influence web architecture has on web vibrations and that the orb web is a structure with the potential to transmit a great deal of information to the spider. The analysis of possible tradeoffs in localization with respect to web architecture is possible using the methods presented in this work, which will aid in understanding the evolution of the orb web as both a snare and sensory extension of the spider. Moving forward, it will be necessary to confirm if these same localization cues are present in the webs of actual orb weaving spiders, or to what extent the frequency response patterns in silk webs differ from our model. It will also be important to assess if and how localization cues vary across different types of orb webs and how the predicted tradeoffs between localization and prey capture may drive diversity in web types across spiders. In addition, we plan to use our artificial web as a platform to develop refined localization approaches that function in real-time.

Acknowledgements A. W. Otto would like to thank Malcolm Rosenthal for useful discussions during the preparation of this paper. This work was supported by NSF Physics of Living Systems under grant 1504428.

References

1. Masters, W.M.: Vibrations in the orbwebs of Nuctenea sclopetaria (Araneidae): I. Transmission through the web. Behav. Ecol. Sociobiol. **15**, 207–215 (1984). https://doi.org/10.1007/BF00292977
2. Masters, W.M.: Vibrations in the orbwebs of Nuctenea sclopetaria (Araneidae): II. Prey and wind signals and the spider's response threshold. Behav. Ecol. Sociobiol. **15**, 217–223 (1984). https://doi.org/10.1007/BF00292978
3. Landolfa, M.A., Barth, F.G.: Vibrations in the orb web of the spider Nephila clavipes: cues for discrimination and orientation. J. Comp. Physiol. A. (1996). https://doi.org/10.1007/BF00192316
4. Tietsch, V., Alencastre, J., Witte, H., Torres, F.G.: Exploring the shock response of spider webs. J. Mech. Behav. Biomed. Mater. **56**, 1–5 (2016). https://doi.org/10.1016/j.jmbbm.2015.11.007
5. Mortimer, B., Soler, A., Siviour, C.R., et al.: Tuning the instrument: sonic properties in the spider's web. J. R. Soc. Interface. **13**, 20160341 (2016). https://doi.org/10.1098/rsif.2016.0341
6. Yu, H., Yang, J., Sun, Y.: Energy absorption of spider orb webs during prey capture: a mechanical analysis. J. Bionic Eng. **12**, 453–463 (2015). https://doi.org/10.1016/S1672-6529(14)60136-0
7. Frohlich, C., Buskirk, R.E.: Transmission and attenuation of vibration in orb spider webs. J. Theor. Biol. **95**, 13–36 (1982). https://doi.org/10.1016/0022-5193(82)90284-3
8. Cranford, S.W., Tarakanova, A., Pugno, N.M., Buehler, M.J.: Nonlinear material behaviour of spider silk yields robust webs. Nature. **482**, 72–76 (2012). https://doi.org/10.1038/nature10739
9. Soler, A., Zaera, R.: The secondary frame in spider orb webs: the detail that makes the difference. Sci. Rep. (2016). https://doi.org/10.1038/srep31265
10. Ko, F.K., Jovicic, J.: Modeling of mechanical properties and structural design of spider web. Biomacromolecules. **5**, 780–785 (2004). https://doi.org/10.1021/bm0345099
11. Lin, L.H., Edmonds, D.T., Vollrath, F.: Structural engineering of an orb-spider's web. Nature. **373**, 146–148 (1995). https://doi.org/10.1038/373146a0
12. Zaera, R., Soler, A., Teus, J.: Uncovering changes in spider orb-web topology owing to aerodynamic effects. J. R. Soc. Interface. **11**, 20140484–20140484 (2014). https://doi.org/10.1098/rsif.2014.0484
13. Morassi, A., Soler, A., Zaera, R.: A continuum membrane model for small deformations of a spider orb-web. Mech. Syst. Signal Process. **93**, 610–633 (2017). https://doi.org/10.1016/j.ymssp.2017.02.018
14. Wirth, E., Barth, F.: Forces in the spider orb web. J. Comp. Physiol. A. (1992). https://doi.org/10.1007/BF00223966
15. Alam, M.S., Wahab, M.A., Jenkins, C.H.: Mechanics in naturally compliant structures. Mech. Mater. **39**, 145–160 (2007). https://doi.org/10.1016/j.mechmat.2006.04.005
16. Qin, Z., Compton, B.G., Lewis, J.A., Buehler, M.J.: Structural optimization of 3D-printed synthetic spider webs for high strength. Nat. Commun. **6**, 7038 (2015). https://doi.org/10.1038/ncomms8038
17. Zheng, L., Behrooz, M., Gordaninejad, F.: A bioinspired adaptive spider web. Bioinspir. Biomim. **12**, 016012 (2017). https://doi.org/10.1088/1748-3190/12/1/016012
18. Klerk, D.D., Rixen, D.J., Voormeeren, S.N.: General framework for dynamic substructuring: history, review and classification of techniques. AIAA J. **46**, 1169–1181 (2008). https://doi.org/10.2514/1.33274
19. Bampton, M.C.C., Craig Jr., R.R.: Coupling of substructures for dynamic analyses. AIAA J. **6**, 1313–1319 (1968). https://doi.org/10.2514/3.4741
20. Hurty, W.C.: Dynamic analysis of structural systems using component modes. AIAA J. **3**, 678–685 (1965). https://doi.org/10.2514/3.2947
21. Gosline, J., Lillie, M., Carrington, E., et al.: Elastic proteins: biological roles and mechanical properties. Philos. Trans. R. Soc. B Biol. Sci. **357**, 121–132 (2002). https://doi.org/10.1098/rstb.2001.1022
22. Buchholdt, H.A.: In: Telford, T. (ed.) An Introduction to Cable Roof Structures, 2nd edn. ASCE Press, Reston (1999)
23. Bishop, R.E.D., Johnson, D.C.: The mechanics of vibration, Paperback re-issue, 1. paperback Ed. Cambridge Univ. Press, Cambridge (2011)

Chapter 13
Recent Advances to Estimation of Fixed-Interface Modal Models Using Dynamic Substructuring

Mathew S. Allen and Randall L. Mayes

Abstract In 2010, Allen & Mayes proposed to estimate the fixed-interface modes of a structure by measuring the modes of the structure bolted to a fixture and then applying constraints to the fixture using the transmission simulator method. While the method proved useful, and has indeed been used in studies since that point, a few peculiarities were noted. First, in some cases the estimated fixed-base natural frequencies were observed to converge very slowly to the true values (in simulated experiments) as the number of constraints was increased. To formulate these constraints, prior studies used only the free-interface modes of the fixture or the measured modes of the assembly. This work extends that to consider other sets of constraints, showing improved results. Furthermore, in some prior studies it has been observed that there were errors of more than 10% in the natural frequencies even when the fixture motion was hundreds of times smaller than the motion of the structure of interest (and so it had presumably been removed). This work explores this phenomenon, seeking to use the strain energy in the fixture, to the extent that it can be estimated using a test-analysis model for the fixture, as a metric to predict frequency error. The proposed methods are explored by applying them to simulated measurements from a beam and from the NASA Space Launch System coupled to the Mobile Launcher.

Keywords Component mode synthesis · Hurty/Craig-Bampton · Boundary conditions · Residual flexibility · Residual stiffness

13.1 Introduction

The problem outlined in Fig. 13.1 is a variant on a classical substructuring problem, and so various methods could be used to address it. The method to be used here was developed in about 2010 and was validated through both simulations and experiments as published in [1] and as will be briefly reviewed subsequently. In most of the literature on substructuring the goal is to assemble two substructures by applying a set of constraints (enforce compatibility) and enforcing force equilibrium. The proposed application is somewhat simpler because we seek only to apply a set of constraints to one system to estimate the fixed base response. As reviewed in [2], there are two main classes of approaches: Modal Substructuring or Component Mode Synthesis (CMS) and Frequency Based Substructuring (FBS).

This paper focuses on the Modal Substructuring (CMS) family of approaches, which were pioneered by Martinez & Carne [3], Ewins [4] and others. Early attempts proved to be extremely challenging because of the difficulty of estimating rotational responses at the interfaces; the methods are very sensitive to small errors in the measurements near the connection points. One important advance in this area occurred recently with the discovery that the substructuring could be improved by using a basis of shapes to average over several sensors near the interface [5, 6] and adding additional sensors away from the interface [6, 7]. In light of these findings, Allen & Mayes developed the Transmission Simulator (TS) method [6, 8, 9] and began to report the first successful case studies in which an experimentally measured substructure was coupled to a finite element model through a continuous interface. These ideas were later used to develop a procedure to estimate the fixed-interface modes of a structure from measurements of the structure on an assembly in [1]. In that study the method was

M. S. Allen (✉)
Department of Engineering Physics, University of Wisconsin, Madison, WI, USA
e-mail: msallen@engr.wisc.edu

R. L. Mayes
Structural Dynamics Department, Sandia National Laboratories, Albuquerque, NM, USA
e-mail: randy.mayes@awe.co.uk

A. Linderholt et al. (eds.), *Dynamics of Coupled Structures, Volume 4*, Conference Proceedings of the Society for Experimental Mechanics Series, https://doi.org/10.1007/978-3-319-74654-8_13

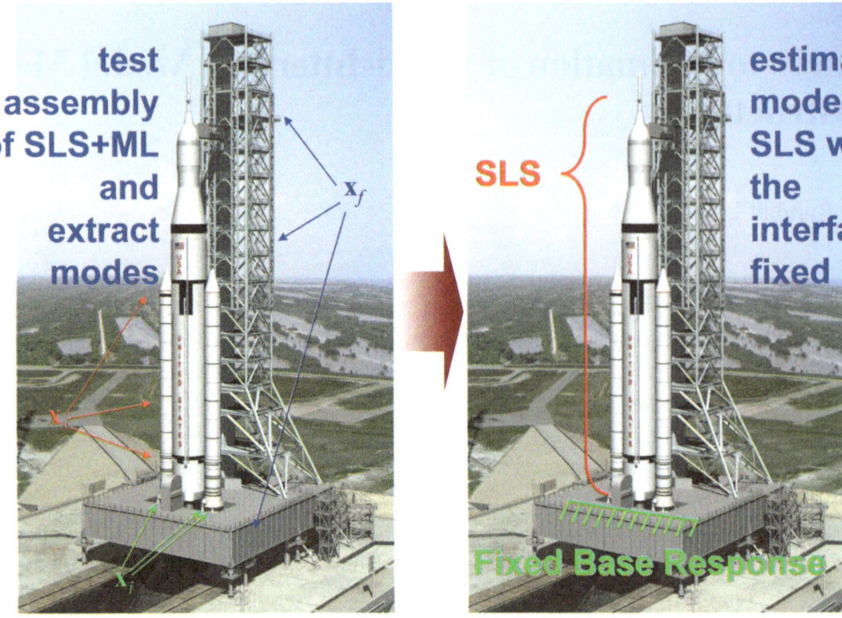

Fig. 13.1 Schematic of one potential application of the proposed method. A modal test could be performed on the SLS when it is assembled to the ML and the modes of this assembly could be used to estimate the modes of the SLS with the interface fixed

demonstrated on two test structures, which are shown in Fig. 13.1, and some results from those studies will be discussed subsequently. This is the procedure that will be used in this paper. Note, the TS is sometimes called a "Substitute" in the works of Rixen et al. [10].

The FBS family of approaches are also known by a variety of names: Impedance Coupling [11, 12], Admittance Modeling [13, 14], Receptance Coupling, Norton-Thevenin Equivalent Circuit modeling, FRF Structural Modification [15, 16], etc... All of these approaches use measured frequency responses at various points on the structure (and in rare cases time domain impulse responses) to estimate the FRFs of an assembly from FRFs of each component separately. At a minimum, the input and output FRFs must be measured at each of the connection DOF, and this can lead to difficulties for multi-point connections since the FRF matrix becomes large, potentially ill-conditioned, and slight errors in reciprocity can produce anomalies in the results [13]. The challenge is understanding to what level the residual mass and stiffness from out of band modes must be captured; the anti-resonances in the FRF matrix are very sensitive to these terms. Recently, these challenges have been addressed by defining a set of basis vectors that span the space of base motion and using those to improve the conditioning in the force and response locations near the base. For example see the substructuring work done by de Klerk, Rixen et al. [17, 18] and the work by Napolitano & Yoder [19, 20]. This work focuses entirely on the modal substructuring approach, but the FBS alternative is important to keep in mind since it may prove advantageous in some situations.

As explained in the abstract, this work explores two variants on the substructuring approach proposed in [1] and applies them to two structures, a simple beam and a test-analysis model of the NASA Space Launch System (SLS).

13.2 Theoretical Development

A modal test [4] can be performed on a structure to identify a linear time invariant model of a structure. The resulting modal model is comprised of N modes of vibration, whose shapes are measured at a certain set of measurement points. For the purposes of this work, we presume that a test has been performed on a structure of interest which is connected to a fixture or transmission simulator [21] so the equation of motion for the fixture + structure is given by

$$\begin{gathered}[\mathbf{I}]\,\ddot{\mathbf{q}} + \left[{}^{\backslash}2\omega_r\zeta_{r\backslash}\right]\dot{\mathbf{q}} + \left[{}^{\backslash}\omega_r^{\,2}{}_{\backslash}\right]\mathbf{q} = \begin{bmatrix}\boldsymbol{\varphi}_f \\ \boldsymbol{\varphi}_s\end{bmatrix}^{\mathrm{T}}\{\mathbf{F}\} \\ \mathbf{x} = \begin{bmatrix}\boldsymbol{\varphi}_f \\ \boldsymbol{\varphi}_s\end{bmatrix}\mathbf{q}\end{gathered} \tag{13.1}$$

where ω_r is the rth natural frequency of the assembly, ζ_r is the corresponding damping ratio and the matrices $\boldsymbol{\varphi}_f$ and $\boldsymbol{\varphi}_s$ contain the mass-normalized mode shapes at the measurement points on the fixture, $\mathbf{x}_f$, and substructure, $\mathbf{x}_s$, respectively.

The goal is to estimate the motion of the substructure with the interface between the fixture and substructure fixed, and this can be done if the motion of the fixture can be driven to zero. To do this, we presume that the motion of the fixture can be described by a set of shapes, $\boldsymbol{\varphi}_f^{\text{fixt}}$.

$$\mathbf{x}_f = \boldsymbol{\varphi}_f^{\text{fixt}}\mathbf{q}^{\text{fixt}} \tag{13.2}$$

The following subsections will discuss various strategies for choosing the basis $\boldsymbol{\varphi}_f^{\text{fixt}}$.

If the fixture were truly rigid and perfectly fixed to ground, then the motion of the fixture, $\mathbf{x}_f$ would be zero, but all fixtures have finite stiffness, and so some motion will always be present. The most straightforward remedy would be to estimate both the displacement and rotation at the points where the structure joins the fixture and then to apply constraints to force that motion to be zero, i.e. $\mathbf{x}_i = 0$, where $\mathbf{x}_i$ are the DOF at the interface (see Fig. 13.1). Using this approach, the number of constraints to be applied is equal to the number of interface DOF, so if the fixture + substructure has N modes, and there are N_i interface DOF, then the resulting model for the fixed-interface substructure will have N-N_i modes. While this method is straightforward, it was unfortunately illustrated in [1], that this method produces complicated results, with many spurious modes remaining in the model, and it seems to lack robustness when the measurements near the interface are noisy. Instead, in this work we prefer to use a global set of constraints applied over all DOF $\mathbf{x}_f$. If it is feasible to take measurements at or near the interface DOF, then those can also be included. They are not required; it is only necessary to use measurement locations which cause the fixture mode shape matrix to be easily invertible.

If the basis chosen, $\boldsymbol{\varphi}_f^{\text{fixt}}$, is full rank, then the fixture motions can be estimated from the measured motions, $\mathbf{x}_f$, as follows,

$$\mathbf{q}^{\text{fixt}} \approx \left(\boldsymbol{\varphi}_f^{\text{fixt}}\right)^{\dagger}\mathbf{x}_f = \left(\boldsymbol{\varphi}_f^{\text{fixt}}\right)^{\dagger}\boldsymbol{\varphi}_f\mathbf{q}, \tag{13.3}$$

where $\mathbf{q}$ denotes the modal coordinates of the assembled fixture + structure and $()^{\dagger}$ denotes the Moore-Penrose pseudoinverse. The desired model would have negligible motion of the fixture, or in other words,

$$\mathbf{q}^{\text{fixt}} = 0. \tag{13.4}$$

These constraints can be applied using standard Ritz or substructuring methods [6, 8, 22]. In terms of the modal coordinates of the fixture + structure, the constraint equations are

$$\left(\boldsymbol{\varphi}_f^{\text{fixt}}\right)^{+}\boldsymbol{\varphi}_f\mathbf{q} = 0, \tag{13.5}$$

either of which constitutes a set of N_{fixt} constraint equations. The matrix multiplying $\mathbf{q}$ is the matrix $[a]$ in the text by Ginsberg [22], or $\mathbf{B}$ in the review by De Klerk, Rixen and Voormeeren [2]. The procedure described in either of those works can be used to enforce these constraints and hence to estimate the modes of the fixture + structure with the fixture motion nullified. The "ritzscomb" Matlab® routine, which is freely available from the first author or on the Matlab Central File Exchange, implements the method in [22] and was used to perform the calculations for this work. For completeness, the basic operations are summarized below, using the notation from [2].

First a matrix $\mathbf{L}$ is found that resides in the null space of $\mathbf{B} = \left(\boldsymbol{\varphi}_f^{\text{fixt}}\right)^{+}\boldsymbol{\varphi}_f$. Then, applying the coordinate change $\mathbf{q} = \mathbf{L}\mathbf{q}_u$ and pre-multipying by $\mathbf{L}^{\text{T}}$ to obtain a weak form of the equations (i.e. one that ignores any forces that are not within the span of the basis), the equations of motion can be written as follows,

$$\widehat{\mathbf{M}}\ddot{\mathbf{q}}_u + \widehat{\mathbf{C}}\dot{\mathbf{q}}_u + \widehat{\mathbf{K}}\mathbf{q}_u = \mathbf{L}^{\text{T}}\left[\boldsymbol{\varphi}_{\text{f}}^{\text{T}}\ \ \boldsymbol{\varphi}_{\text{s}}^{\text{T}}\right]\{\mathbf{F}\} \tag{13.6}$$

where $\widehat{\mathbf{K}} = \mathbf{L}^{\text{T}}\left[\backslash\omega_r^2\backslash\right]\mathbf{L}$, etc…

One key challenge is to find a basis set, $\boldsymbol{\varphi}_f^{\text{fixt}}$, that effectively constrains the fixture to ground using the smallest possible set of shapes, in order to maximize robustness and the number of modes that can be estimated. The key method presented in [8] is reviewed below, as well as a few proposed alternatives.

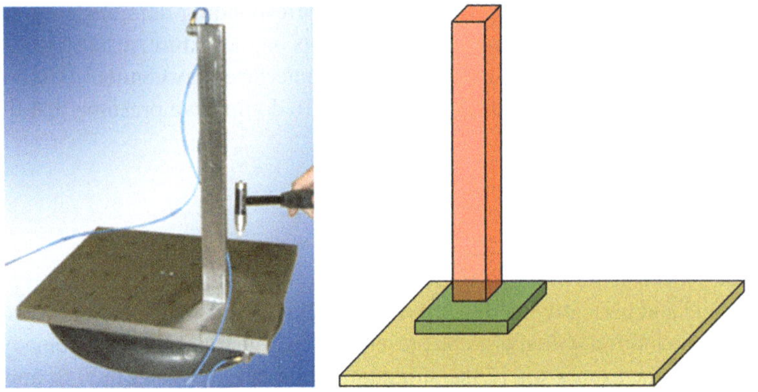

Fig. 13.2 (left) Photograph of a 0.5″ thick by 1.0″ wide steel beam mounted on a 0.625″ thick steel plate. The modes of this system were measured at a grid of points on the plate and used to estimate the fixed-interface modes of the beam. (right) Schematic of a FEM of the structure with the region near the interface made thicker (more rigid)

13.2.1 Modal Constraints (or MCFS Method [8])

Using the approach outlined in [8], a finite element model is created for the fixture, described by $\mathbf{K}^{fixt}$, $\mathbf{M}^{fixt}$, and then the following eigenvalue problem is solved to estimate a set of shapes, $\boldsymbol{\varphi}^{\mathrm{MCFS}}$,

$$\left[\mathbf{K}^{fixt} - \left(\omega^{fixt}\right)^2 \mathbf{M}^{fixt}\right] \boldsymbol{\varphi}^{\mathrm{MCFS}} = 0, \tag{13.7}$$

which are partitioned to the measurement DOF and used as the desired basis, i.e. $\boldsymbol{\varphi}_f^{\mathrm{fixt}} = \boldsymbol{\varphi}_f^{\mathrm{MCFS}}$.

In the study in [8], the authors applied this approach to a plate with a beam attached, as shown in Fig. 13.2. In that study, the authors found that even after constraining out many mode of the free-free plate, the resulting mode shapes of the cantilever beam agreed very well with theory, yet the natural frequencies estimated were in error by between 3 and 10%. The study was repeated with the experimental measurements replaced with modes from a FEM (to eliminate measurement errors) and the natural frequencies had almost the same pattern of errors. This revealed that modal truncation was the culprit, i.e. that it was not possible to completely nullify the motion of the fixture due to an incomplete basis for the fixture + substructure and an incomplete set of shapes $\boldsymbol{\varphi}_f^{\mathrm{fixt}}$. A further study using the FEM revealed that the natural frequency errors decreased to between 0.2 and 3% if the plate was made three times thicker near the connection point (schematic on the right in Fig. 13.2).

13.2.2 Modal Constraints with Rigidized Substructure (MC-R)

In a recent work, it was noted that MCFS shapes don't exercise the interface, and hence this might explain the slow convergence of the MCFS approach that was seen in [8]. If the mass properties of the substructure are known, then a rigid mass can readily be added to the finite element model of the fixture to load the interface and hopefully produce a better basis. The resulting FEM is denoted $\mathbf{K}^{fixt+R}$, $\mathbf{M}^{fixt+R}$, and once again an eigenvalue problem is solved to estimate the motion of the fixture + rigid substructure.

$$\left[\mathbf{K}^{fixt+R} - \left(\omega^{fixt+R}\right)^2 \mathbf{M}^{fixt+R}\right] \boldsymbol{\varphi}^{\mathrm{MCR}} = 0, \tag{13.8}$$

which are partitioned to the measurement DOF and used as the desired basis, i.e. $\boldsymbol{\varphi}_f^{\mathrm{fixt}} = \boldsymbol{\varphi}_f^{\mathrm{MCR}}$.

It is important to note that the constraints in Eqs. (13.7) and (13.8) only enforce zero motion at the fixture measurement points if the number of measurement points equals the number of fixture modes, but ideally one would actually use more measurement points on the fixture than there are active modes in the structure in order to average out noise and measurement errors. As a result, the motions of the physical measurement points will typically not be exactly zero after applying the constraints. In the best case the residual motion would be due only to measurement noise, but experience has shown that there is typically some residual motion in the fixture that is physical, since one is seeking to constrain an infinite dimensional system with a finite number of constraints. Fortunately, one can readily observe the fixture motions after applying the constraints to see whether the constraints were effective in enforcing a rigid boundary condition, and this was demonstrated in [8]. In this work we take this a step further, seeking to use the FEM of the fixture and the measured fixture motions to obtain better estimates of the natural frequencies of the substructure on a rigid foundation.

13.2.3 Natural Frequency Correction Scheme

In the example cited in Fig. 13.2, even after using the best available set of constraints, the natural frequencies estimated continued to show errors. It was noted that even though the residual motions of the fixture (the plate in Fig. 13.2) were small, less than 1% of the motion of the structure of interest (the vertical beam), because the stiffness of the plate is very high, it is possible that exceedingly small motions of the fixture have an important contribution to the strain energy in each mode, and hence to the natural frequencies. In this work, we propose to use Test Analysis Models (TAMs) derived from finite element models of the fixture to estimate the error in the natural frequencies due to the residual flexibility that remains in the fixture. Consider an undamped structure that is vibrating in the rth mode, so that $\mathbf{x}(t) = \mathrm{Re}\left(\boldsymbol{\varphi}_r e^{i\omega_r t}\right)$. The peak potential and kinetic energies must be equal, and the kinetic energy, $KE_{true} = \frac{1}{2}\dot{\mathbf{x}}^{\mathrm{T}}\mathbf{M}\dot{\mathbf{x}}$ can then be rewritten to find that the maximum kinetic energy is $\max(KE) = \omega_r^2 \times \frac{1}{2}\boldsymbol{\varphi}_r^{\mathrm{T}}\mathbf{M}\boldsymbol{\varphi}_r$. Similarly, the maximum potential energy is $\max(PE) = \frac{1}{2}\boldsymbol{\varphi}_r^{\mathrm{T}}\mathbf{K}\boldsymbol{\varphi}_r$, and equating these two we obtain the following equation that relates the rth natural frequency to the energy in the structure.

$$\omega_r^2 = \frac{PE}{KE^*} = \frac{\frac{1}{2}\boldsymbol{\varphi}_r^{\mathrm{T}}\mathbf{K}\boldsymbol{\varphi}_r}{\frac{1}{2}\boldsymbol{\varphi}_r^{\mathrm{T}}\mathbf{M}\boldsymbol{\varphi}_r}, \tag{13.9}$$

where the notation KE^* reminds the reader that this is not the true kinetic energy, but an energy like quantity.

The equation above can be expanded in a Taylor series to obtain the desired relationship,

$$2\omega_r \frac{\partial \omega_r}{\partial PE} = \frac{1}{KE^*}, \tag{13.10}$$

where we have assumed that the kinetic energy does not change significantly. This is in line with our experience, where the mode shape of the structure as a whole is nearly unchanged, so that the kinetic energy would not change significantly. The potential energy can then be decomposed by separating the mode shape into DOF on the structure of interest, denoted $\boldsymbol{\varphi}_{Sr}$, and those on the fixture or transmission simulator, $\boldsymbol{\varphi}_{Fr}$. We have measurements of the natural frequency and mode shape for a case where the fixture motion is non-negligible, $\omega_r|_{meas}$ and we desire to know the natural frequency that would be obtained if the fixture motion were zero $\omega_r|_0$. Expanding about the latter case, the relationship becomes the following.

$$\omega_r|_{meas} = \omega_r|_0 + \frac{1}{2\omega_r KE^*}\left(PE_S - PE_{S,0} + PE_F - PE_{F,0} + PE_J - PE_{J,0}\right), \tag{13.11}$$

where

$$PE_S = \frac{1}{2}\boldsymbol{\varphi}_{Sr,meas}{}^{\mathrm{T}}\mathbf{K}\boldsymbol{\varphi}_{Sr,meas}, \tag{13.12}$$

and so forth for the other sensor sets and for the measured case, $()|_{meas,}$ and the case where the fixture motion is zero $()|_0$.

Before proceeding it is helpful to illustrate using a simple example. Consider a spring attached to ground via two springs in series, k_s representing the stiffness of the system of interest and k_f representing the stiffness of a fixture. Then let the mass and k_s be unity as the stiffness of the fixture is varied between 1 and 100 (i.e. 1x to 100x stiffer than the system of interest). The natural frequency of the system was then computed and is shown in Fig. 13.3. The line labeled "Measured" corresponds to the natural frequency that would be measured on the elastic fixture, while the line labeled "True" is the frequency that we would obtain if the fixture was infinitely stiff. The measured frequency can be seen to decrease as the stiffness of the boundary decreases. The regime of interest are those cases where the natural frequency shows a small (i.e. ~10%) error.

The second pane in Fig. 13.3 shows the energy in each of the springs as a function of their relative stiffness. As the fixture becomes stiffer (with Ks/Kf approaching 0), less and less energy is stored in the fixture and the energy in the system approaches that which it would have on a rigid foundation. If the energies $\Delta PE_S = PE_S - PE_{S,0}$ and $\Delta PE_F = PE_F - PE_{F,0} = PE_F - 0$ were known, then one could use Eq. (13.11) to correct the measured natural frequency, as shown by the dashed black line in Fig. 13.3a. Unfortunately, while we expect that ΔPE_F could be estimated using a high quality ROM of the fixture or transmission simulator, it seems less likely that ΔPE_S could be reliably estimated.

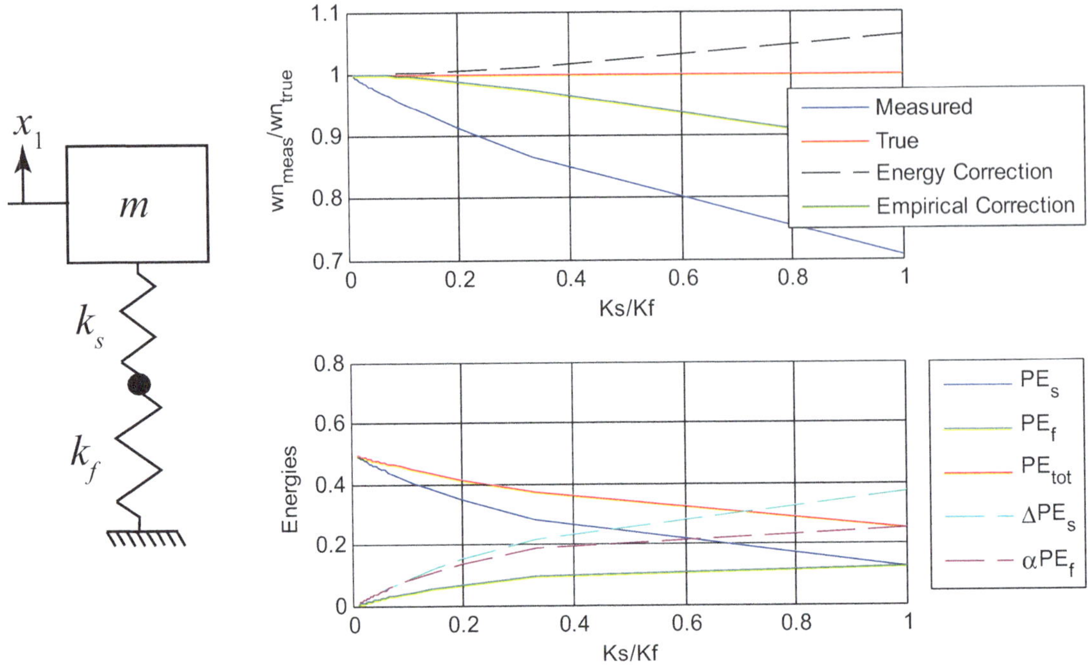

Fig. 13.3 (**a**) Ratio between the true natural frequency (i.e. that measured on an infinitely stiff foundation) and that measured on a foundation with stiffness k_f. (**b**) The potential energy (PE) in each of the springs k_s and k_f as a function of the ratio k_s/k_f. For the magenta curve, $\alpha = 2.0$

It appears that the only viable option is to use some empirical means to estimate the other energy quantities from ΔPE_F. For example, in the simple case shown, ΔPE_S is approximately equal to $-\alpha PE_F$, with $\alpha = 2.0$, as shown with the magenta line in Fig. 13.3. Using this assumption, the empirical natural frequency correction scheme becomes the following

$$\omega_r|_{meas} = \omega_r|_0 + \frac{1}{2\omega_r KE^*}\left((1-\alpha)\frac{1}{2}\boldsymbol{\varphi}_{Fr,meas}{}^{\mathrm{T}}\mathbf{K}\boldsymbol{\varphi}_{Fr,meas}\right), \tag{13.13}$$

and the result for this simple system, shown in Fig. 13.3a, can be seen to work quite well up to $k_s/k_f = 0.2$, or to the point where the error in the natural frequency is about 10%. In the absence of other information, we shall presume that this empirical factor, $\alpha = 2.0$, is reasonable for structures in general and evaluate the method for other structures.

13.3 Numerical Results

These two concepts, using mass-loaded shapes to enhance the constraint basis (Sect. 13.2.2), and the natural frequency correction scheme (Sect 13.2.3), were tested on two structures to evaluate their performance. The results are discussed in the following subsections.

13.3.1 Free-Free Beam

The first structure considered is an assembly of two free-free beams, shown below. The beam on the left is the "fixture" or transmission simulator, and the goal is to estimate the modes of the beam on the right when it is perfectly fixed. Two cases are simulated using one-dimensional finite element models. For all of the cases the beams have the following properties: height, h = 0.75 in, width, b = 1 in, Young's Modulus, E = 30.0e6 psi, density, ρ = 0.28 lb./in^3. For Case 1, the lengths of the beams are L_A = 12 in, L_B = 24 in, while Case 2 represents a more challenging situation in which Beam A is much more flexible relative to Beam B, with L_A = 20 in, L_B = 16 in. For all of the cases considered, we use all modes of A and B up to 6400 Hz, in other words we presume that the modal test on each component is able to extract all of the modes from 0 to 6400 Hz, including the rigid body modes (Fig. 13.4).

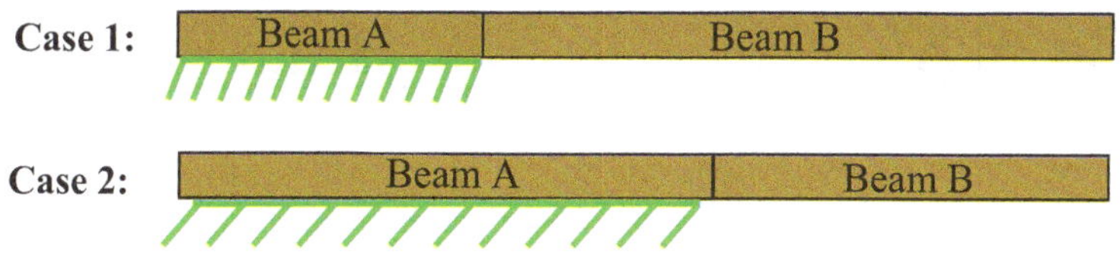

Fig. 13.4 Schematic of the beam system studied in this section. Two cases will be considered, Case 1 (top) and Case 2 (bottom), which differ in the relative lengths of Beams A and B

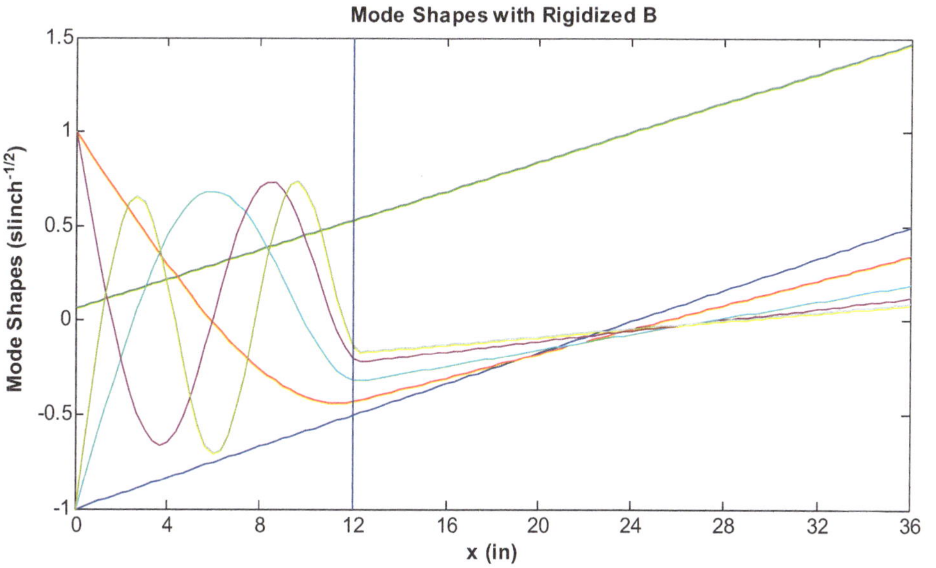

Fig. 13.5 Case 1: Mode shapes of the Transmission Simulator (Beam A) when Beam B is replaced by a rigid mass. This is the basis of shapes used to form the constraints in Eq. (13.5)

The first step is to compute the mode shapes of the transmission simulator when a rigid mass is used in place of the structure of interest. Figure 13.5 shows the resulting mode shapes for Case 1; these shapes are used in Eq. (13.5) to eliminate the motion of the transmission simulator. One can see that the rigid mass enriches the modal basis, as the resulting shapes show asymmetry and involve deformation near the interface.

The standard modal substructuring approach is then used to enforce these constraints, and the natural frequencies and mode shapes of the system with Beam A fixed are then computed. For Case 1, the resulting shapes are compared with the true fixed-interface shapes in Fig. 13.6 and the natural frequencies are compared in Table 13.1. For this case there were 12 modes of the assembly and 5 constraint shapes below 6400 Hz, and so this modal basis produces estimates for the first 7 modes of beam B. A Test-Analysis-Model (TAM) was created of Beams A and B using the displacement of each node (the rotations were statically reduced out), and this was used to estimate the potential energy in Beam A when it deformed into each mode, and the kinetic energy of Beam B (Fig. 13.7).

The results show that adding a Beam B to the transmission simulator enriches the basis, reducing the errors in the natural frequencies noticeably, although not dramatically. When using a free-free transmission simulator, the error, show in the third column, range between −3% and −7% In contrast, when a rigid Beam B is used to enrich the basis, the errors drop to −2% to 5%. The empirical natural frequency correction scheme improves the results further, most notably reducing the error in the first natural frequency from −4.9% to −2.3%. However, for some of the higher natural frequencies the scheme over-corrects the frequencies. When considering this, one should note that when using substructuring, they typical rule of thumb states that the natural frequencies should be accurate up to 50–66% of the frequency range used for each subcomponent, or 3200–4300 Hz in this case. However, here we obtain excellent results up to 80% of the frequency range. In our experience, if the fixture is quite rigid (i.e. well approximated as a rigid mass) then the results may be excellent in the entire frequency range. In contrast, if the fixture is very flexible then the results may be poor even below 50% of the measured frequency range.

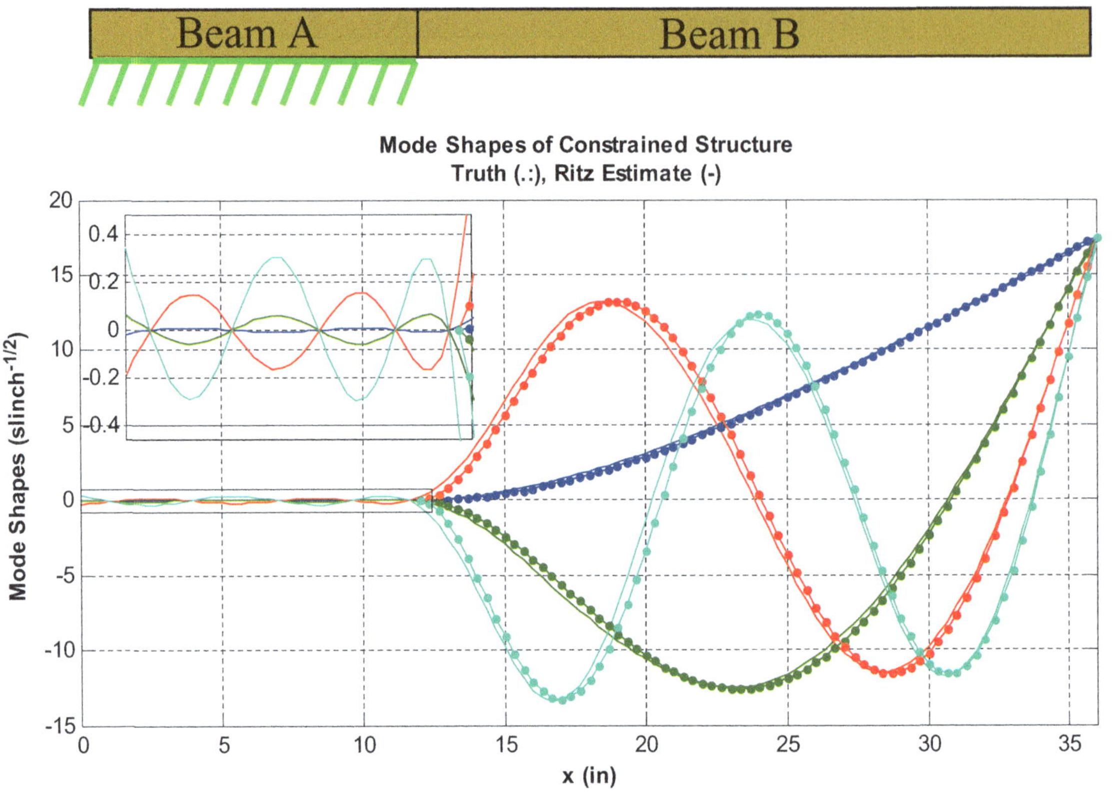

Fig. 13.6 **Case 1**: Mode shapes of Beam B with the Fixed-Interface fixed. **Solid lines** show the true fixed-interface shapes of the FEM, while the **dotted lines** show those estimated by Modal Substructuring using the MC-R method in Sect. 13.2.2

Table 13.1 **Case 1**: Natural Frequencies estimated by substructuring for two cases: (A = Free-Free) when the modes of the transmission simulator are computed with the interface free, (A + Rigid B) when the system of interest is replaced with a rigid mass, and (A + Rigid B, Corrected) the latter case with the frequencies corrected using the empirical formula

B Freq. True	B Freq. CMS A=Free-Free	% Error	B Freq. CMS A+Rigid B	% Error	B Freq. CMS A+Rigid B, Corrected	% Error
42.8	39.9	-6.7	40.7	-4.9	41.8	-2.3
268.2	251.5	-6.2	256.1	-4.5	262.1	-2.3
751.0	707.5	-5.8	720.1	-4.1	738.7	-1.6
1471.6	1393.2	-5.3	1417.2	-3.7	1464.9	-0.5
2432.7	2314.7	-4.9	2353.6	-3.3	2456.9	1.0
3634.1	3478.3	-4.3	3534.5	-2.7	3723.9	2.5
5075.7	4903.9	-3.4	4975.5	-2.0	5255.9	3.6

Next consider the case where Beam A is longer, i.e. the fixture is more flexible relative to the structure of interest. In this case there are once again 12 modes of the assembly below 6400 Hz, but now we have 7 constraint shapes in that frequency range. As might be expected, the more flexible fixture produces errors in the natural frequencies that are much larger, ranging between -11% and -13%. In this case using the enhanced basis for Beam A reduces these errors to -7% to -9%. Table 13.2 also shows the natural frequencies corrected using the empirical scheme with $\alpha = 2$. While the correction scheme works very well for the first mode, it increases the errors for the other modes, dramatically for the highest frequency modes. In the case of those modes, it may be that the strain energy is so large that the Taylor series expansion is no longer valid.

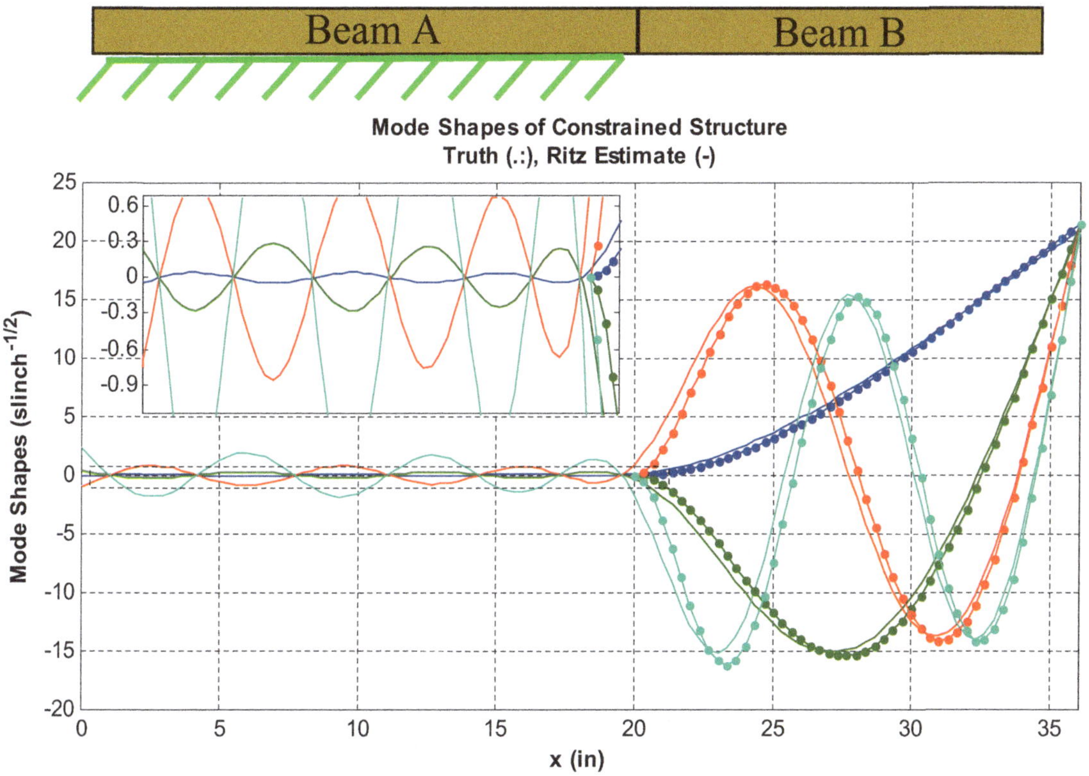

Fig. 13.7 **Case 2**: Mode shapes of Beam B with the Fixed-Interface fixed. **Solid lines** show the true fixed-interface shapes of the FEM, while the **dotted line**s show those estimated by Modal Substructuring using the MC-R method in Sect. 13.2.2

Table 13.2 **Case 2**: Natural Frequencies estimated by substructuring for two cases: (A = Free-Free) when the modes of the transmission simulator are computed with the interface free, (A + Rigid B) when the system of interest is replaced with a rigid mass, and (A + Rigid B, Corrected) the frequencies corrected using the empirical formula

B Freq. True	B Freq. CMS A=Free-Free	% Error	B Freq. CMS A+Rigid B	% Error	B Freq. CMS A+Rigid B, Corrected	% Error
96.3	83.8	-13.0	88.1	-8.5	95.0	-1.4
603.5	530.0	-12.2	555.9	-7.9	680.0	12.7
1689.7	1493.1	-11.6	1565.2	-7.4	2912.6	72.4
3311.2	2936.1	-11.3	3078.6	-7.0	9963.9	200.9
5473.7	4841.9	-11.5	5066.5	-7.4	35127.1	541.7

13.3.2 NASA Space Launch System on Mobile Launcher

The methods were further evaluated by testing them on the new Space Launch System (SLS) that NASA is developing. While legacy vehicles, such as the Saturn V and the Space Shuttle, were tested in free-free conditions in order to update the corresponding finite element models, NASA no longer has a facility with the capability to lift and hang even the initial 70 t payload model so that it could be tested free free. This initial model is expected to weigh approximately 5.5 million pounds, or the equivalent of 7.5 fully-loaded 747 jets.[1] Instead, plans are being developed to test the SLS while it is mounted on the Mobile Launch (ML) platform, within the Vehicle Assembly Building (VAB). This test is internally called the Integrated

[1] https://www.nasa.gov/pdf/588413main_SLS_Fun_Facts.pdf, accessed October 2017.

Modal Test (IMT). Even though the SLS is more massive than the SLS, the ML has significant low frequency dynamics, so it will become necessary to also model the ML in order to correlate to this test. This is a significant undertaking, since the ML is a complicated frame structure with thousands of bolted structural members, extensive cabling systems, etc . . .

One possible alternative would be to use the measured modes of the SLS assembled to the ML (SLS + ML) in order to predict the modes that the SLS would have on a truly rigid foundation (i.e. a fixed interface). This section uses test analysis models of the SLS and ML to explore this possibility. Statically reduced test-analysis models (TAMs) were provided with a few thousand degrees of freedom, and which were used to estimate the motion of the (SLS + ML) at 625 measurement DOF. The measurement DOF were chosen to distinguish between the (SLS + ML) modes, as measured by the MAC and Cross-Orthogonality. The model also contains 24DOF at the interface between the SLS and ML (note that the restraining arm towards the top of the tower was not connected to the SLS). These DOF consist of translations at 8 posts, four at the base of each Solid Rocket Booster (SRB). These 24DOF were fixed in the model to estimate the true fixed based modes of the SLS, and that model will be used as a truth model.

Two case studies will be presented here. In the first the TAMs will be used to estimate the free-free modes of the ML, and the same fixed-interface method that was presented in [1] is used, which is here denoted ML-F since the ML has a free-interface. For the second case, the modes of the ML will be computed with the SLS replaced with a rigid mass, as proposed in Sect. 13.2.2, and the results will be denoted ML-R. In both cases, we presume that the first 149 modes of the SLS + ML are extracted from the modal test, corresponding to all modes in a certain frequency range. Then 84 modes of the ML-F or ML-R are used as constraints, corresponding to all modes of the ML in approximately the same frequency range. The ML-F actually had about 10% more modes in the frequency range of interest, but the same number of constraints was used to obtain a more straightforward comparison.

Table 13.3 shows the errors in the natural frequencies and the corresponding MAC values for three cases. First, it is interesting to consider the errors that would be obtained if the SLS + ML modes were compared directly with the modes of the SLS-Fixed. If this is done, the natural frequencies have errors ranging between −12.4% and 1.4% over the first ten modes, and the MAC values range between 0.92 and 1.00, so the ML does form a fairly effective fixed-interface for the SLS. However, it is important to note that SLS + ML contains many additional modes and this table only shows those SLS + ML modes that best matched the true SLS-Fixed modes. The modes estimated by CMS reduce these frequency

Table 13.3 Errors in the natural frequencies estimated by substructuring for the SLS. Shaded modes are the primary modes of interest

	Measured SLS+ML		84 Constraints: Modes of ML-R (i.e. SLS Rigidized)			ML-R Corrected	84 Constraints: Modes of ML-F (i.e. SLS Removed)		
SLS Mode #	Freq. % Error	MAC	CMS Mode #	Freq. % Error	MAC	Freq. % Error	CMS Mode #	Freq. % Error	MAC
1	-12.4	1.00	1	-11.0	1.00	2.0	1	-2.6	1.00
2	-9.4	1.00	2	-5.1	1.00	-3.2	2	-6.0	1.00
3	-8.7	0.92	4	-1.7	0.98	-1.3	4	0.3	1.00
4	-6.8	0.95	3	-6.0	0.87	-4.0	3	-5.9	0.99
5	-2.1	0.95	5	-0.5	1.00	-0.5	5	-0.7	1.00
6	-3.4	1.00	6	-2.6	0.98	-2.5	6	-3.3	0.99
7	1.4	0.99	7	-2.9	0.98	-0.6	7	0.5	0.93
8	-1.6	0.94	8	-3.3	0.97	-2.9	8	-3.2	0.79
9	-2.2	0.97	9	-2.3	0.98	-0.6	9	0.6	0.96
10	-1.9	0.95	10	-0.1	1.00	0.0	10	-0.4	1.00
11	-2.1	0.97	11	-2.2	0.99	-0.4	11	-1.9	0.99
12	-2.2	0.99	12	-2.8	0.97	-2.4	12	-2.3	0.95
13	-6.3	0.98	13	-5.4	0.82	-2.8	13	-5.8	0.95
14	-1.6	0.92	14	-1.8	0.87	8.5	14	-2.1	0.96
15	-6.5	0.76	15	-1.1	0.64	18.8	15	-1.1	0.70
16	-35.2	0.45	16	-1.2	0.80	5.5	16	6.3	0.43
17	-0.1	0.99	18	0.3	0.98	22.9	18	-0.3	0.98
18	0.8	0.39	20	-0.5	0.32	6.0	20	-1.2	0.27
19	-1.5	0.87	19	-4.8	0.67	-0.8	19	-9.0	0.43
20	-6.9	0.70	17	-8.7	0.13	9.7	17	-7.7	0.24

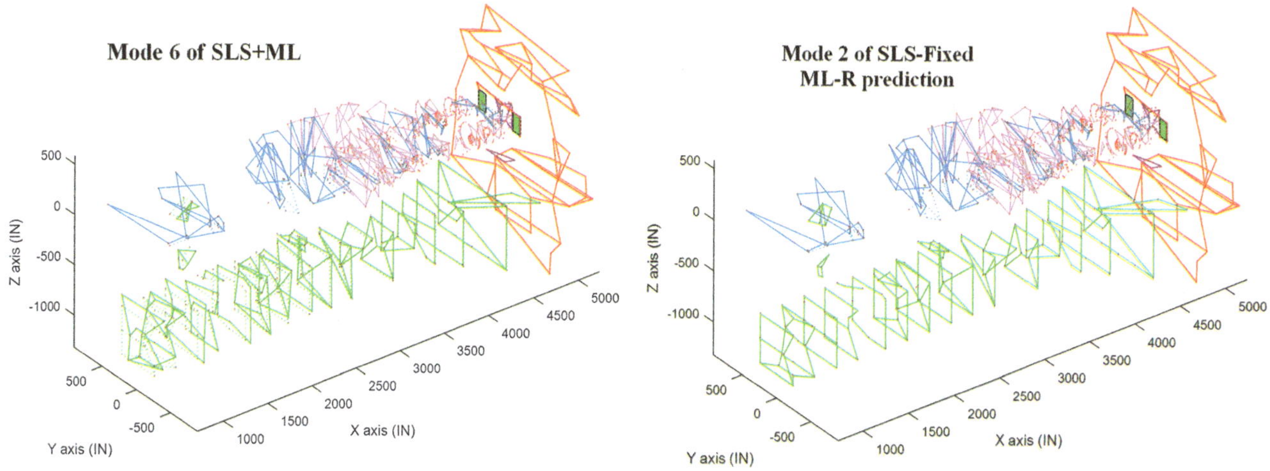

Fig. 13.8 Mode shapes of Mode 2 from (left) the measurements on the SLS + ML, i.e. the first column in Table 13.3, and (right) from substructuring using the ML rigidized, i.e. ML-R

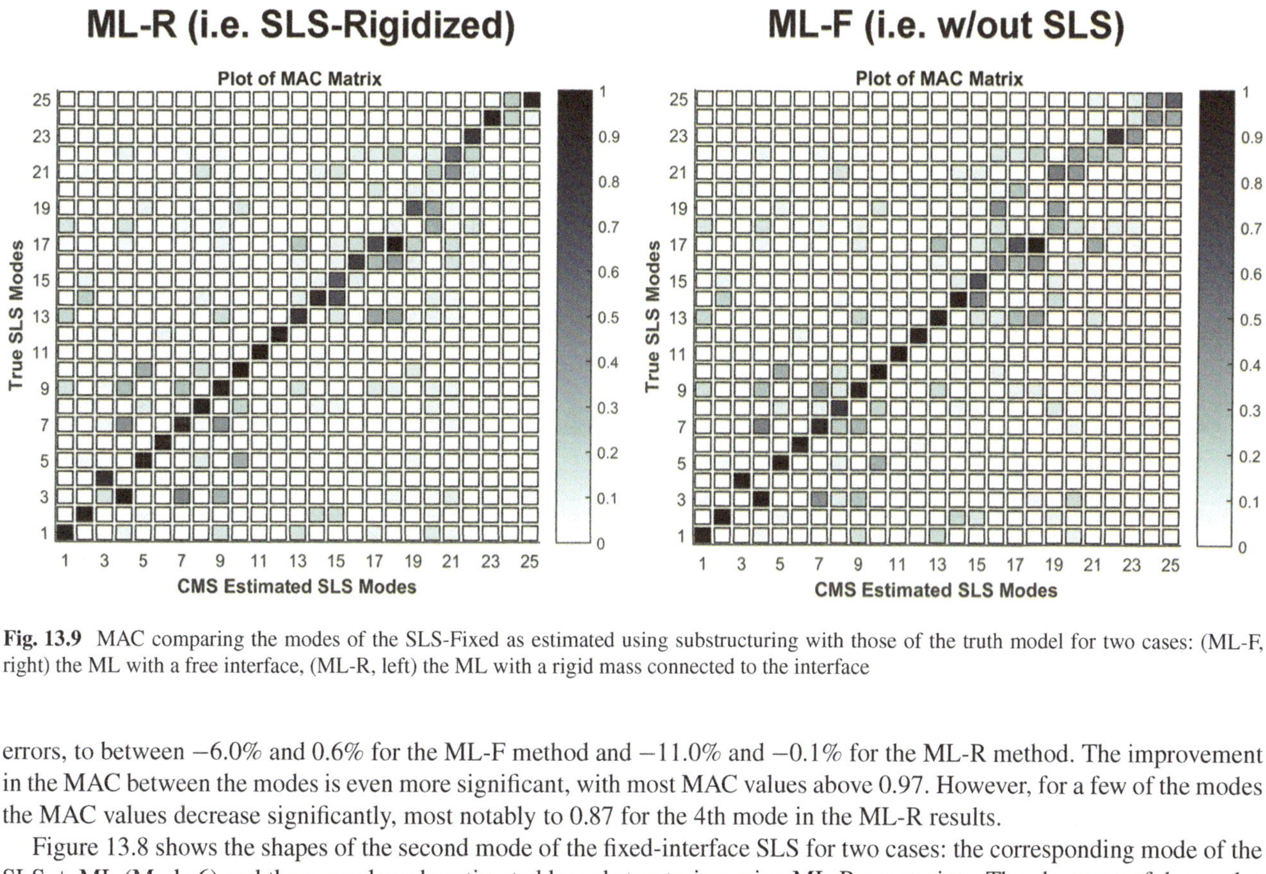

Fig. 13.9 MAC comparing the modes of the SLS-Fixed as estimated using substructuring with those of the truth model for two cases: (ML-F, right) the ML with a free interface, (ML-R, left) the ML with a rigid mass connected to the interface

errors, to between −6.0% and 0.6% for the ML-F method and −11.0% and −0.1% for the ML-R method. The improvement in the MAC between the modes is even more significant, with most MAC values above 0.97. However, for a few of the modes the MAC values decrease significantly, most notably to 0.87 for the 4th mode in the ML-R results.

Figure 13.8 shows the shapes of the second mode of the fixed-interface SLS for two cases: the corresponding mode of the SLS + ML (Mode 6) and the second mode estimated by substructuring using ML-R constraints. The character of the modes is difficult to discern from these static plots, but they do show that the ML base and tower show significant motion for the SLS + ML mode, even if Table 13.3 shows that its frequency and shape are fairly accurate. In contrast, the modes estimated from substructuring show no visible motion of the ML tower or base.

Figure 13.9 shows the MAC between the modes estimated by substructuring, with the true modes of the SLS with a fixed interface (SLS-Fixed). The MAC plots show that in both ML-F and ML-R cases there is a good correspondence between the true modes and those estimated by substructuring. However, the ML-R results show better agreement for the higher frequency modes, while the ML-F results seem to still contain several spurious modes beyond about the 16th.

It is also informative to consider the drive point FRFs at a few points as would be obtained by the various models. Drive points near the top of the SLS were used that were approximately in the axial and the two lateral directions, and the FRFs were reconstructed using the modes from the ML-F and ML-R method and are shown in Fig. 13.10. The FRF show that the

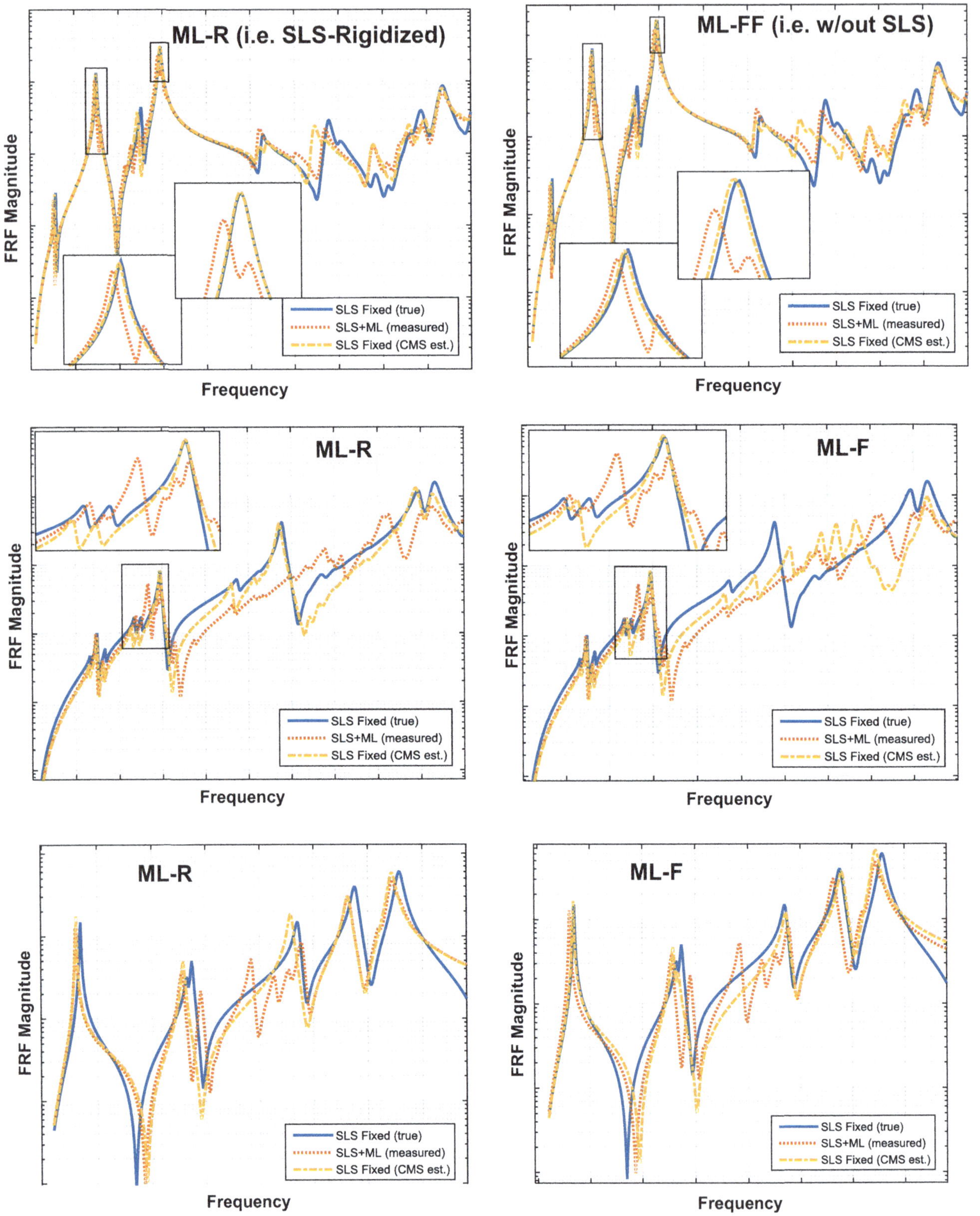

Fig. 13.10 Drive-point frequency response functions for the SLS-Fixed as estimated using substructuring with those of the truth model for two cases: (ML-F, right) and (ML-R, left). The FRFs are also shown for the SLS + ML, or the FRFs that would be obtained in the test of the coupled system if no post processing is performed. (top) – Lateral direction, (middle) – Axial direction, (bottom) – Bending towards/away from ML tower

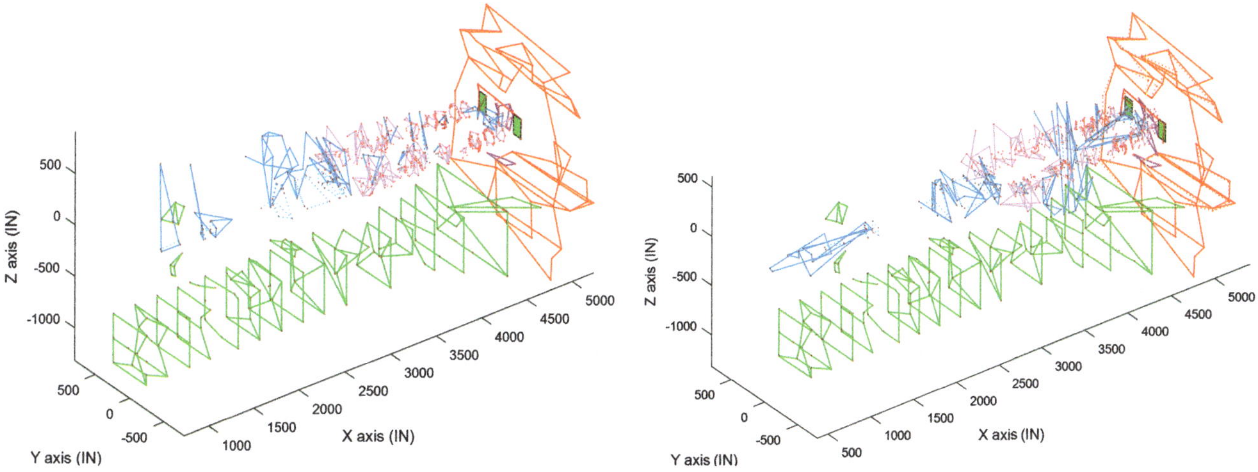

Fig. 13.11 Mode shapes of Modes 1 (left) and 16 (right) for the ML-R case. The character of the modes is difficult to discern from these static plots, but they do show that the ML base and tower shows no visible motion for Mode 1, while it shows some residual vibration for the axial Mode 16

ML-R method is far superior in estimating the high frequency axial modes (middle row), although it is less accurate for the first bending mode in the in-out direction and for the higher frequency in-out modes.

As was done in the previous section, we can visualize the residual motion in the fixture (i.e. the ML) to understand how effective the constraints were. Figure 13.11 shows the SLS + ML for two modes estimated using the ML-R approach. The residual motion in the ML is not visible, but it is still presumed to be the cause of the frequency errors seen in Table 13.3. In order to evaluate this, the natural frequency correction scheme was evaluated for this system using $\alpha = 2$ as in the previous section and using the mode shapes estimated by the ML-R method. The method proves effective at correcting the natural frequencies of the first few modes, reducing the errors in those frequencies from between -11.0% to -0.5% to -4% to 2.0%.

13.4 Conclusions

Two enhancements have been presented to the method proposed by Allen, Gindlin & Mayes [1] that estimates the fixed-interface modes of a structure that is connected to a flexible fixture. The first is a small modification to the modal basis used to define the constraints that are applied to the fixture; local deformations near the interface are captured by adding a rigid mass to the fixture before computing its modes. This was found to improve the estimates of the structure's fixed-interface natural frequencies to a certain extent, although not dramatically. We also explored the use of schemes to correct the natural frequencies estimated for the fixed-interface structure, based on the strain energy remaining in the fixture, as estimated by a Test-Analysis Model. The theoretical development revealed that, because the change in Potential Energy in the structure of interest could not be estimated accurately, it was necessary to assume that it was proportional to the energy in the fixture, and this makes the method empirical and calls its general applicability into question. However, for the cases studied to date the empirical method seemed to provide good results so long as the deformation in the fixture was not too large. A general statement that continues to hold with these and other investigations since the publication of [1] is that the fewer modes the fixture has in the measured bandwidth, the better the substructuring approach works to estimate the fixed base result.

Acknowledgements This research is based on work that was originally supported by Sandia National Laboratories. Sandia National Laboratories is a multi-mission laboratory managed and operated by National Technology and Engineering Solutions of Sandia, LLC., a wholly owned subsidiary of Honeywell International, Inc., for the U.S. Department of Energy's National Nuclear Security Administration under contract DE-NA-0003525. The authors also gratefully acknowledge the support of the National Aeronautics and Space Administration (NASA) and the helpful comments provided by members of the Loads and Dynamics Technical Discipline Team of the NASA Engineering & Safety Center (NESC).

References

1. Allen, M.S., Gindlin, H.M., Mayes, R.L.: Experimental modal substructuring to estimate fixed-base modes from tests on a flexible fixture. J. Sound Vib. **330**, 4413–4428 (2011)
2. de Klerk, D., Rixen, D.J., Voormeeren, S.N.: General framework for dynamic substructuring: history, review, and classification of techniques. AIAA J. **46**, 1169–1181 (2008)
3. Martinez, D.R., Carne, T.G., Gregory, D.L., Miller, A.K.: Combined experimental/analytical modeling using component mode synthesis, AIAA/ASME/ASCE/AHS structures, Structural Dynamics & Materials Conference, Palm Springs, CA, USA (1984)
4. Ewins, D.J.: Modal Testing: Theory, Practice and Application. Research Studies Press, Baldock (2000)
5. Corus, M., Balmes, E., Nicolas, O.: Using model reduction and data expansion techniques to improve SDM. Mech. Syst. Signal Process. **20**, 1067–1089 (2006)
6. Allen, M.S., Mayes, R.L.: Comparison of FRF and modal methods for combining experimental and analytical substructures. In: 25th International Modal Analysis Conference (IMAC XXV), Orlando, Florida (2007)
7. Sjovall, P., Abrahamsson, T.: Substructure system identification from coupled system test data. Mech. Syst. Signal Process. **22**, 15–33 (2008)
8. Allen, M.S., Mayes, R.L., Bergman, E.J.: Experimental modal substructuring to couple and uncouple substructures with flexible fixtures and multi-point connections. J. Sound Vib. **329**, 4891–4906 (2010)
9. Mayes, R.L., Hunter, P.S., Simmermacher, T.W., Allen, M.S.: Combining experimental and analytical substructures with multiple connections. In: 26th International Modal Analysis Conference (IMAC XXVI), Orlando, Florida (2008)
10. de Klerk, D., Rixen, D.J.: Component transfer path analysis method with compensation for test bench dynamics. Mech. Syst. Signal Process. **24**, 1693–1710 (2010)
11. Urgueira, A.P.V.: Dynamic analysis of coupled structures using experimental data, PhD Thesis, Imperial College of Science, Technology and Medicine, University of London, London (1989)
12. Imregun, M., Robb, D.A., Ewins, D.J.: Structural modification and coupling dynamic analysis using measured FRF data. In: 5th International Modal Analysis Conference (IMAC V), London, England (1987)
13. Carne, T.G., Dohrmann, C.R.: Improving experimental frequency response function matrices for admittance modeling. In: 24th International Modal Analysis Conference (IMAC XXIV), St. Louis, Missouri (2006)
14. Martinez, D.R., Miller, A.K., Carne, T.G.: Combined experimental and analytical modeling of shell/payload structures. In: The Joint ASCE/ASME Mechanics Conference, Albuquerque, NM (1985)
15. Mottershead, J.E., Kyprianou, A., Ouyang, H.: Structural modification. Part 1: rotational receptances. J. Sound Vib. **284**, 249–265 (2005)
16. Kyprianou, A., Mottershead, J.E., Ouyang, H.: Structural modification. Part 2: assignment of natural frequencies and antiresonances by an added beam. J. Sound Vib. **284**, 267–281 (2005)
17. de Klerk, D.: Dynamic response characterization of complex systems through operational identification and dynamic substructuring, PhD Thesis: Delft University of Technology (2009)
18. de Klerk, D., Rixen, D.J., Voormeeren, S.N., Pasteuning, F.: Solving the RDoF problem in experimental dynamic substructuring. In: 26th International Modal Analysis Conference (IMAC XXVI), Orlando, Florida (2008)
19. Napolitano, K., Yoder, N., Fixed base FRF using boundary measurements as references – analytical derivation. In: 30th International Modal Analysis Conference (IMAC XXX), Jacksonville, FL (2012)
20. Yoder, N., Napolitano, K.: Fixed base FRF using boundary measurements as references – experimental results. In: 30th International Modal Analysis Conference (IMAC XXX), Jacksonville, FL (2012)
21. Mayes, R.L., Arviso, M.: Design studies for the transmission simulator method of experimental dynamic substructuring, International Seminar on Modal Analysis (ISMA2010), Lueven, Belgium (2010)
22. Ginsberg, J.H.: Mechanical and Structural Vibrations, 1st edn. John Wiley and Sons, New York (2001)

Chapter 14
On the Problem of Describing the Coupling Interface Between Sub-structures: An Experimental Test for 'Completeness'

J. W. R. Meggitt, A. T. Moorhouse, and A. S. Elliott

Abstract The implementation of both vibration source characterisation and sub-structure coupling/decoupling procedures rely on the complete description of a coupling interface, that is, the inclusion of coupling forces in all significant degrees of freedom (DoFs). However, it is not straight-forward to establish which DoFs are required in the description. E.g. is it necessary to include moments and/or in-plane forces? This is an important question as an incomplete description will lead to an erroneous representation of the dynamics. However, there are currently no methods of quantifying the completeness of an interface description. In this paper an experimental procedure is described for the assessment of interface completeness. Based on the theoretical blocking of DoF subsets, a relation is presented that allows for the contribution of an unknown DoF to be established. Further, a coherence style criterion is proposed to estimate the completeness of a given interface description. This criterion may be used to check whether sufficient coupling DoFs have been included in both source characterisation and sub-structure coupling/decoupling procedures. Numerical and experimental examples are provided to illustrate the concept.

Keywords Interface · Completeness · Characterisation · Sub-structure · Coupling

14.1 Introduction

Interface dynamics play an essential role in the coupling of a built up structures. The *complete* description of an interface (i.e. the correct number, position and orientation of coupling DoFs) is key when attempting to experimentally characterise a structural assembly and/or its constituent components. This paper is concerned with the development of an 'Interface Completeness Criterion' that quantitatively assesses the degree to which an interface has been correctly described.[1]

In general, the interface DoFs of a point connected structure are able to move in 6 coordinate DoFs. In theory, a *complete* interface description must account for all of these. Experimentally, however, this is typically not possible, nor is it often warranted.[2] In any practical scenario access to the coupling interface is likely to be restricted. This complicates the measurement of coupling DoFs. Furthermore, the experimental challenges associated with the measurement of rotational and in-plane DoFs has led to their near constant neglect. For this reason, incomplete interface descriptions are routinely encountered in practice. Whilst considerable research has focused on the importance of rotational DoFs [1–5], in-plane DoFs have received far less attention. According to Moorhouse and Elliott [6], however, their importance in the coupling of structural elements should not be understated. With this in mind, the ability to assess the importance of a particular DoF, perhaps unknown, would clearly be advantageous.

In the analysis of a continuous interface it is standard practice to approximate the interface in terms of a finite number of point-like DoFs so as to acquire an approximate description of its dynamics [7, 8]. Such an approximation is often based on

[1]Although we will refer to the proposed quantity as a criterion, given its definition, the authors are undecided as to whether 'coefficient' may be a more appropriate term.

[2]Whilst a complete interface description is required from a theoretical basis, it is often the case that their exists a subset of DoFs that are of particular importance, and themselves provide a satisfactory description of the interface. In such a case the remaining DoFs may not need to be accounted for.

J. W. R. Meggitt (✉) · A. T. Moorhouse · A. S. Elliott
Acoustics Research Centre, University of Salford, Greater Manchester, UK
e-mail: j.w.r.meggitt1@salford.ac.uk

A. Linderholt et al. (eds.), *Dynamics of Coupled Structures, Volume 4*, Conference Proceedings of the Society for Experimental Mechanics Series, https://doi.org/10.1007/978-3-319-74654-8_14

a points per wavelength argument, although the completeness of this description remains unclear. Alternatively, a continuous interface may be approximated in terms of a truncated orthonormal series expansion [9]. Similarly, however, it is unclear the degree to which this expansion suitably describes the interface.

In the analysis of both discrete and continuous interfaces, an indication as to whether sufficient DoFs have been accounted for would clearly benefit the experimenter, and aid in avoiding the potential problems associated with an incomplete interface description. Based on the mathematical blocking of known interface DoFs, theory is developed that allows for the contribution of unknown interface DoFs to be established. This contribution is used as part of a coherence style criterion to assess the completeness of an interface description. The proposed criterion may be used to indicate missing DoFs, weigh up the potential benefits of including additional DoFs, or provide a level of uncertainty in a given interface description.

The remainder of this paper will be organised as follows: In Sect. 14.2 the interface completeness problem will be demonstrated mathematically by considering the independent characterisation of a vibration source. Having introduced the problem, Sect. 14.3 will describe the relevant theory, before Sect. 14.4 outlines the proposed Interface Completeness Criterion (ICC). Section 14.5 will provide a numerical and experimental case study. Lastly, Sect. 14.6 will present some concluding remarks.

14.2 The Interface Completeness Problem

The interface completeness problem is illustrated here with reference to the inverse determination of blocked forces using the in-situ approach presented in [10–12]. A similar problem is encountered in the coupling (or decoupling) of sub-structures through dynamic sub-structure coupling procedures.

It was shown by Moorhouse et al. [13] that the blocked force, which represents an independent property of a structural source, may be acquired through an inverse procedure similar to that used in operational force identification and standard transfer path analysis (TPA) procedures [14]. The relation of note is given by,

$$\mathbf{v_b} = \mathbf{Y_{bc}}\bar{\mathbf{f}}_\mathbf{c} \tag{14.1}$$

where $\mathbf{Y_{bc}} \in \mathbb{C}^{N\times N}$ is the measured mobility matrix of a coupled assembly, $\mathbf{v_b} \in \mathbb{C}^N$ is a measured operational velocity vector, and $\bar{\mathbf{f}}_\mathbf{c} \in \mathbb{C}^N$ is the vector of unknown blocked forces. Here subscripts $\{b\}$ and $\{c\}$ represent *complete sets* of remote receiver and coupling interface DoFs, respectively. Variables are given in the frequency domain, with explicit notation omitted for clarity.

To illustrate the interface problem we begin by considering the in-situ blocked force relation in the form,

$$\mathbf{v_b} = \mathbf{Y_{bc_i}}\bar{\mathbf{f}}_{\mathbf{c_i}} + \mathbf{Y_{bc_j}}\bar{\mathbf{f}}_{\mathbf{c_j}}. \tag{14.2}$$

where the velocity contributions from known and unknown DoFs are separated. Here, $\{c_i \subseteq c\}$ represents the set of coupling interface DoFs that are *known* and considered measurable, whilst $\{c_j \subseteq c \mid c_j \nsubseteq c_i\}$ represents the set of interface DoFs that are *unknown*, or known but can not be measured. An illustrative example of such an assembly is given in Fig. 14.1a.

In the determination of the blocked force one pre-multiplies the operational velocity vector by the inverse of a mobility matrix pertaining to the *known* DoFs, $\mathbf{Y_{bc_i}}$. Pre-multiplication of Eq. 14.2 by $\mathbf{Y}_{\mathbf{bc_i}}^{-1}$ thus yields,

$$\mathbf{Y}_{\mathbf{bc_i}}^{-1}\mathbf{v_b} = \bar{\bar{\mathbf{f}}}_{\mathbf{c_i}} = \bar{\mathbf{f}}_{\mathbf{c_i}} + \mathbf{Y}_{\mathbf{bc_i}}^{-1}\mathbf{Y_{bc_j}}\bar{\mathbf{f}}_{\mathbf{c_j}}. \tag{14.3}$$

The acquired blocked force, $\bar{\bar{\mathbf{f}}}_{\mathbf{c_i}}$, although correct in its own respect,[3] is not the *true* blocked force at the DoFs c_i, $\bar{\mathbf{f}}_{\mathbf{c_i}}$. The neglect of the DoFs c_j has resulted in the additional term, $\mathbf{Y}_{\mathbf{bc_i}}^{-1}\mathbf{Y_{bc_j}}\bar{\mathbf{f}}_{\mathbf{c_j}}$. This term is a property of the coupled assembly and thus the acquired blocked force is no longer an independent property of the source. This loss of independence may have severe implications when the acquired blocked force is transferred to a secondary assembly and used to predict the operational response, for example in the construction of a Virtual Acoustic Prototype [15]. This is demonstrated below.

[3]The blocked force $\bar{\bar{\mathbf{f}}}_{\mathbf{c_i}}$ represents the reaction forces due to a source that is blocked in c_i DoFs but unrestrained in the remaining c_j DoFs. It is therefore by definition a blocked force, albeit not the true blocked force.

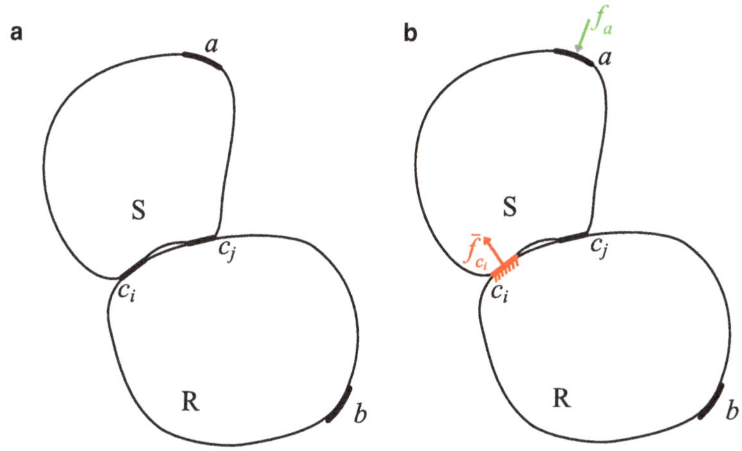

Fig. 14.1 Diagrammatic illustration of the constrained and unconstrained assemblies corresponding to Eqs. 14.7 and 14.8, respectively. (**a**) Unconstrained assembly. (**b**) Constrained assembly

We are interested in the prediction of the remote receiver velocity, $\mathbf{v}'_{\mathrm{b}}$, on a secondary assembly (denoted by the prime symbol, $'$). The true velocity, that we are aiming to predict, may be expressed in terms of the true blocked force as,

$$\mathbf{v}'_{\mathrm{b}} = \mathbf{Y}'_{\mathrm{bc_i}}\bar{\mathbf{f}}_{\mathrm{c_i}} + \mathbf{Y}'_{\mathrm{bc_j}}\bar{\mathbf{f}}_{\mathrm{c_j}}. \tag{14.4}$$

To predict the remote velocity the acquired blocked force, $\bar{\bar{\mathbf{f}}}_{\mathrm{c_i}}$, is pre-multiplied by the coupled transfer mobility of the secondary assembly that corresponds to the known and measurable DoFs, $\mathbf{Y}'_{\mathrm{bc_i}}$.

$$\tilde{\mathbf{v}}'_{\mathrm{b}} = \mathbf{Y}'_{\mathrm{bc_i}}\bar{\bar{\mathbf{f}}}_{\mathrm{c_i}} = \mathbf{Y}'_{\mathrm{bc_i}}(\bar{\mathbf{f}}_{\mathrm{c_i}} + \mathbf{Y}^{-1}_{\mathrm{bc_i}}\mathbf{Y}_{\mathrm{bc_j}}\bar{\mathbf{f}}_{\mathrm{c_j}}) \tag{14.5}$$

Expanding the above we arrive at,

$$\tilde{\mathbf{v}}'_{\mathrm{b}} = \mathbf{Y}'_{\mathrm{bc_i}}\bar{\mathbf{f}}'_{\mathrm{c_i}} + \mathbf{Y}'_{\mathrm{bc_i}}\mathbf{Y}^{-1}_{\mathrm{bc_i}}\mathbf{Y}_{\mathrm{bc_j}}\bar{\mathbf{f}}_{\mathrm{c_j}}. \tag{14.6}$$

From the above it is clear that the true velocity, $\mathbf{v}'_{\mathrm{b}}$, and the predicted velocity, $\tilde{\mathbf{v}}'_{\mathrm{b}}$, are not equal. Hence the blocked force is no longer transferable. Here, the unknown blocked force, $\bar{\mathbf{f}}_{\mathrm{c_j}}$, contributes to the predicted velocity response through the propagating transfer function, $\mathbf{Y}'_{\mathrm{bc_i}}\mathbf{Y}^{-1}_{\mathrm{bc_i}}\mathbf{Y}_{\mathrm{bc_j}}$. So, whilst the response prediction still accounts for the unknown blocked force, its contribution to the response of a secondary assembly is incorrect. This discrepancy is clearly a consequence of an incomplete interface description. Had all coupling DoFs been accounted for the true blocked force would have been acquired, and subsequently transferred to the secondary assembly with no problems.[4] The uncertainty introduced as a result of an incomplete interface description is referred to here as model uncertainty.

In this paper we are interested in establishing an experimental procedure to minimize, or identify the severity/cause of, this source of model uncertainty through the development of an Interface Completeness Criterion (ICC).

14.3 Theoretical Development

In this section we will develop the theory required to formulate the proposed Interface Completeness Criterion.

Consider the assembly depicted in Fig. 14.1a, where a source and receiver sub-structure are coupled via two *sets* of interface DoFs. Denoted c_i and c_j, these DoFs form subsets of the complete coupling interface c, such that $\{c_i \subseteq c\}$ and

[4] A similar problem is encountered in the application of sub-structure coupling and decoupling procedures. The neglect of coupling interface DoFs results in an erroneous representation of the assembly/substructure dynamics.

$\{c_j \subseteq c \mid c_j \not\subseteq c_i\}$. Also included are two sets of remote measurement DoFs on the source and receiver. These are referred to as a and b, respectively. The equations that govern the behaviour of the coupled source-receiver assembly (as depicted in Fig. 14.1a) are given generally by,

$$\begin{pmatrix} \mathbf{v_a} \\ \mathbf{v_{c_i}} \\ \mathbf{v_{c_j}} \\ \mathbf{v_b} \end{pmatrix} = \begin{bmatrix} \mathbf{Y_{aa}} & \mathbf{Y_{ac_i}} & \mathbf{Y_{ac_j}} & \mathbf{Y_{ab}} \\ \mathbf{Y_{c_ia}} & \mathbf{Y_{c_ic_i}} & \mathbf{Y_{cc_j}} & \mathbf{Y_{c_ib}} \\ \mathbf{Y_{c_ja}} & \mathbf{Y_{c_jc_i}} & \mathbf{Y_{c_jc_j}} & \mathbf{Y_{c_jb}} \\ \mathbf{Y_{ba}} & \mathbf{Y_{bc_i}} & \mathbf{Y_{bc_j}} & \mathbf{Y_{bb}} \end{bmatrix} \begin{pmatrix} \mathbf{f_a} \\ \mathbf{f_{c_i}} \\ \mathbf{f_{c_j}} \\ \mathbf{f_b} \end{pmatrix}. \tag{14.7}$$

We begin by considering the case where only two external forces are applied, i.e. $\mathbf{f_{c_j}} = \mathbf{f_b} = 0$. The first is an arbitrary force at the remote source DoFs a, $\mathbf{f_a}$. The second is a constraint force at the coupling interface DoFs c_i, $\bar{\mathbf{f}}_{\mathbf{c_i}}$. This second force is the blocking force required to constrain the velocity $\mathbf{v_{c_i}}$ to $\mathbf{0}$. The constrained assembly may be represented by the following set of equations,

$$\begin{pmatrix} \mathbf{v_a} \\ \mathbf{0} \\ \mathbf{v_{c_j}} \\ \mathbf{v_b} \end{pmatrix} = \begin{bmatrix} \mathbf{Y_{aa}} & \mathbf{Y_{ac_i}} & \mathbf{Y_{ac_j}} & \mathbf{Y_{ab}} \\ \mathbf{Y_{c_ia}} & \mathbf{Y_{c_ic_i}} & \mathbf{Y_{c_ic_j}} & \mathbf{Y_{c_ib}} \\ \mathbf{Y_{c_ja}} & \mathbf{Y_{c_jc_i}} & \mathbf{Y_{c_jc_j}} & \mathbf{Y_{c_jb}} \\ \mathbf{Y_{ba}} & \mathbf{Y_{bc_i}} & \mathbf{Y_{bc_j}} & \mathbf{Y_{bb}} \end{bmatrix} \begin{pmatrix} \mathbf{f_a} \\ \bar{\mathbf{f}}_{\mathbf{c_i}} \\ \mathbf{0} \\ \mathbf{0} \end{pmatrix}. \tag{14.8}$$

The above describes an assembly that is excited by an arbitrary force at a, whilst the coupling interface, c, is *partially* constrained. This constrained assembly is shown diagrammatically in Fig. 14.1b. Taking the second line of Eq. 14.8,

$$\mathbf{0} = \mathbf{Y_{c_ia}f_a} + \mathbf{Y_{c_ic_i}}\bar{\mathbf{f}}_{\mathbf{c_i}}. \tag{14.9}$$

and solving for the blocked force, we arrive at,

$$\bar{\mathbf{f}}_{\mathbf{c_i}} = -\mathbf{Y}_{\mathbf{c_ic_i}}^{-1}\mathbf{Y_{c_ia}f_a}. \tag{14.10}$$

The above relation is a force transmissibility that relates the externally applied force at a, to the resultant blocked force at c_i. Taking the 4th line of Eq. 14.8, whilst introducing the notation $\mathbf{v}_\mathbf{b}^{(\mathrm{c_j})}$ to describe the velocity of the constrained assembly, i.e. the contribution through the coupling DoFs c_j only,

$$\mathbf{v}_\mathbf{b}^{(\mathrm{c_j})} = \mathbf{Y_{ba}f_a} + \mathbf{Y_{bc_i}}\bar{\mathbf{f}}_{\mathbf{c_i}}. \tag{14.11}$$

The first RHS term describes the velocity at b due an applied force at a on the *unconstrained* assembly, i.e. $\mathbf{v}_\mathbf{b}^{(\mathrm{c})} = \mathbf{Y_{ba}f_a}$. The second RHS term describes an added contribution to this velocity due to the blocking force required to constrain the interface c_i. The blocked force may be eliminated by expressing it in terms $\mathbf{f_a}$, as per Eq. 14.10,

$$\mathbf{v}_\mathbf{b}^{(\mathrm{c_j})} = \mathbf{Y_{ba}f_a} - \mathbf{Y_{bc_i}}\mathbf{Y}_{\mathbf{c_ic_i}}^{-1}\mathbf{Y_{c_ia}f_a} \tag{14.12}$$

or equivalently,

$$\mathbf{v}_\mathbf{b}^{(\mathrm{c_j})} = \left[\mathbf{Y_{ba}} - \mathbf{Y_{bc_i}}\mathbf{Y}_{\mathbf{c_ic_i}}^{-1}\mathbf{Y_{c_ia}}\right]\mathbf{f_a}. \tag{14.13}$$

Equation 14.13 relates an externally applied force at a to the resultant velocity at b on the constrained assembly depicted in Fig. 14.1b.[5] The bracketed mobility term therefore represents the transfer mobility through the constrained assembly. Consequently, let us define,

$$\mathbf{Y}_\mathbf{ba}^{(\mathrm{c_j})} \triangleq \left[\mathbf{Y_{ba}} - \mathbf{Y_{bc_i}}\mathbf{Y}_{\mathbf{c_ic_i}}^{-1}\mathbf{Y_{c_ia}}\right] \tag{14.14}$$

as the transfer mobility from a to b, through the DoFs c_j, whilst the DoFs c_i are blocked.

[5] Although not discussed in any further detail here, for $|c_j| = 1$, Eq. 14.13 may be considered the velocity at b due to the contribution of a single *path*, i.e. whilst all other paths are blocked. This is a similar concept to that used by Margans [17] in the formulation of the GTDT TPA procedure [18].

Noting that the left hand term in the bracket of Eq. 14.13 is the transfer mobility of the unconstrained assembly, let us redefine this as,

$$\mathbf{Y}_{\mathbf{ba}}^{(\mathbf{c})} \triangleq \mathbf{Y}_{\mathbf{ba}} \tag{14.15}$$

that is, the transfer mobility from a to b through all coupling DoFs (c_i and c_j). Lastly, the right hand term in the bracket of Eq. 14.13 can be seen to form a round trip identity [16] for a mobility similar to $\mathbf{Y}_{\mathbf{ba}}$. It is interesting to note that for the case whereby only a single set of coupling DoFs exist, i.e. $|c_j| = 0$, this round trip identity is equal exactly to the unconstrained mobility, $\mathbf{Y}_{\mathbf{ba}}$ [16].[6] For the constrained interface considered, however, the mobility product $\mathbf{Y}_{\mathbf{bc_i}}\mathbf{Y}_{\mathbf{c_ic_i}}^{-1}\mathbf{Y}_{\mathbf{c_ia}}$ corresponds to a transfer mobility between a and b whilst accounting for only the coupling DoFs c_i. As such, let us define,

$$\mathbf{Y}_{\mathbf{ba}}^{(\mathbf{c_i})} \triangleq \mathbf{Y}_{\mathbf{bc_i}}\mathbf{Y}_{\mathbf{c_ic_i}}^{-1}\mathbf{Y}_{\mathbf{c_ia}}. \tag{14.16}$$

Using the above definitions Eq. 14.14 may be rewritten as,

$$\mathbf{Y}_{\mathbf{ba}}^{(\mathbf{c_j})} = \mathbf{Y}_{\mathbf{ba}}^{(\mathbf{c})} - \mathbf{Y}_{\mathbf{ba}}^{(\mathbf{c_i})} \tag{14.17}$$

or alternatively,

$$\mathbf{Y}_{\mathbf{ba}}^{(\mathbf{c})} = \mathbf{Y}_{\mathbf{ba}}^{(\mathbf{c_i})} + \mathbf{Y}_{\mathbf{ba}}^{(\mathbf{c_j})}. \tag{14.18}$$

Equation 14.18 states that the transfer mobility from a to b through the unconstrained interface, $\mathbf{Y}_{\mathbf{ba}}^{(\mathbf{c})}$, may be represented as the sum of two mobility terms. These correspond to; the transfer mobility from a to b whilst the DoFs c_i are constrained, $\mathbf{Y}_{\mathbf{ba}}^{(\mathbf{c_j})}$; and transfer mobility from a to b whilst the DoFs c_j are in some way neglected, $\mathbf{Y}_{\mathbf{ba}}^{(\mathbf{c_i})}$.

Experimentally we are able to measure $\mathbf{Y}_{\mathbf{ba}}^{(\mathbf{c})}$ directly, simply through the excitation of a, and response measurement of b, since the physical assembly is not constrained. Similarly, we are able to compute $\mathbf{Y}_{\mathbf{ba}}^{(\mathbf{c_i})}$ through the direct measurement of its associated mobilities, $\mathbf{Y}_{\mathbf{bc_i}}$, $\mathbf{Y}_{\mathbf{c_ic_i}}^{-1}$, and $\mathbf{Y}_{\mathbf{c_ia}}$. We are therefore able to calculate the contribution of the DoFs c_j without having access to them directly.

We note here that if the DoF subset c_i contains all DoFs, $c_i = c$, then $|c_j| = 0$ and $\mathbf{Y}_{\mathbf{ba}}^{(\mathbf{c})} = \mathbf{Y}_{\mathbf{ba}}^{(\mathbf{c_i})}$. Experimentally this would mean that all of the interface DoFs have been included and the interface description is *complete*.[7] This final remark will be used in the formulation of an Interface Completeness Criterion. This criteria may be used to provide a quantitative assessment of an interface's 'completeness', that is, whether sufficient DoFs have been accounted for.

14.4 Interface Completeness Criterion

In the field of modal analysis there exist a number of criteria to assess confidence in acquired modal parameters. For example, the modal assurance criterion (MAC), the coordinate modal assurance criterion (COMAC), and the frequency response assurance criterion (FRAC), to name but a few [19].

The aim here is to establish a similar criterion for the 'completeness' of an interface description. Here 'completeness' refers to the degree to which the coupling interface DoFs are accounted for by the coupling interface mobility matrix, $\mathbf{Y}_{\mathbf{cc}}$. A complete interface is defined as one whose DoFs are completely described by $\mathbf{Y}_{\mathbf{cc}}$.

A suitable criteria should be bound between 0 and 1 such that the degree of completeness may be assessed with ease. An assessment is proposed in the same form as the well established modal assurance criterion (MAC). The proposed criteria will be referred to as the Interface Completeness Criterion (ICC).

[6] Consequently, $\mathbf{Y}_{\mathbf{ba}}^{(\mathbf{c_j})} = 0$ since the DoF subset c_j does not exist.

[7] Alternatively, if one considers Eq. 14.14, a complete interface description would block entirely the coupling interface, resulting in a velocity of 0 at the remote source DoFs b.

Using an expression identical in form to the MAC, the ICC yields a frequency dependent scalar value bound between 0 and 1, and is defined as,

$$\mathrm{ICC}_{b\mathbf{a}} = \frac{\left|\mathbf{Y}^{\mathbf{c}}_{\mathrm{ba}}\left(\mathbf{Y}^{\mathbf{c}_i}_{\mathrm{ba}}\right)^{H}\right|^2}{\mathbf{Y}^{\mathrm{c}}_{\mathrm{ba}}(\mathbf{Y}^{\mathrm{c}}_{\mathrm{ba}})^{H}\mathbf{Y}^{\mathrm{c}_i}_{\mathrm{ba}}(\mathbf{Y}^{\mathrm{c}_i}_{\mathrm{ba}})^{H}} \tag{14.19}$$

where, $\mathbf{Y}^{(\mathbf{c}_i)}_{\mathrm{ba}} = \mathbf{Y}_{\mathrm{bc}_i}\mathbf{Y}^{-1}_{\mathbf{c}_i\mathbf{c}_i}\mathbf{Y}_{\mathbf{c}_i\mathbf{a}}$.

Recalling the closing remark of Sect. 14.3, we note that if all coupling DoFs are accounted for, that is, $\mathbf{Y}^{(\mathbf{c})}_{\mathrm{ba}} = \mathbf{Y}^{(\mathbf{c}_i)}_{\mathrm{ba}}$, the $\mathrm{ICC}_{b\mathbf{a}}$ is equal to one. Similarly, if none of the coupling DoFs are accounted for, that is, $\mathbf{Y}^{(\mathbf{c}_i)}_{\mathrm{ba}} = \mathbf{0}$, the $\mathrm{ICC}_{b\mathbf{a}}$ is equal to zero. The $\mathrm{ICC}_{b\mathbf{a}}$ is thus related to the degree of completeness, and consequently, the degree of model uncertainty present in a given interface description.

We note that the $\mathrm{ICC}_{b\mathbf{a}}$ describes the degree to which an interface is complete through the blocking of a transfer function measured between a set of source side DoFs, a, and a *single* receiver side DoF, b. As such, a different $\mathrm{ICC}_{b\mathbf{a}}$ is obtained for each remote receiver DoF considered. Furthermore, it should be noted that whilst the above criterion, in theory, yields 1 for a complete interface description this is likely unachievable in an experimental scenario. The measured and predicted mobilities, $\mathbf{Y}^{(\mathbf{c})}_{\mathrm{ba}}$ and $\mathbf{Y}^{(\mathbf{c}_i)}_{\mathrm{ba}}$, respectively, will be contaminated to some extent by measurement uncertainty and perfect agreement is unlikely. The question of experimental error is further addressed in the case study of Sect. 14.5.2, although a more thorough investigation regarding the sensitivity of the ICC is considered beyond the scope of this work.

14.5 Case Studies

In this section the application of the ICC will be demonstrated as part of two case studies. In the first, the discretisation of a continuous interface between two coupled plates will be considered numerically. In the second, an experimental study will investigate the completeness of an interface between two sections of steel rod.

14.5.1 Numerical: Continuous Interface

In this study the Interface Completeness Criterion (ICC) is used too assess the completeness of a discrete approximation to a continuous interface. The interface is considered to be an arbitrary line that separates two sides of a simply supported rectangular plate, as illustrated in Fig. 14.2. Each side of the interface is considered a separate plate, and are referred as source (plate 1) and receiver (plate 2). The coupled plate assembly is modelled analytically using a truncated modal summation. Both translational z, and x y rotational DoFs, α and β, are included. The geometric and material properties of the model are presented in Table 14.1. Five source side DoFs (a) are considered on plate 1 and a single receiver side DoF (b) on plate 2. The coupling interface is represented by an increasing number of positional DoFs, as illustrated in Fig. 14.3. All positional DoFs, unless otherwise specified, are made up of three coordinate DoFs, relating to z, α and β. The aim of this study is to illustrate the use of the ICC in quantitatively assessing the discretisation of a continuous interface. Interfaces of this type are often encountered in practical scenarios and are typically represented by a series of discrete point like DoFs. The degree to which this approximation accounts for the continuous nature of the interface is of interest. The ICC between the source side DoFs and each of the receiver side DoFs are calculated (these are referred to as ICC_z, ICC_α and ICC_β).

Five interface descriptions are considered here, corresponding to 1, 2, 3, 4 and 5 positional DoFs, evenly spread across the breadth of the plate, as illustrated through Fig. 14.3a–e. The corresponding ICCs are shown in Fig. 14.4, from top to bottom, respectively. The ICCs pertaining to translational and rotational (α and β) receiver DoFs are shown in blue, orange

Table 14.1 Geometry of the coupled plate assembly, plate 1 and plate 2. The material properties of the three plates were; Young's modulus $E = 200 \times 10^9$, density $\rho = 9000$, Poison's ratio $\mu = 0.3$, and loss factor $\eta = 0.1$

Plate	Dimensions
Coupled	$1 \times 0.8 \times 0.005$
Plate 1	$0.35 \times 0.8 \times 0.005$
Plate 2	$0.65 \times 0.8 \times 0.005$

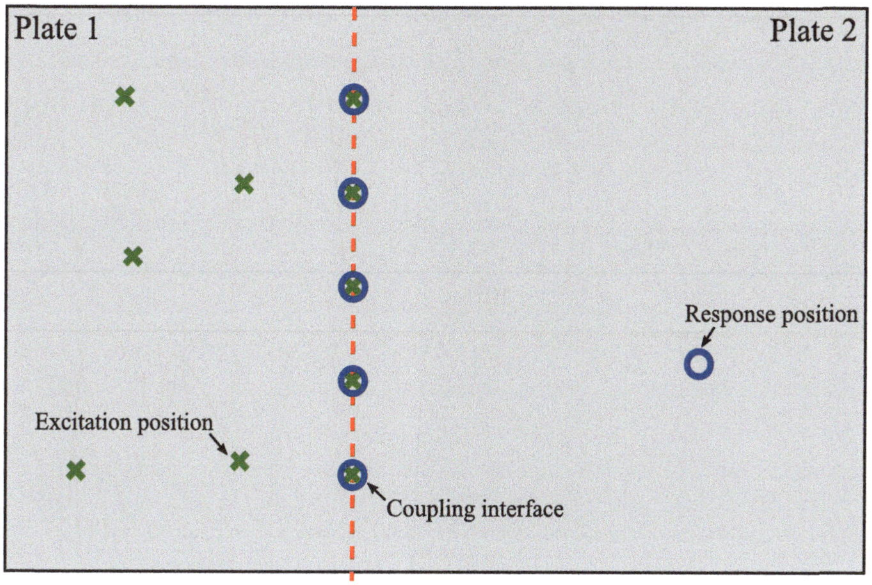

Fig. 14.2 Diagrammatic representation of the continuous interface numerical study. Two plates are coupled via a continuous interface (dashed red line) in the translational z and x/y rotational DoFs, α and β

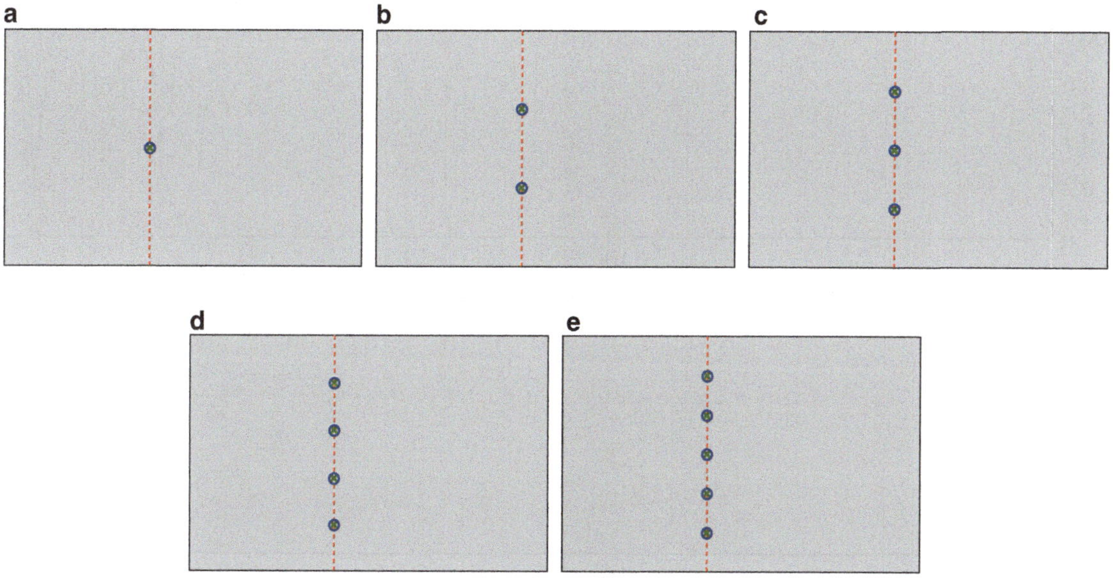

Fig. 14.3 Diagrammatic representation of the continuous interface numerical study. Two plates are coupled via a continuous interface which is approximated through varying degrees of discretisation. (**a**) 1 positional DoF. (**b**) 2 positional DoFs. (**c**) 3 positional DoFs. (**d**) 4 positional DoFs. (**e**) 5 positional DoFs

and yellow, respectively. Figure 14.4 clearly illustrates that the ICC converges to one as the number of DoFs used to describe the interface is increased. Furthermore, the results highlight the difference between ICC_z, ICC_α and ICC_β. Although sharing an overall trend, clear differences can be observed. This suggests that a single ICC may not sufficiently describe the interface completeness, and that multiple ICCs may be required for a more thorough insight. That said, the experimental effort required to include additional response DoFs is minimal as no further excitations are required.

Shown in Fig. 14.5 are the ICCs corresponding to an interface description made up of 10 positional DoFs. In Fig. 14.5a the coupling interface description includes both translational and rotational coordinate DoFs. In Fig. 14.5b the coupling interface description includes only translational coordinate DoFs, i.e. the rotational DoFs were neglected. As one might expect, the neglect of rotational coupling DoFs has resulted in a worsening of the ICC. Whilst this is an intuitively obvious result, it highlights the importance of rotational DoFs in the description of a continuous interface and, furthermore, the ability of the ICC to quantify it.

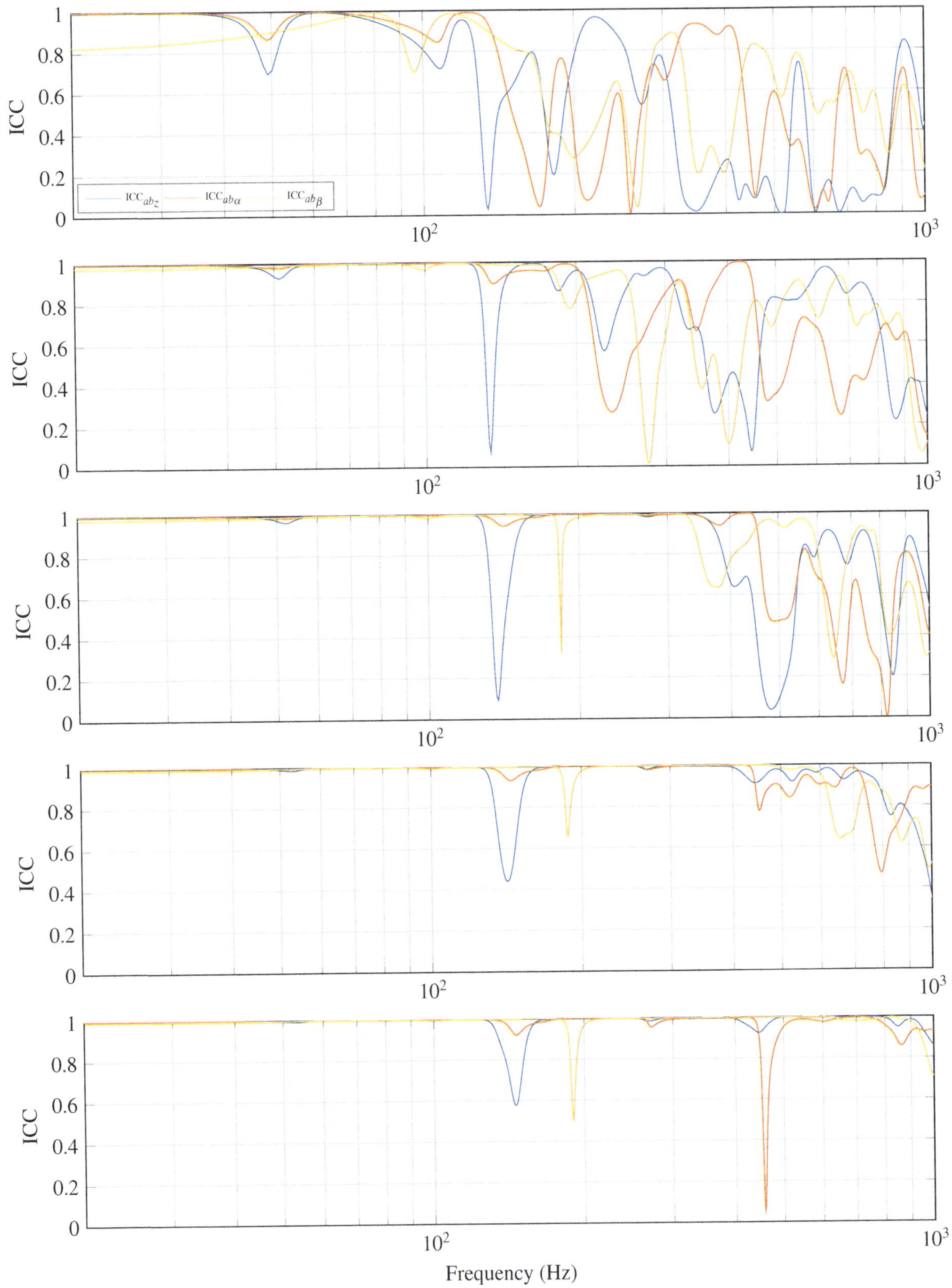

Fig. 14.4 Interface Completeness Criteria (ICC) for a continuous interface approximated using five different levels of discretisation (see Fig. 14.3). From top to bottom, the interface is represented by 1, 2, 3, 4 and 5 positional DoFs, each of which includes translational z and x/y rotational DoFs, α and β. The ICC is calculated using 15 source side positional DoFs (including z, α and β at each) and a single receiver side DoF (including z, α and β). An ICC is presented for each receiver DoF

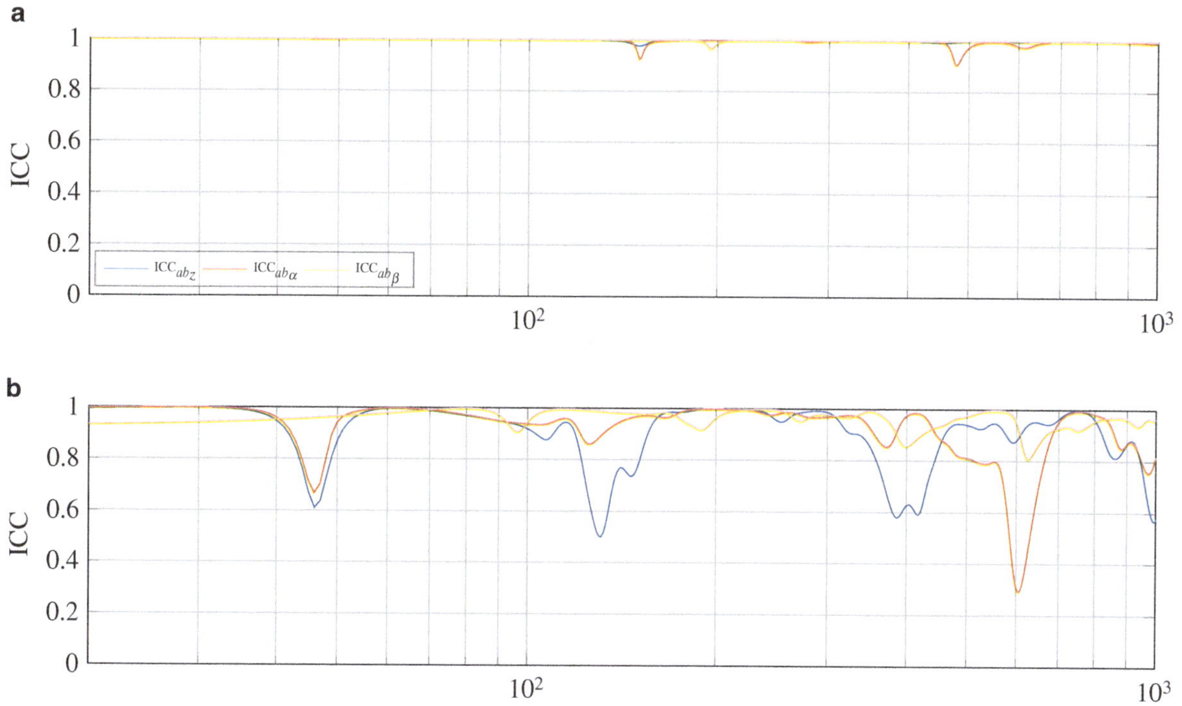

Fig. 14.5 Interface Completeness Criteria (ICC) for acontinuous interface approximated using 10 positional DoFs, both with (**a**) and without (**b**) rotational coupling. The ICC is calculated using 15 source side positional DoFs (including z, α and β at each) and a single receiver side DoF (including z, α and β). An ICC is presented for each receiver DoF. (**a**) 10 positional DoFs. (**b**) 10 positional DoFs (translation z only, neglecting α and β)

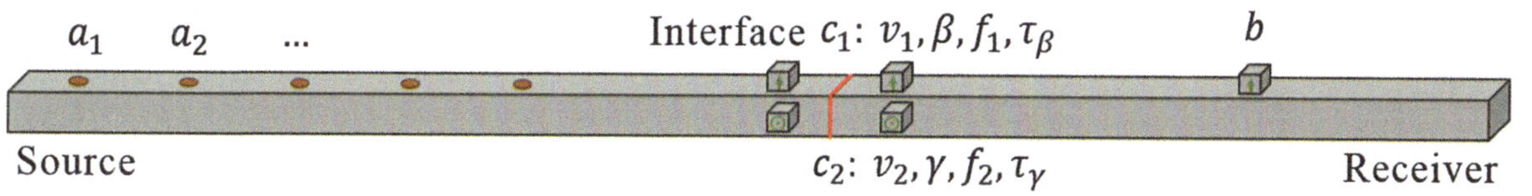

Fig. 14.6 Diagrammatic representation of the experimental study. Two rods are coupled via a single discrete interface (red line) in translational y and z, and y/z rotational DoFs, β and γ

Lastly, it is perhaps worth noting the improvement in the ICC between Figs. 14.4 (bottom) and 14.5a due to the extra 5 positional DoFs.

14.5.2 Experimental: Point Connected Interface

The experimental case study presented here concerns the description of an interface between two rigidly coupled rods. A diagrammatic illustration of the study is given in Fig. 14.6. The coupling interface, shown in red, was considered as a single positional DoF, and therefore described, generally, by the 6 coordinate DoFs, x, y, z, α, β and γ.

The aim of this study was to use the ICC to access various interface descriptions consisting of the DoFs; y, z, β and γ. The in-plane DoF x (along the length of the rod) and its associated rotation α were neglected due to the difficulties associated with their measurement.

The translational, rotational and cross mobilities associated with the retained DoFs considered were approximated using the finite difference method [20], which itself imposes a limit on the validity of an interface description over a frequency range related to the sensor spacing (taken here as 2 cm).

Twelve source side DoFs (a) were excited whilst simultaneously measuring the response at the coupling interface (c_i) and the single receiver side DoF (b). This enabled the calculation of $\mathbf{Y}_{\mathrm{ba}}^{(\mathrm{c})}$ and $\mathbf{Y}_{\mathrm{c_i a}}$. The interface DoFs were subsequently excited and the mobilities $\mathbf{Y}_{\mathrm{c_i c_i}}$ and $\mathbf{Y}_{\mathrm{bc_i}}$ measured. Together, the above measurements allowed for the calculation of the mobility

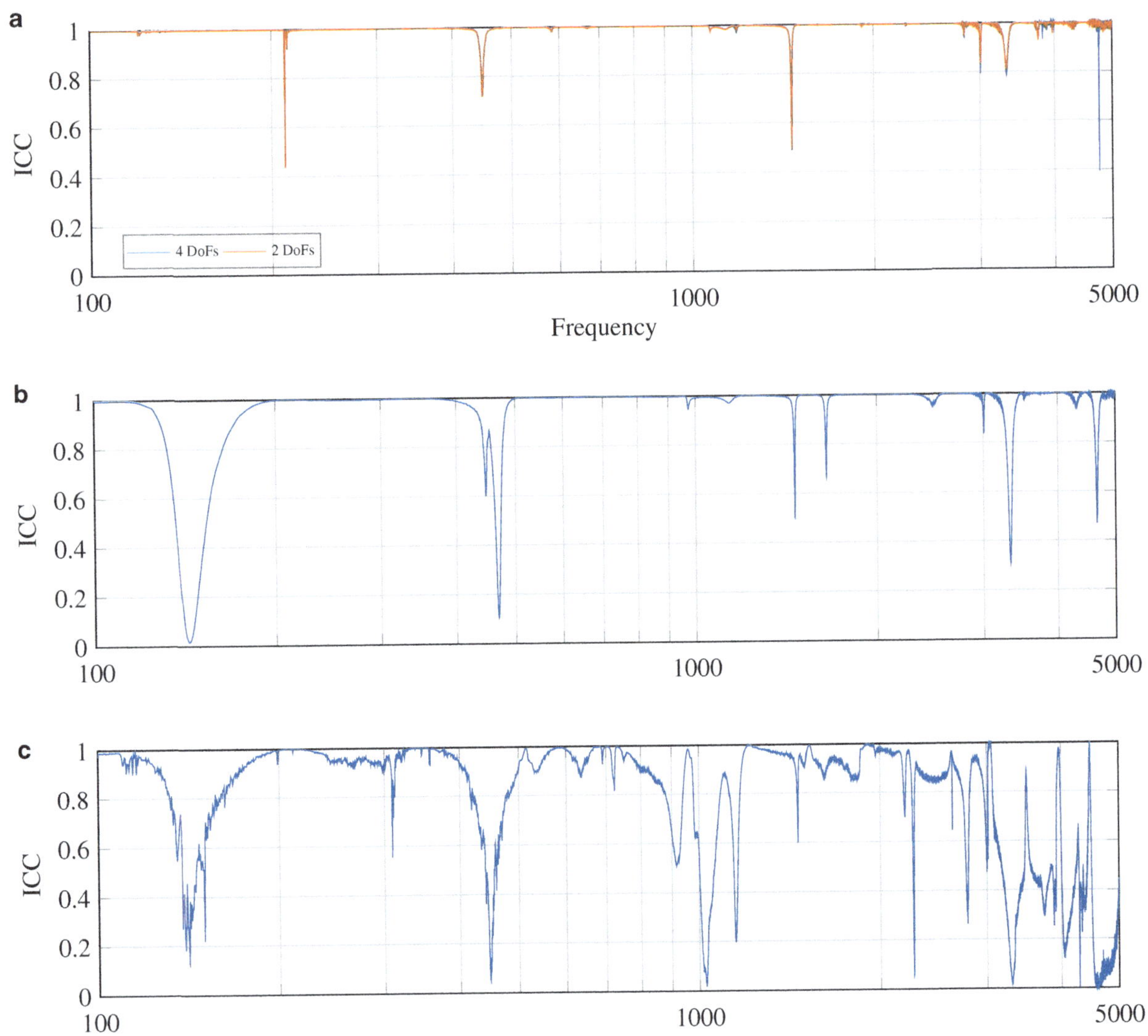

Fig. 14.7 Interface Completeness Criteria (ICC) obtained from the experimental study using 4 different interface descriptions. (**a**) An ICC comparison based on all 4 interface DoFs (z, y, γ, β), and the reduced interface DoFs (z, γ). (**b**) An ICC based on the translational z DoF, i.e. neglecting the rotational DoF, γ. (**c**) An ICC based on the translational and rotational DoFs, y and β

$\mathbf{Y}_{\mathrm{ba}}^{(\mathrm{c_i})} = \mathbf{Y}_{\mathrm{bc_i}}\mathbf{Y}_{\mathrm{c_ic_i}}^{-1}\mathbf{Y}_{\mathrm{c_ia}}$. The ICC was then calculated as per Eq. 14.19. The importance of a particular DoF to the interface description may be assessed by simply neglecting said DoF in the calculation of the ICC.

Shown in Fig. 14.7 are the ICC*s* obtained from the experimental study illustrated in Fig. 14.6 for a number of different interface descriptions.

Presented in Fig. 14.7a are the ICCs associated with a 4 DoF description (blue) and a reduced 2 DoF description (orange), where the DoFs y and β have been neglected. It is interesting to note that the two ICCs are in near perfect agreement. This suggests that the interface DoFs, y and β, have a negligible effect on the coupling of the structure. It should be noted, however, that this result was obtained from an ICC in which the source side excitation was made in the vertical z direction. Had additional excitations been applied in the y direction, coupling in y and β would likely have been of greater importance.

It is important to note that the calculation of the ICC, as per Eq. 14.19, is particularly sensitive to experimental error, notably the error associated with inconsistent excitations in the measurement of the required mobility matrices. Errors of this sort can lead to inconsistencies within the interface mobility matrix, $\mathbf{Y}_{\mathrm{c_ic_i}}$, which may result in large errors following its inversion. This effect is particularly relevant for under-damped systems, where structural anti-resonances are sensitive to excitation position. As an example, consider the sharp notch in the ICCs of Fig. 14.7a in the region of 200 Hz. This is likely a result of inconsistencies in the measured mobilities due to the small amount of damping present in the rod, and not due to an incomplete interface description.

Shown in Fig. 14.7b is the ICC associated with the single translational coupling DoF, z. A comparison against the reduced description in Fig. 14.7a (orange) clearly illustrates a worsening of the ICC due to the now neglected rotational DoF, γ. The effect is particularly pronounced in the region of 150 and 450 Hz. This suggests that in these regions the rotational coupling of the interface is of greater importance.

Lastly, shown in Fig. 14.7c is the ICC associated with the coupling DoFs y and β. As one would expect considering Fig. 14.7a, the ICC suggests that, by themselves, the DoFs y and β provide a very poor interface description.

14.6 Conclusion

This paper has been concerned with the development of an Interface Completeness Criterion (ICC) suitable for the quantitative assessment of the degree to which a coupling interface has been described. The ICC is based on the mathematical blocking of a subset of the coupling interface DoFs. This blocking allows for the separation of known and unknown DoFs, and consequently the formulation of a bound completeness criterion, referred to here as the ICC. The application of the ICC was illustrated by two case studies; the first a numerical simulation concerning the discretisation of a continuous interface, and the second, and experimental study concerning a discrete interface.

Acknowledgements This work was funded through the EPSRC Research Grant EP/P005489/1, Design by Science.

References

1. Gialamas, T., Tsahalis, D., Bregant, L., Dtte, D., Van Der Auweraer, H.: Substructuring by means of FRFs: some investigations on the significance of rotational DOFs. In: Proceedings of the 14th International Modal Analysis Conference – IMAC, pp. 619–625, Jan 1996
2. Ewins, D.J., Liu, W.: The importance assessment of RDOF in FRF coupling analysis. IMAC XVII **2**, 1481–1487 (1999)
3. Duarte, M.L.M., Ewins, D.J.: Some insights into the importance of rotational degrees-of-freedom and residual terms in coupled structure analysis. In: Proceedings of the 13th International Modal Analysis Conference – IMAC, pp. 164–170, May 1995
4. Klerk, D.D., Rixen, D.J., Voormeeren, S.N., Pasteuning, F.: Solving the RDoF problem in experimental dynamic substructuring. In: Proceedings of the 26th International Modal Analysis Conference – IMAC (2008)
5. D'Ambrogio, W., Fregolent, A.: Substructure decoupling without using rotational DoFs: fact or fiction? Mech. Syst. Signal Process. **72–73**, 499–512 (2016)
6. Moorhouse, A.T., Elliot, A.S.: Indirect measurement of frequency response functions applied to the problem of substructure coupling. In: NOVEM: Noise and Vibration – Emerging Methods, pp. 1–7 (2012)
7. Klerk, D.D., Rixen, D.J., Voormeeren, S.N.: General framework for dynamic substructuring: history, review and classification of techniques. AIAA J. **46**(5), 1169–1181 (2008)
8. Patil, N.S., Elliott, A.S., Moorhouse, A.T.: A blocked pressure based transfer path analysis (TPA) method to measure the sound insulation path contributions of a partition subjected to an incident airborne field. In: 24th International Congress on Sound and Vibration, ICSV 2017, pp. 1–8 (2017)
9. Bonhoff, H.A., Petersson, B.A.T.: The influence of cross-order terms in interface mobilities for structure-borne sound source characterization: plate-like structures. J. Sound Vib. **329**(16), 3280–3303 (2007)
10. Elliott, A.S., Meggitt, J.W.R., Moorhouse, A.T.: Blocked forces for the characterisation of structure borne noise. In: Internoise2015, San Fransisco, pp. 5798–5805 (2015)
11. Alber, T., Sturm, M., Moorhouse, A.T.: Independent characterization of structure-borne sound sources using the in-situ blocked force method. In Proceedings of Internoise, pp. 7302–7313 (2016)
12. Lennström, D., Olsson, M., Wullens, F., Nykänen, A.: Validation of the blocked force method for various boundary conditions for automotive source characterization. Appl. Acoust. **102**, 108–119 (2016)
13. Moorhouse, A.T., Elliott, A.S., Evans, T.A.: In situ measurement of the blocked force of structure-borne sound sources. J. Sound Vib. **325**(4–5), 679–685 (2009)
14. van der Seijs, M.V., Klerk, D.D., Rixen, D.J.: General framework for transfer path analysis: history, theory and classification of techniques. Mech. Syst. Signal Process. **68**, 1–28 (2015)
15. Meggitt, J.W.R., Elliott, A.S., Moorhouse, A.T.: Virtual assemblies and their use in the prediction of vibro-acoustic responses. In: Proceedings of the Institute of Acoustics, Warickshire (2016)
16. Moorhouse, A.T., Elliott, A.S.: The "round trip" theory for reconstruction of Green's functions at passive locations. J. Acoust. Soc. Am. **134**(5), 3605–3612 (2013)
17. Magrans, F.X.: Method of measuring transmission paths. J. Sound Vib. **74**(3), 321–330 (1981)

18. Guasch, O., Magrans, F.X.: The global transfer direct transfer method applied to a finite simply supported elastic beam. J. Sound Vib. **276**(1–2), 335–359 (2004)
19. Allemang, R.J.: The modal assurance criterion (MAC): twenty years of use and abuse. In: Proceedings of SPIE – the International Society for Optical Engineering, vol. 4753, pp. 397–405 (2002)
20. Elliott, A.S., Moorhouse, A.T., Pavić, G.: Moment excitation and the measurement of moment mobilities. J. Sound Vib. **331**(1), 2499–2519 (2012)

Chapter 15
Coupling Acoustic-Structure Systems Using Dynamic Substructuring

R. Benjamin Davis and Ryan Schultz

Abstract Dynamic substructuring allows analysts to combine component structural dynamics models into system-level models. For purely structural systems, dynamic substructuring techniques are well established. Here, one such technique is adapted to solve the coupled response of elastic structures in contact with acoustic cavities. The inputs to the process are the component natural frequencies and the free-interface structural and acoustic modes. The approach is well suited for use with analytical, finite element and/or empirical component modes. Unlike alternative acoustic-structure interaction approaches, the present technique avoids the cumbersome calculation of coupling coefficients and can result in real-valued system natural frequencies and modes. The technique is applied to a simple acoustic-structure system and the results from the model are used to gain insight into acoustic-structure interaction phenomena that have been observed in test hardware.

15.1 Introduction

Dynamic substructuring refers to the joining of subsystem models to produce a dynamic model of a built-up system. The term encompasses two primary substructuring techniques—component mode synthesis (CMS) [1–3] and frequency-based substructuring (FBS) [4]. While CMS methods seek to join reduced-order representations of subsystems, FBS methods join subsystems using their respective frequency response functions (FRFs). Both techniques have been studied extensively and a number of variations of each have been developed. Recent studies have considered a CMS approach that joins modal representations of subsystems by identifying a set of generalized coordinates that satisfy the interface constraints [5–7]. This approach is attractive because, unlike traditional CMS methods, it does not require explicit knowledge of subsystem mass and stiffness matrices, which are often cumbersome to extract from commercial finite element software. This approach, which will be referred to here as generalized coordinate assembly component mode synthesis (GCA-CMS) can be used in conjunction with any combination subsystem modes that are calculated by analytical, numerical, or empirical methods.

While the overwhelming thrust of dynamic substructuring research involves the joining of structural subsystems, dynamic substructuring techniques have been used in various ways to solve acoustic-structure interaction (ASI) problems. Here, ASI is defined as the fully coupled system response of structures in contact with acoustic cavities. An early use of CMS techniques in the context of ASI problems is due to Wolf [8], who considers the problem of a flexible rectangular plate coupled to a variable-length rectangular cavity filled with air. Wolf's approach involved calculating the uncoupled natural modes of the plate using an FE model and then coupling the structural modes to analytically-obtained acoustic modes. In the interest of predicting interior noise in passenger vehicles, Sung and Nefske [9–11] implemented and experimentally verified a CMS technique that relies on both structural and acoustic FE models. Their procedure involves solving the *in vacuo* forced response problem and then transforming the vehicle response into acoustic excitations of the cavity. The overall cavity response is then found using a separate acoustic FE model. This approach—though reasonable for its intended application—does not constitute a fully coupled analysis because it does not account for any influence that the fluid might have on the dynamics of the structure. A more recent study has adapted the classical CMS methods of Craig and Bampton and Rubin for coupling structural finite element models (FEMs) to acoustic boundary element models [12].

Dowell's acoustoelasticity formulation [13] is a popular approach to solving ASI problems. While most implementations of this approach have involved analytically-derived modes, examples of using the approach with FEM modes do exist [14].

R. Benjamin Davis
College of Engineering, University of Georgia, Athens, GA, USA
e-mail: ben.davis@uga.edu

R. Schultz (✉)
Structural Dynamics Department, Sandia National Laboratories, Albuquerque, NM, USA
e-mail: rschult@sandia.gov

A. Linderholt et al. (eds.), *Dynamics of Coupled Structures, Volume 4*, Conference Proceedings of the Society for Experimental Mechanics Series, https://doi.org/10.1007/978-3-319-74654-8_15

In a sense, Dowell's formulation can be interpreted as a CMS approach where rigid wall acoustic modes are combined with *in vacuo* structural modes to model the coupled system. The formulation requires the calculation of coupling coefficients between each pair of subsystem modes. Calculating these coupling coefficients involves computing an area integral over the product of the structural and acoustic component mode pair. Computing these coefficients numerically with FEM-generated modes can be non-trivial, especially for axi-symmetric structures and cavities where the acoustic and structural component modes can set up at an arbitrary azimuthal angle and need to be "clocked" in order to calculate the coupling coefficients. Dowell's formulation results in a set of gyroscopically coupled system equations that have complex-valued frequencies and modes. Relative to classical modal analysis which operates on real-valued natural frequencies and modes, it is considerably more involved to perform a frequency response analysis using complex frequencies and modes.

Here we present an adaptation of the GCA-CMS method for use with ASI systems. This approach is well suited for use with any combination of analytical, FEM, or empirical component modes. An undamped application of this method results in real-valued natural frequencies and modes and does not require the calculation of coupling coefficients between individual component mode pairs. The GCA-CMS method is used to examine the coupling of an acoustic-structure mode pair in a rectangular enclosure backed by a flexible plate. The method is found to offer a convenient framework for considering the effects of varying acoustic subsystem damping.

15.2 Theory

The disjoint system of equations is found by stacking the subsystem modal matrices along the diagonals of the system matrices

$$\begin{bmatrix} [I_A] & [0] \\ [0] & [I_S] \end{bmatrix} \begin{Bmatrix} \ddot{q}_A \\ \ddot{q}_S \end{Bmatrix} + \begin{bmatrix} [\omega_A^2] & [0] \\ [0] & [\omega_S^2] \end{bmatrix} \begin{Bmatrix} q_A \\ q_S \end{Bmatrix} = \begin{Bmatrix} 0 \\ 0 \end{Bmatrix}. \tag{15.1}$$

Here, $[I]$ is the identity matrix and $[\omega]$ represents a diagonal matrix of the subsystem natural frequencies. We are considering a single acoustic subsystem, A, and a single structural subsystem, S, though an arbitrary number of subsystem modal matrices can be concatenated in the same manner. We are also considering the case of undamped free vibration. Subsystem modal damping and external force terms can easily be included in Eq. (15.1), if desired. The transformation from physical coordinates, u, to generalized coordinates is given by

$$\begin{Bmatrix} u_A \\ u_S \end{Bmatrix} = \begin{bmatrix} [\nabla\Phi_A] & [0] \\ [0] & [\Phi_S] \end{bmatrix} \begin{Bmatrix} q_A \\ q_S \end{Bmatrix}, \tag{15.2}$$

where $[\nabla\Phi_A]$ represents a matrix containing the gradients of the normalized acoustic subsystem modes, and $[\Phi_S]$ is the mass normalized structural mode matrix. The gradient of the pressure mode shape enables the enforcement of displacement constraints. The relationship between pressure, p, and fluid particle displacement for harmonic responses is given by [15]

$$u_A = \frac{\nabla p}{\rho_f \omega^2}, \tag{15.3}$$

where ρ_f is the fluid density and ω is the oscillatory frequency. The mass normalized structural mode matrix can be easily output from FE software, but obtaining $[\nabla\Phi_A]$ typically requires that the calculation of the gradient and the normalization of the modes take place outside of the FE software. The acoustic modes are normalized according to

$$\nabla\Phi_{A_j} = \frac{\nabla\phi_{A_j}}{\sqrt{m_j}}, \tag{15.4}$$

where ϕ_{A_j} are non-normalized acoustic mode shapes with units of pressure. Here we will calculate acoustic subsystem mode shapes assuming the acoustic-structure interface to be a pressure release (i.e., $p = 0$) surface. This enables the application of effective constraints by ensuring that the gradients of the mode shapes at the interface are non-zero. The modal mass, m_j, is given by

$$m_j = \rho_f \int_V \nabla\phi_j \cdot \nabla\phi_j dV. \tag{15.5}$$

The equations of motion in Eq. (15.1) now need to be joined by enforcing the desired constraints in terms of physical coordinates. The constraint equations take the form of

$$[a]\begin{Bmatrix} u_A \\ u_S \end{Bmatrix} = \begin{Bmatrix} 0 \\ 0 \end{Bmatrix}, \tag{15.6}$$

where $[a]$ is a boolean constraint matrix in which each element is either zero, one, or minus one. Inserting Eq. (15.6) into Eq. (15.2) gives a set of constraint equations in terms of the generalized coordinates

$$[a]\begin{bmatrix} [\nabla\Phi_A] & [0] \\ [0] & [\Phi_S] \end{bmatrix}\begin{Bmatrix} q_A \\ q_S \end{Bmatrix} = [\hat{a}]\begin{Bmatrix} q_A \\ q_S \end{Bmatrix} = \begin{Bmatrix} 0 \\ 0 \end{Bmatrix}. \tag{15.7}$$

In the constrained system, the constraint forces need to be included in the equations of motion. Constraint forces are proportional to the rows of the constraint coefficient matrix $[\hat{a}]$, where the constants of proportionality, λ, are Lagrange multipliers [16]. The constraint equation in Eq. (15.7) and the equations of motion in Eq. (15.1) (with constraint forces), are the complete equations of motion for the system

$$\begin{aligned} \begin{bmatrix} [I_A] & [0] \\ [0] & [I_S] \end{bmatrix}\begin{Bmatrix} \ddot{q}_A \\ \ddot{q}_S \end{Bmatrix} + \begin{bmatrix} [\omega_A^2] & [0] \\ [0] & [\omega_S^2] \end{bmatrix}\begin{Bmatrix} q_A \\ q_S \end{Bmatrix} &= [\hat{a}]^T\{\lambda\}, \\ [\hat{a}]\begin{Bmatrix} q_A \\ q_S \end{Bmatrix} &= \begin{Bmatrix} 0 \\ 0 \end{Bmatrix}. \end{aligned} \tag{15.8}$$

We can now proceed with the solution of Eq. (15.8) which is a set of algebraic-differential equations. We wish to find a set of unconstrained degrees of freedom, denoted ξ. This set is related to the complete set of generalized coordinates by

$$\begin{Bmatrix} q_A \\ q_S \end{Bmatrix} = [B]\{\xi\}, \tag{15.9}$$

where $[B]$ is a transformation matrix that eliminates the constrained degrees of freedom. Using Eq. (15.9) and the constraint equations in Eq. (15.8), gives

$$[\hat{a}]\,[B]\,\{\xi\} = \{0\}. \tag{15.10}$$

For Eq. (15.10) to hold for non-trivial values of ξ, $[\hat{a}]\,[B] = [0]$. In linear algebra, $[B]$ is called the nullspace (or kernel) of $[\hat{a}]$.

Substituting Eq. (15.9) into Eqs. (15.8) and premultiplying both sides of the first equation by $[B]^T$ gives

$$\begin{aligned} [B]^T\begin{bmatrix} [I_A] & [0] \\ [0] & [I_S] \end{bmatrix}[B]\{\ddot{\xi}\} + [B]^T\begin{bmatrix} [\omega_A^2] & [0] \\ [0] & [\omega_S^2] \end{bmatrix}[B]\{\xi\} &= \cancelto{0}{[B]^T[\hat{a}]^T}\{\lambda\}, \\ [\hat{a}]\,[B]\,\{\xi\} &= \{0\}, \end{aligned} \tag{15.11}$$

where $[B]^T\,[\hat{a}]^T = 0$ since $[\hat{a}]\,[B] = [0]$. The natural frequencies and modes of the system can now be calculated via the eigenvalue problem $\left(\left[\hat{K}\right] - \omega^2\left[\hat{M}\right]\right)\{\phi\} = \{0\}$ where

$$\begin{aligned} \left[\hat{M}\right] &= [B]^T\begin{bmatrix} [I_A] & [0] \\ [0] & [I_S] \end{bmatrix}[B], \\ \left[\hat{K}\right] &= [B]^T\begin{bmatrix} [\omega_A^2] & [0] \\ [0] & [\omega_S^2] \end{bmatrix}[B]. \end{aligned} \tag{15.12}$$

15.3 Rectangular Enclosure Example

Consider the commonly studied ASI system of a rectangular acoustic enclosure with a flexible plate on one end. Figure 15.1 is a diagram of the geometry of interest. On the end of the enclosure coinciding with the XY plane is an aluminum (E=70 GPa, $\rho = 2700\,\mathrm{kg/m^3}$, $\nu = 0.3$) plate that is simply supported on all four edges. The box is filled with air ($\rho_0 = 1.21\,\mathrm{kg/m^3}$, $c_0 = 343\,\mathrm{m/s}$) and shares a width of $W = 0.91$ m and a height of $H = 0.31$ m with the plate. A plate thickness of 4 mm is chosen to promote strong acoustic-structure coupling between the (1,1) (i.e., the "drum head") mode of the plate, and the first longitudinal acoustic mode of the cavity, which is denoted in terms of mode indices as (0,0,1). The (1,1) mode of the structure *in vacuo* is found to have a frequency of 115.6 Hz and the acoustic natural frequency of the (0,0,1) mode is 112.5 Hz, assuming entirely rigid walls. This near-coincidence of the uncoupled natural frequencies is known to lead to a strongly coupled responses in acoustic-structure systems [17].

To generate the basis modes for the GCA-CMS approach, FE models of the *in vacuo* structure and the uncoupled acoustic enclosure are created using the commercial finite element software ANSYS. The acoustic model treats the fluid-structure interface (i.e., the XY plane) as a pressure release surface. The first 30 structural natural frequencies and modes as well as the first 100 acoustic natural frequencies and modes are output from the model. By default, ANSYS outputs mass normalized structural modes that can be used directly in the GCA-CMS process. It is necessary, however, to numerically calculate the gradient of the acoustic modes and re-normalize them according to Eq. (15.4).

Note that every DOF in the structural model is on the fluid-structure interface and that enforcing compatibility at every interface node would highly over-constrain the problem. For the purposes of this study, we enforce a single constraint equation that sets the sum of the structural displacements equal to the sum of the acoustic displacements. This ensures compatibility in an average sense and is appropriate for the consideration of coupling between the (1,1) and (0,0,1) modes, though it will likely fail to capture the coupled behavior of other mode pairs. The determination of effective constraint equations to use in conjunction with the GCA-CMS method and ASI problems is a topic of future work.

15.3.1 Accuracy of the GCA-CMS Approach

The accuracy of the GCA-CMS approach is verified by comparing its output to that of a fully coupled acoustic-structure FEM. The coupling of a structural mode and an acoustic mode that have nearly equivalent frequencies manifests in two coupled modes with relatively separated natural frequencies. The behavior is very similar to that of a tuned-mass-damper system, where the acoustic cavity acts like a parasitic mass, effectively splitting a single structural resonance into two acoustic-structure resonances. Table 15.1 lists the coupled natural frequencies calculated via the GCA-CMS approach and the acoustic-

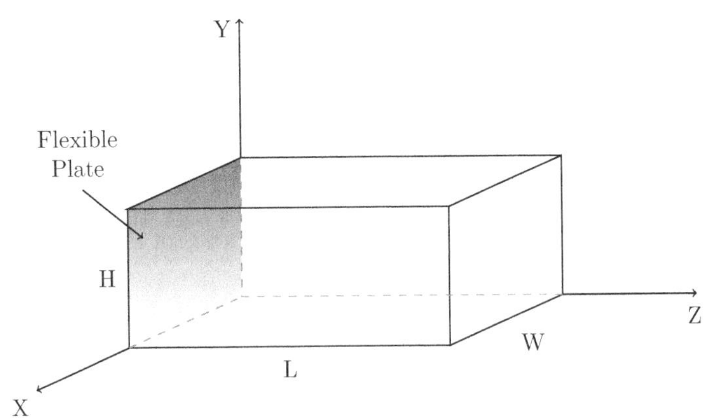

Fig. 15.1 Schematic of a flexible plate backed by a rectangular enclosure

Table 15.1 Comparison of coupled natural frequencies as calculated by FEM and GCA-CMS

	f_{c_1} (Hz)	f_{c_2} (Hz)
FEM	105.6	122.5
GCA-CMS	107.2	123.4

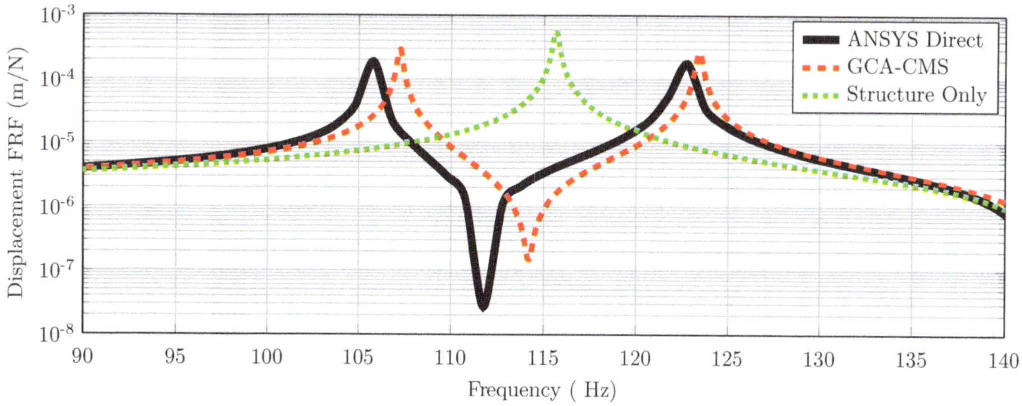

Fig. 15.2 Comparison of direct FEM, GCA-CMS, and *in vacuo* FRFs

structure FEM. The results of the two methods agree within 1.5% for the first mode and 0.7% for the second. Note that the coupled natural frequencies are approximately 7.5% above and below the frequency of the *in vacuo* (1,1) mode of the structure. Unlike other ASI CMS approaches that use rigid wall acoustic modes as basis functions, the GCA-CMS models the coupled behavior with pressure release modes. Due to the quarter-wave nature of the pressure release mode shapes, the first acoustic natural frequency of the cavity with the fluid-structure interface treated as a pressure release surface is one half of the corresponding rigid wall natural frequency. Thus, even though the acoustic basis modes used in the GCA-CMS approach significantly differ from the coupled mode in both frequency and shape, the method can still produce accurate coupled solutions.

To further assess the accuracy of the GCA-CMS method, FRFs are compared in Fig. 15.2 for both the GCA-CMS method and a direct harmonic response analysis of the acoustic-structure FEM. An FRF of the *in vacuo* structure is also shown. In all cases, the force is applied to the $X = 0.45W$, $Y = 0.20H$ point on the plate, and the response is calculated at the $X = 0.35W$, $Y = 0.55H$ point. A viscous damping ratio of 0.1% is applied at the system level. When modal damping is applied at the system level, using the GCA-CMS approach is especially convenient because the method results in real-valued natural frequencies and modes which can then be used in a classical modal analysis. Alternative ASI formulations result in unsymmetric system matrices that require a complex modal analysis.

In addition to showing that the GCA-CMS results compare favorably with the direct FEM solution, Fig. 15.2 illustrates the effects of acoustic-structure coupling on structural response. The coupling effects manifest as two resonant peaks that occur at frequencies above and below the *in vacuo* resonance peak. At these coupled resonances, energy resides in both the fluid and the structure, and consequently the response of the structure at resonance is lower than it would be *in vacuo*. In this particular case, the amplitude response of the coupled peaks is roughly half that of the *in vacuo* peak.

15.3.2 Effects of Acoustic Subsystem Damping

Given that test-analysis correlation typically involves *in vacuo* structural analysis models, it is clear that the effects of acoustic-structure coupling can potentially complicate correlation efforts. To remove the confounding effects of the acoustic fluid in a test, a possible solution is to introduce acoustic damping (e.g., foam) to the cavity. To understand how this might affect system response a numerical experiment is devised in which the viscous damping ratio of the acoustic subsystem is changed while the resulting system eigenvalues, resonant frequencies, and the corresponding response amplitudes are calculated. While acoustic damping is more accurately modeled by invoking complex-valued frequency dependent impedance boundary conditions, modeling the fluid dissipation effects as a single viscous damping ratio is a convenient way to qualitatively assess the system behavior in response to acoustic damping.

With the introduction of subsystem damping, the coupled system frequencies and modes become complex-valued. To calculate the system response, a complex modal analysis procedure is used as described by Davis[18]. In the numerical experiment, the structural subsystem damping is held fixed at 0.001 while the acoustic damping is varied from zero to 0.25. Figure 15.3 shows how the imaginary part of the first two system eigenvalues change with acoustic damping. The figure also shows the resonant frequencies of the FRF (with the same drive and response points specified above), and their amplitudes against acoustic damping. All eigenvalues and resonant frequency values are normalized by the corresponding

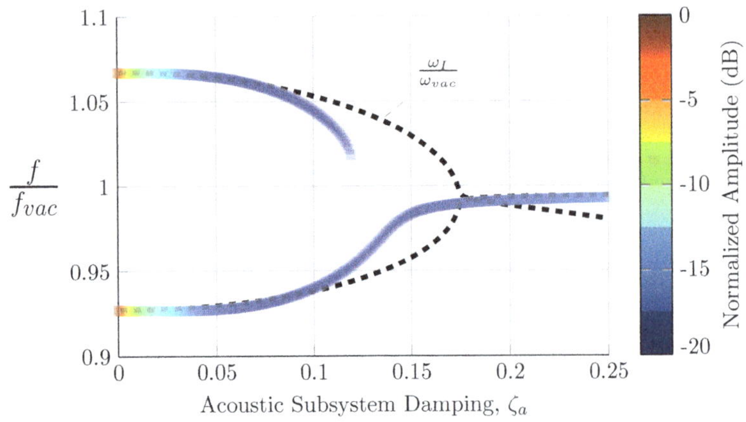

Fig. 15.3 Normalized resonant frequencies and amplitudes vs. acoustic subsystem damping ratio. The normalized imaginary parts of the system eigenvalues are also shown

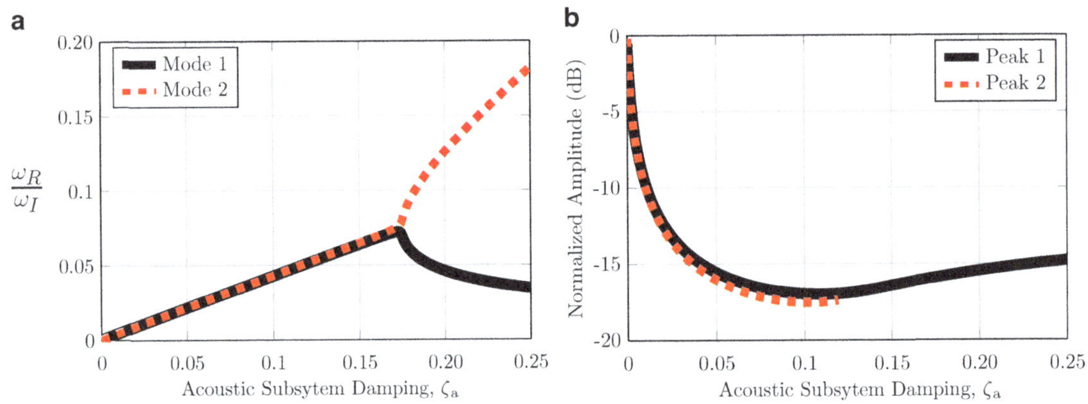

Fig. 15.4 (**a**) Ratio of real and imaginary parts of system eigenvalues and (**b**) normalized amplitude response versus acoustic subsystem damping

value for the *in vacuo* (1,1) mode. The structural response amplitudes are shown on a dB scale that is calculated according to $10 \times log(A_{res}/A_{vac})$, where A_{res} is the FRF amplitude at resonance and A_{vac} is the amplitude at resonance of the *in vacuo* structure with 0.1% damping.

For low levels of acoustic damping, the resonant frequencies closely follow the curve representing the imaginary portion of the eigenvalue. Near $\zeta_a = 0.05$, the first resonant frequency deviates slightly below the lower λ_I curve. In a modal test, this behavior can manifest as coupled resonances that become more separated as acoustic damping is added. As ζ_a increases further, the imaginary parts of the two eigenvalues begin to approach each other. Near $\zeta_a = 0.10$, the two resonant frequencies begin to deviate significantly from the λ_I curves and approach each other more rapidly, until the second resonance peak abruptly disappears near $\zeta_a = 0.12$. The λ_I curves cross near $\zeta_a = 0.17$. The first resonance persists at higher values of ζ_a and eventually follows the higher of the two λ_I curves. The amplitude response at resonance drops sharply with small initial increases in damping. However, for acoustic damping ratios beyond $\zeta_a \approx 0.10$, resonant response amplitudes appear to increase somewhat.

More insight into the coupled system response can be obtained by considering the real part of the system eigenvalues. The ratio of the real and imaginary parts of the eigenvalues is equivalent to the modal viscous damping ratio of the coupled system. In Fig. 15.4a this ratio is shown against the acoustic subsystem damping ratio for the two coupled acoustic-structure modes. Again, the structural subsystem damping is fixed at $\zeta_s = 0.001$. The results indicate that for $\zeta_a < 0.17$, the total damping $(\zeta_s + \zeta_a)$ is effectively split between the two modes. This occurs even when the vast majority of the damping resides in the fluid. Because the modes are well coupled, energy is exchanged and dissipated efficiently between the modes. Near $\zeta_a = 0.17$, a critical point is reached at which the real parts of the system eigenvalues begin to diverge. As acoustic damping increases, one mode becomes much more heavily damped while the other becomes more lightly damped. In the response, this results in a single resonance peak that actually increases in amplitude as acoustic damping is increased. Another interesting feature of the system behavior is that beyond the critical point, the total system damping appears to be noticeably less than the total subsystem damping.

Figure 15.4b shows the normalized amplitude of the plate at resonance versus acoustic subsystem damping. It is observed that the amplitudes of the two resonances are roughly equivalent for low levels of acoustic damping. Near $\zeta_a = 0.05$, the amplitude of the second resonance begins to drop relative to the first, and near $\zeta_a = 0.10$ both resonances reach a minimum. Beyond this point, increasing acoustic damping actually increases the resonant response amplitudes. The abrupt disappearance of the second resonance can be clearly seen near $\zeta_a = 0.12$, and beyond this damping level, a single resonance that increases in amplitude with increases in acoustic damping persists.

Many of the qualitative observations made in this numerical experiment involving acoustic damping can help explain some recent experimental findings. In experimental studies, Schultz and Pacini [19] found that adding foam to the interior of a cylindrical shell can eventually prompt the disappearance of one of the acoustic-structure resonances. They also found that the frequency of the remaining single resonance does not correspond with the frequency of the *in vacuo* structural resonance. Additionally, they saw that increasing the amount of acoustic absorbing material had a non-obvious influence on the apparent structural damping. Each of these observations is consistent with the present numerical results.

15.4 Conclusions

A component mode synthesis method that assembles the system in terms of generalized coordinates is adapted for use with acoustic-structure systems. This approach, termed the GCA-CMS method, has the benefit of not requiring explicit knowledge of subsystem mass and stiffness matrices. Unlike alternative ASI methods, an undamped application of the GCA-CMS method results in real-valued natural frequencies and modes. A third advantage is that it does not require the calculation of coupling coefficients between individual component mode pairs.

The GCA-CMS method is used to calculate the system frequency and response behavior due to a (1,1) plate mode coupling with a (0,0,1) acoustic cavity mode. The subsystem frequencies and modes are obtained from finite element models. An FRF obtained from the GCA-CMS method agrees well with an FRF calculated using a direct response analysis of the an ASI FEM. Next, a complex modal analysis procedure is used to calculate structural response for varying levels of acoustic subsystem damping. The system eigenvalues, resonant frequencies, and amplitudes at resonance are shown as a function of acoustic subsystem damping. It is observed that at some critical value of subsystem damping, the imaginary parts of the coupled system eigenvalues cross while the real parts diverge. This helps explain empirical observations where, at sufficient levels of acoustic damping, the two coupled resonances routinely observed in ASI systems become a single resonance. Interestingly, the response analysis indicates that this transition from two resonances to one occurs abruptly at levels of damping that are well below the level predicted by the eigenvalues alone.

References

1. Craig, R.R., Bampton, M.C.C.: Coupling of substructures for dynamic analysis. AIAA J. **6**(7), 1313–1319 (1968)
2. MacNeal, R.H.: A hybrid method of component mode synthesis. J. Comput. Struct. **1**(4), 581–601 (1971)
3. Rubin, S.: Improved component-mode representation for structural dynamic analysis. AIAA J. **13**(8), 995–1006 (1975)
4. Jetmundsen, B.: On frequency domain methodologies for structural modification and subsystem synthesis. Ph.D. thesis, Rensselaer Polytechnic Institute, Troy (1986)
5. de Klerk, D., Rixen, D.J., Voormeeren, S. N.: General framework for dynamic substructuring: history, review, and classification of techniques. AIAA J. **46**(5), 1169–1181 (2008)
6. Allen, M.S., Kammer, D.C., Mayes, R.L.: Uncertainty in experimental/analytical substructuring predictions: a review with illustrative examples. In: ISMA2010—International Conference on Noise and Vibration Engineering, Leuven (2010)
7. Davis, R.B., Fulcher, C.W.: Framework for optimizing the vibroacoustic response of component loaded panels. AIAA J. **53**(1), 265–272 (2014)
8. Wolf, J.A., Jr.: Modal synthesis for combined structural-acoustic systems. AIAA J. **15**, 743–745 (1977)
9. Nefske, D.J., Sung, S.H.: Automobile interior noise prediction using a coupled structural-acoustic finite element model. In: Proceeding of the 11th International Congress on Acoustics, Paris, vol. 5, pp. 465–468 (1983)
10. Sung, S.H., Nefske, D.J.: A coupled structural-acoustic finite element model for vehicle interior noise analysis. J. Vib. Stress Reliab. Des. **106**, 314–318 (1984)
11. Sung, S.H., Nefske, D.J.: Component mode synthesis of a vehicle structural-acoustic system. AIAA J. **24**(6), 1021–1026 (1986)
12. Junge, M., Brunner, D., Becker, J., Gaul, L.: Interface-reduction for the Craig–Bampton and Rubin method applied to FE–BE coupling with a large fluid–structure interface. Int. J. Numer. Methods Eng. **77**(12), 1731–1752 (2009)
13. Dowell, E.H., Gorman, G.F., III, Smith, D.A.: Acoustoelasticity: general theory, acoustic natural modes and forced response to sinusoidal excitation including comparisons with experiment. J. Sound Vib. **52**(4), 519–542 (1977)
14. Davis, R.B., Virgin, L.N., Brown, A.M.: An efficient modal analysis method for structures coupled to fluid-filled cavities. In: 49th AIAA/ASME/ASCE/AHS/ASC Structures, Structural Dynamics, and Materials Conference, p. 1849 (2008)

15. Pierce, A.D.: Acoustics: An Introduction to its Physical Principles and Applications. Acoustical Society of America, New York (1989)
16. Ginsberg, J.H.: Mechanical and Structural Vibrations: Theory and Applications. Wiley, New York (2001)
17. Davis, R.: A simplified approach for predicting interaction between flexible structures and acoustic enclosures. J. Fluids and Struct. **70**, 276–294 (2017)
18. Davis, R.B.: Techniques to assess acoustic-structure interaction in liquid rocket engines. Ph.D. thesis, Duke University, Durham (2008)
19. Schultz, R., Pacini, B.: Mitigation of structural-acoustic mode coupling in a modal test of a hollow structure. In: Maio, D.D., Castellini, P. (eds.) Rotating Machinery, Hybrid Test Methods, Vibro-Acoustics & Laser Vibrometry, vol. 8, pp. 71–84. Springer, Cham (2017)

Chapter 16
Dynamic Substructuring Applied to the Decoupling of Acoustic-Structure Systems

Ryan Schultz and R. Benjamin Davis

Abstract Dynamic substructuring allows analysts to combine component structural dynamics models into a system-level model. An analogous process can be used to subtract off a component model from the system model. Here, this decoupling approach is adapted for use with acoustic-structure systems. This has applications to model validation where structural models are correlated to test data. When acoustic cavities are present, coupling with the acoustic subsystem can confound the test response and inhibit the model validation effort. Here, the Transmission Simulator method of Component Mode Synthesis is applied to extract the structure-only response from a coupled structural-acoustic system using the acoustic cavity modes as the subtracted component. This approach is demonstrated on a simple plate-box system using component modes from analytical and numerical models.

Keywords Acoustoelasticity · Component mode synthesis · Structural acoustics · Transmission simulator · Substructuring

16.1 Introduction

Often, modal test results are used to correlate to or update finite element (FE) models of structures. Thus, it is important that the model matches the test conditions (and vice versa). One aspect seldom considered is the possibility that the air inside hollow structures could be modifying the structural modes measured in test. This modification of the structural response is due to coupling of the structure and acoustic modes, dubbed acoustoelasticity. For a system to exhibit acoustoelastic coupling, the modes of the hollow structure need to match closely to the modes of the internal air cavity in both shape and frequency [1]. When the modes are very similar in shape and close in frequency, the result is a split of the structural mode into two peaks, akin to a tuned absorber. While it is possible to prevent the acoustoelastic coupling in test by modifying the boundary conditions or adding damping to the air cavity, at times the problem of acoustoelastic coupling is not known until after the test [2]. As such, a method for decoupling the structure from the air is desired.

Component mode synthesis (CMS) has long been used to couple modal models of structures [3]. Similarly, related methods, such as the Transmission Simulator method, have been developed to remove components from system models. The objective of this work is to first apply CMS to couple a structural-acoustic plate-box system to get a coupled system with acoustoelastic coupling traits. Next, the Transmission Simulator method will be applied to the coupled system to decouple, that is, remove the acoustic component and result in a structure-only model. Comparisons will be made against direct numerical simulation of a finite element model of the plate-box system.

Sandia National Laboratories is a multimission laboratory managed and operated by National Technology and Engineering Solutions of Sandia, LLC., a wholly owned subsidiary of Honeywell International, Inc., for the U.S. Department of Energy's National Nuclear Security Administration under contract DE-NA0003525.

R. Schultz (✉)
Structural Dynamics Department, Sandia National Laboratories, Albuquerque, NM, USA
e-mail: rschult@sandia.gov

R. Benjamin Davis
College of Engineering, University of Georgia, Athens, GA, USA
e-mail: ben.davis@uga.edu

A. Linderholt et al. (eds.), *Dynamics of Coupled Structures, Volume 4*, Conference Proceedings of the Society for Experimental Mechanics Series, https://doi.org/10.1007/978-3-319-74654-8_16

16.2 Review of Component Mode Synthesis Methods

CMS is used to connect modal models of components to form a system model with accurate dynamics. Typically, the size of the component modal models is greatly reduced compared to the system models built with full degree of freedom (DOF) component models, bringing efficiency to system modeling. Additionally, this component assembly approach provides useful modularity, allowing for the insertion or removal of components. While many different CMS methods exist, here we consider the simple case of isolated component modes. That is, the component mode basis will be formed solely with the component normal modes without additional constraint or supplemental modes.

16.2.1 Coupling Dynamic Systems with Component Mode Synthesis

Consider two components, A and B, joined together to create a system, C. The CMS method used to couple $A + B = C$ is as follows [4]:

1. Obtain component modes and form the component shape, modal mass and modal stiffness matrices. Truncate or select the modes of interest. Modes can come from numerical simulation or experimental measurements.

$$\text{Component A} : \omega_A, \Phi_A \tag{16.1}$$

$$\text{Component B} : \omega_B, \Phi_B$$

2. Write constraint equations in matrix form which link the connected DOF in each component. This is typically called a boolean constraint matrix, S.

$$S \begin{Bmatrix} x_A \\ x_B \end{Bmatrix} = \{0\} \tag{16.2}$$

3. Cast this constraint matrix into modal space by multiplying by a block diagonal matrix of the component mode shape matrices.

$$\overline{S} = S \begin{bmatrix} \Phi_A & 0 \\ 0 & \Phi_B \end{bmatrix} \tag{16.3}$$

4. Generate a transformation matrix, L, which will connect the modal DOF of the components based on the modal constraint matrix. This is done by taking the null space of the modal constraint matrix.

$$L = null\left(\overline{S}\right) \tag{16.4}$$

5. Use the transformation matrix to form the coupled system modal mass and stiffness matrices using the diagonally stacked component modal mass and stiffness matrices.

$$\overline{M} = L^T \overline{M}_U L, \quad \overline{K} = L^T \overline{K}_U L \tag{16.5}$$

6. Solve for the coupled system modes then transform back to physical space using the uncoupled component modes and the transformation matrix.

$$\Phi_C = \Phi_U L \overline{\Phi}_C \tag{16.6}$$

The choice of component modes affects the system modes; enough of the right component modes are needed to properly span the space. Several of the CMS approaches supplement or alter the free component normal modes with additional modes to enrich the basis. In particular, the Transmission Simulator method uses an additional component to provide more realistic connection boundary conditions on the components. An interesting aspect of the Transmission Simulator method is that the Transmission Simulator component is removed from the system via a subtraction operation as described in the next section.

16.2.2 Decoupling Dynamic Systems with the Transmission Simulator Method

To remove the Transmission Simulator component from the system assembly, a procedure similar to the coupling process is used; however here there is a subtraction of component modes. This decoupling of a component from a system can notionally be considered as: C – B = A, with B being be the Transmission Simulator and A being the desired, extracted component.

The process for removing a Transmission Simulator component from a coupled system is as follows [4, 5]:

1. Obtain system and Transmission Simulator component modes and form the shape, modal mass and modal stiffness matrices. Truncate or select the modes of interest.

$$\text{Coupled System} : \omega_C, \Phi_C \tag{16.7}$$

$$\text{Transmission Simulator Component} : \omega_B, \Phi_B$$

2. Create the modal space constraint matrix using a softened form of the constraint equations which uses a pseudo-inverse of the Transmission Simulator component modes. Φ_B^+ is the pseudo-inverse of the B component shape matrix.

$$\overline{S} = \left[\Phi_B^+\right]\left[\Phi_C \quad -\Phi_B\right] \tag{16.8}$$

3. Generate a transformation matrix, *L*, which will connected the modal DOF of the components based on the modal constraint matrix. This is done by taking the null space of the modal constraint matrix.

$$L = null\left(\overline{S}\right) \tag{16.9}$$

4. Use the transformation matrix to form the uncoupled system modal mass and stiffness matrices using the diagonally stacked system and component modal mass and stiffness matrices. This time, however, there is a subtraction of the *B* component modal matrices:

$$\overline{M}_U = \begin{bmatrix} \overline{M}_C & 0 \\ 0 & -\overline{M}_B \end{bmatrix}, \quad \overline{K}_U = \begin{bmatrix} \overline{K}_C & 0 \\ 0 & -\overline{K}_B \end{bmatrix} \tag{16.10}$$

$$\overline{M} = L^T \overline{M}_U L, \quad \overline{K} = L^T \overline{K}_U L \tag{16.11}$$

5. Solve for the decoupled system modes then transform back to physical space using the uncoupled component modes and the transformation matrix.

$$\Phi_A = \Phi_U L \overline{\Phi}_A \tag{16.12}$$

16.3 Application of Component Mode Synthesis to Couple and Decouple an Acoustoelastic System

Typically, the Transmission Simulator is used to provide a better set of basis modes for the A component. For this acoustoelastic system decoupling problem, we propose to use the same method but treat the Transmission Simulator component, B, as the entrained air in a hollow structure, A, for which we have coupled modes from system C. More concisely, we want to achieve A using modes of C and B. In this work, an example acoustoelastic system is created to explore the use of CMS and Transmission Simulator approaches to the coupling and decoupling of acoustoelastic systems. Below is a diagram of the coupling of components to form an acoustoelastic system and the decoupling of an acoustoelastic system to leave an uncoupled component (Fig. 16.1).

Ideally, the decoupling process would be attempted using coupled system modes from a numerical simulation. However, obtaining coupled structural-acoustic system modes from FE packages can be tricky. Coupled modes from a FE simulation was tried here but the results were not trustworthy and seemed to indicate an issue with scaling of the mode shapes coming out of the FE package. As such, for this work the decoupling process is trialed using coupled system modes obtained by coupling together structure and acoustic components with the CMS process described above. This is not especially satisfying since the notional math is simply: (A + B) – B = A, however until a robust method for obtaining useful coupled structural-acoustic system modes is found, this is an option.

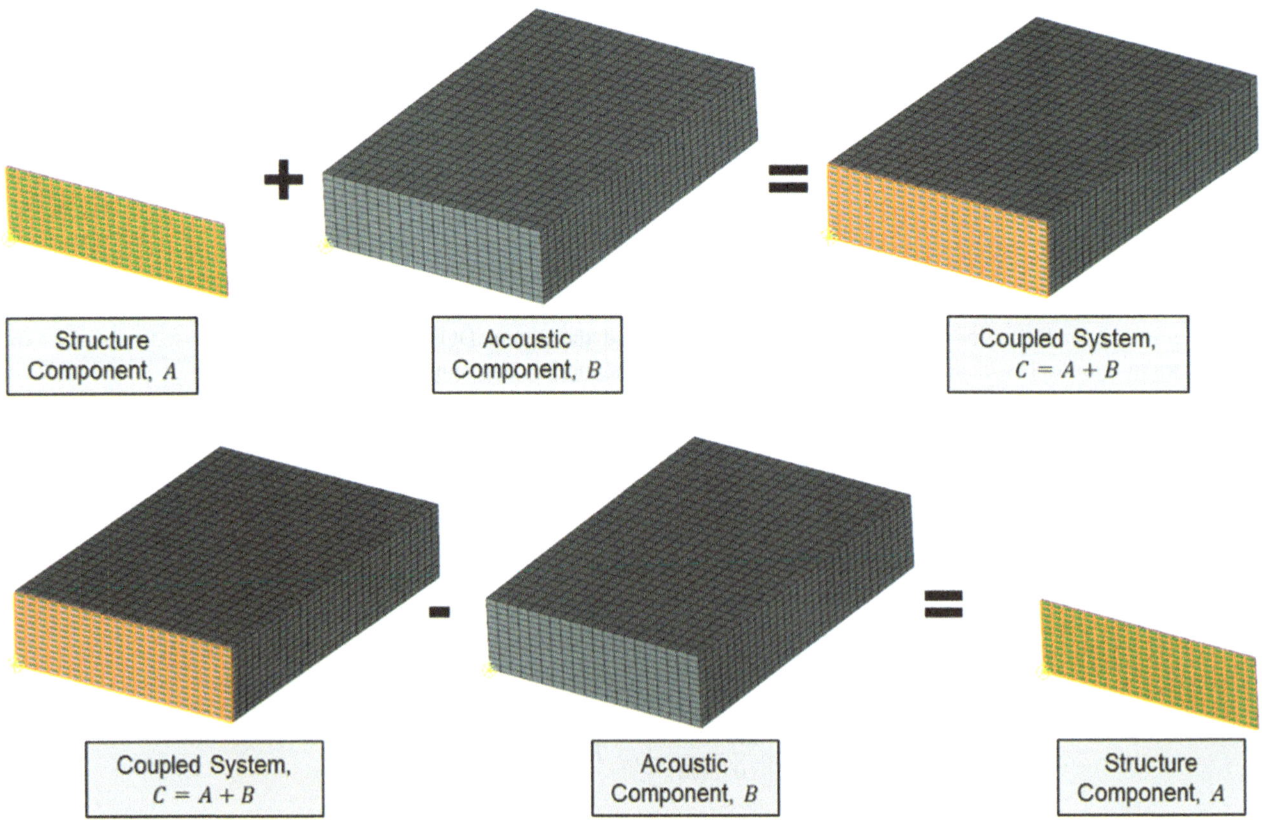

Fig. 16.1 Coupling of the structural-acoustic plate-box system (*top*) and decoupling of the system (*bottom*)

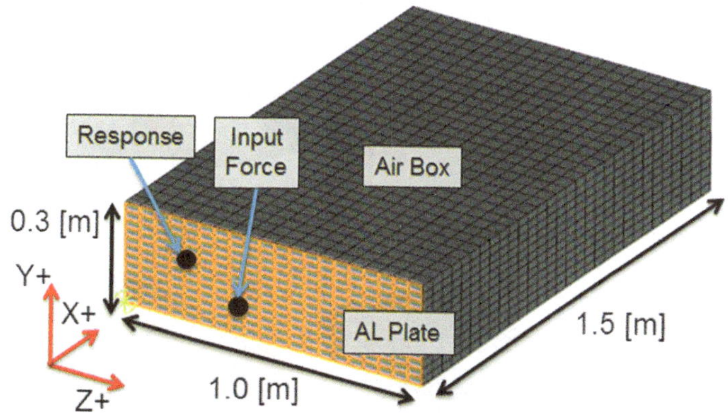

Fig. 16.2 Dimensions of the plate-box system and locations of the input and response points

16.3.1 Acoustoelastic Systems

Acoustoelastic systems are comprised of a structure entraining an acoustic fluid volume and have structural modes with similar frequencies and shapes to the hollow cavity acoustic modes. This mode similarity allows for a coupling of the acoustic and structural components which changes the response of the structure. Depending on mode shape similarity and the proximity of the natural frequencies, the coupling effect can be large or small [6].

16.3.2 Example System: Plate-Box

A simple acoustoelastic system is a box of air with an elastic plate on one side, an example used in the acoustoelasticity literature [7]. A finite element (FE) model of such a system was developed with dimensions to align the first mode of a 4 [mm] thick simply supported aluminum plate with the first mode of a rigid-walled, air-filled box. A diagram showing the mesh and dimensions is shown below. The modulus of elasticity of the plate was tuned to closely align the structure and acoustic mode frequencies. The mesh is a regular grid with 31 nodes in X, 13 nodes in Y, and 21 nodes in Z, providing 273 nodes on the wetted surface (Fig. 16.2).

16.3.3 Structure Component Modes

The plate-only (*in vacuo*) modes were computed using Sierra/SD, part of the Sierra Mechanics finite element package [8]. Plate-only shapes and frequencies are shown below alongside the rigid-walled acoustic modes.

16.3.4 Acoustic Component Modes

The acoustic component modes were first computed with rigid-wall boundary conditions for the purposes of visualizing the acoustic pressure mode shapes, shown below alongside the plate modes. In the figure below, it can be seen that the first plate and acoustic modes are compatible; the shapes and frequencies are similar. No other mode pair in the frequency range shows the same similarity, thus we expect only one coupled mode split peak in the response (Fig. 16.3).

Rigid-walled boundary conditions cannot be used in this CMS process because there is zero acoustic particle velocity at the plate-box wetted surface; a condition that will result in unusable acoustic mode shapes in the B component basis. Thus, a different set of acoustic modes was needed. Using a pressure release boundary condition at the wetted surface and rigid walls elsewhere provides modes with non-zero particle velocity at the wetted surface. However, computing acoustic modes with any type of absorbing boundary condition can be challenging using most finite element packages. For this simple geometry and idealistic boundary conditions, the acoustic modes can be computed analytically as [9]:

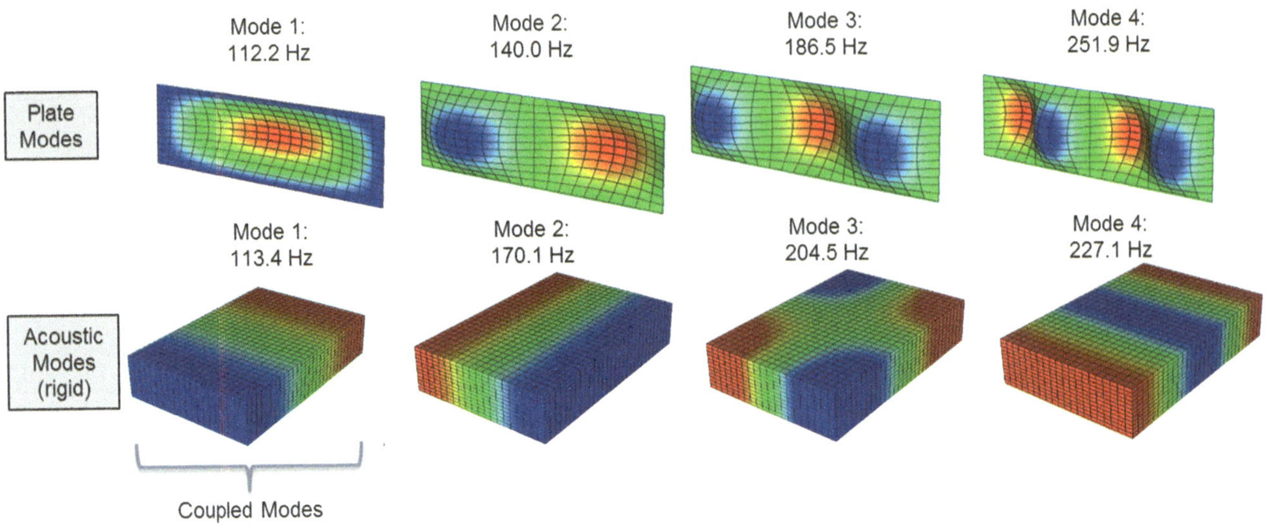

Fig. 16.3 Component modes of the plate (*top*) and rigid-walled box (*bottom*)

$$p\left(x, y, z, n_x, n_y, n_z\right) = A_{scale}\sin\left(\frac{(2n_x + 1)\,\pi x}{2L_x}\right)\cos\left(\frac{n_y \pi y}{L_y}\right)\cos\left(\frac{n_z \pi z}{L_z}\right) \tag{16.13}$$

$$f_{n_x,n_y,n_z} = \frac{c_0}{2\pi}\sqrt{\left(\frac{(2n_x + 1)\,\pi}{2L_x}\right)^2 + \left(\frac{n_y \pi}{L_y}\right)^2 + \left(\frac{n_z \pi}{L_z}\right)^2} \tag{16.14}$$

where $p(x, y, z, n_x, n_y, n_z)$ is the pressure mode shape as a function of x, y, z space and n_x, n_y, n_z mode orders. The box dimensions are L_x by L_y by L_z. The shape scaling, A_{scale} is 1.0 to unit scale the shapes. f_{n_x,n_y,n_z} is the mode natural frequency and c_0 is the acoustic sound speed. The first mode of the box is order (1,0,0), meaning it is a quarter sine in X and constant in Y and Z.

At the wetted surface, the plate DOF are constrained to the acoustic DOF in terms of displacements, not pressures. Thus, the pressure mode shapes had to be converted to a displacement-like quantity. Acoustic particle velocity is proportional to the gradient of pressure [10]:

$$\vec{u} \propto \vec{\nabla} p \tag{16.15}$$

This gradient was computed numerically by generating the pressure shape in three dimensions and then computing the gradient of that grid of shape values. The acoustic modes need to be mass normalized to be useful in a CMS process so a modal mass was determined using a volume integral of the square of the acoustic pressure gradient mode shape:

$$m_j = \rho_0 \int_V \vec{\nabla}\Phi_{\mathrm{p,j}} \cdot \vec{\nabla}\Phi_{\mathrm{p,j}} dV \tag{16.16}$$

16.3.5 Variations of Component Modes

The acoustic modes were computed at the fine mesh grid shown above. However, this is an impractically large number of acoustic DOF so subsets of these modes were also considered. Next, only the acoustic DOF at the wetted surface were retained in both the coupled system (C) and acoustic component (B) modes. Finally, the acoustic DOF were removed entirely from the C coupled system DOF, leaving only the plate DOF to be constrained to the B component acoustic DOF. Here, there are no B component DOF in the C coupled modes, and so the pseudo-inverse of the partitioned C shapes used to form the modal constraint matrix have to be replaced by the A DOF in the C shapes. This issue may cause some of the errors in the extracted B component modes observed in Fig. 16.7b.

16.3.6 Numerical Simulations of Uncoupled and Coupled Systems

Direct frequency response simulations were performed using the plate-only and plate and box coupled system model using Sierra/SD. These frequency response function (FRF) curves will be the basis of comparison for the response predictions made using the CMS coupled and decoupled systems.

A node was chosen for the input DOF and another for the output DOF, shown below. Then, a unit force linear spectrum was input over a broad frequency range and the response extracted at the output DOF. As this is a direct simulation, proportional damping was used. 0.5% damping was desired for the plate and 1.0% for the air. The proportional damping terms were determined to achieve nearly flat damping in the range of 50–500 Hz, but there is some overshoot in the damping at high frequencies.

16.4 Results

16.4.1 Direct Simulation for In-Vacuo Structural Component and Coupled System Response

FRFs from direct simulation using Sierra/SD for the plate-only and the coupled plate-box system are shown below. As expected, there is strong evidence of acoustoelastic coupling near the first mode of the plate. As these shapes are compatible and the frequencies are very close, there is a dramatic split peak observed in the coupled system FRF. The CMS coupled system predictions will be compared against the coupled system FRF from this simulation and the CMS decoupled plate predictions will be compared against the plate FRF (Fig. 16.4).

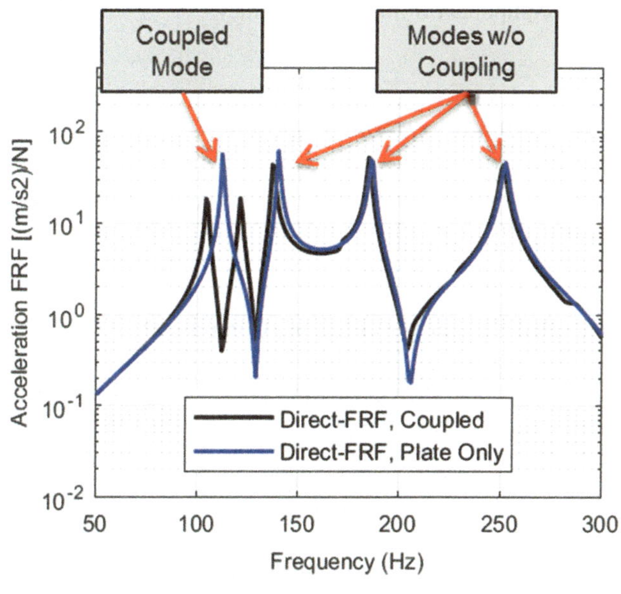

Fig. 16.4 FRFs from direct simulation of the plate component and the coupled plate-box system

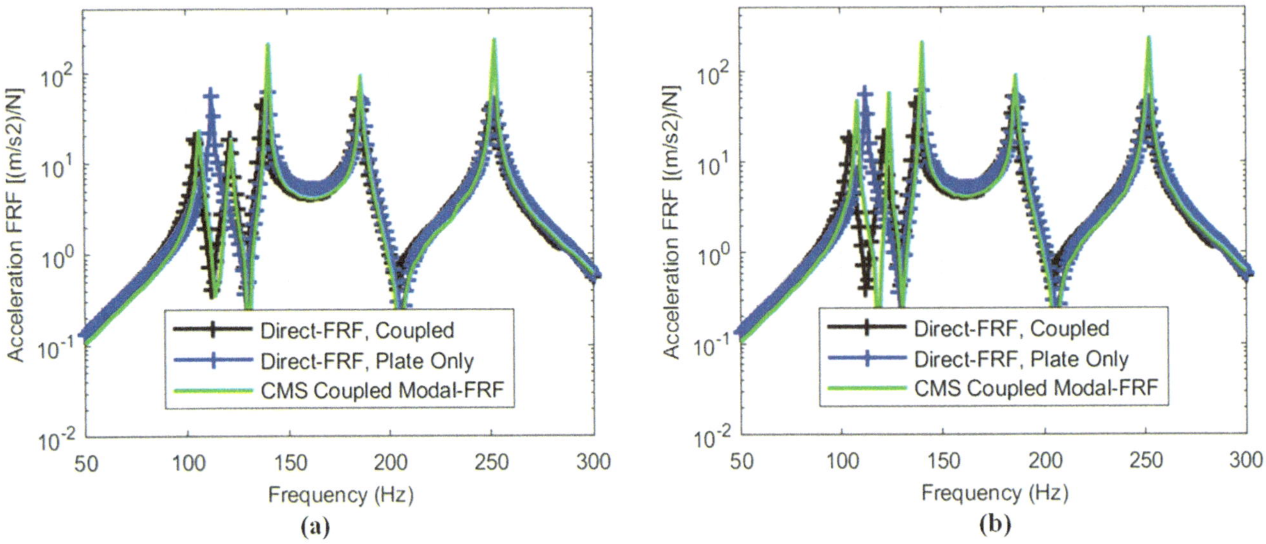

Fig. 16.5 FRFs of the CMS-coupled plate-box system compared with the direct simulation using many modes (50 structure, 100 acoustic) in the plate and box basis (**a**) and fewer modes (30 structure, 10 acoustic) in the basis (**b**)

16.4.2 Coupled Structural-Acoustic System Response

The FRFs for the CMS coupled system are compared with the coupled system response predicted with direct simulation in the plot below. The CMS-coupled system agrees very well with the direct FRF simulation, capturing the coupled modes effect well. There is some error in the peaks at the higher frequency modes, but this may be due to the damping models being different; in the CMS mode-based predictions, the CMS system FRF is computed with modal damping whereas the direct FRF simulations use proportional damping.

As expected for CMS problems, the number of modes in the components affects the results, as seen in Fig. 16.5b. Introducing more acoustic modes into component B improves accuracy.

16.4.3 Decoupled Structural-Acoustic System Response

Predictions of response for the decoupled plate response are shown below for various sets of basis modes and degrees of freedom in the coupled system modes.

16.4.3.1 All Acoustic DOF in Coupled System & Subtracted Component

Figure 16.6 shows the FRF for the coupled and then decoupled system using all the plate DOF and all the acoustic DOF in components C and B. Of course, it is unrealistic to obtain this DOF count using experimentally-derived structure and acoustic shapes. In the next subsections, cases with a reduced DOF count are considered.

16.4.3.2 Reduced Count Acoustic DOF in Coupled System & Subtracted Component

Next, shown in Fig. 16.7a, only the acoustic DOF on the wetted surface in components C and B are retained in the basis modes. This is a drastic reduction in the DOF count, from 8463 acoustic DOF to 273 acoustic DOF. The decoupling procedure is again successful at removing the coupled modes peak at mode 1. Good results were obtained using 30 basis modes for component C and 20 modes for B.

Entirely removing the acoustic DOF from the C component modes was also considered, Fig. 16.7b. Here, only the 273 displacement DOF were retained in the C modes and then the 273 acoustic DOF were retained in the B modes. This is a

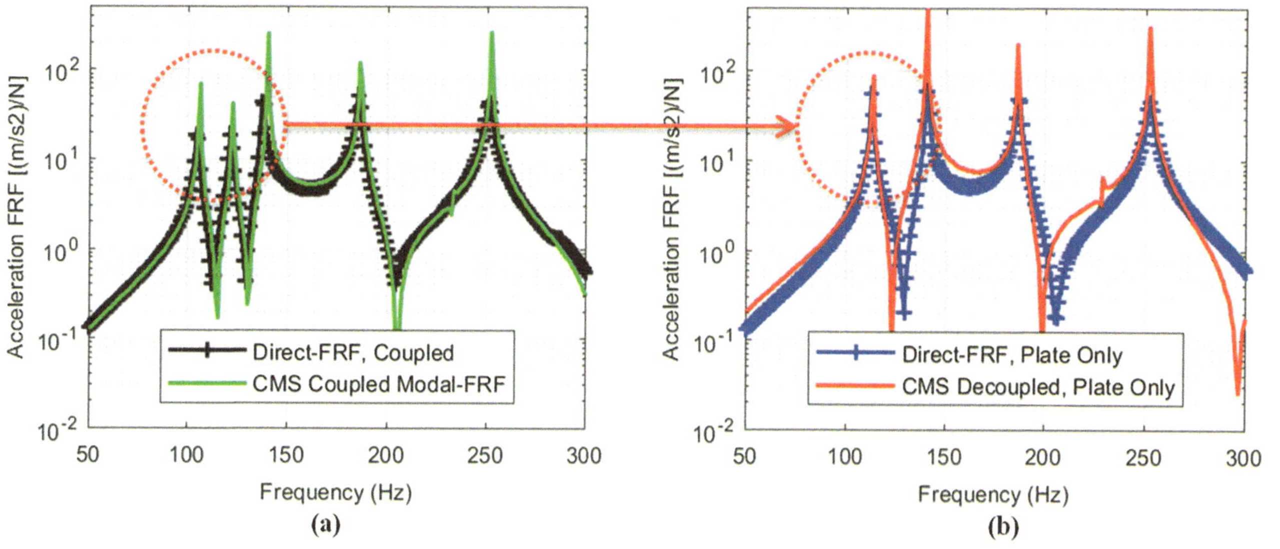

Fig. 16.6 FRFs of the CMS-coupled plate (**a**) which is then decoupled using the Transmission Simulator method (**b**)

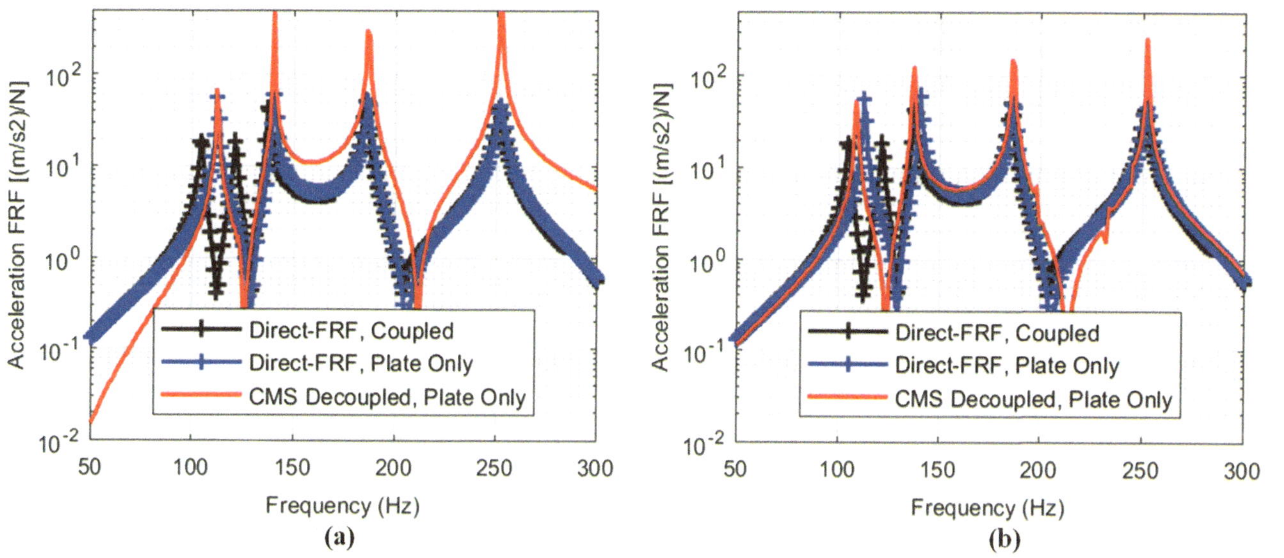

Fig. 16.7 FRFs of the decoupled CMS system using (**a**) only DOFs on the wetted surface (no interior acoustic DOFs in B) and (**b**) using only displacement DOFs in component C (no acoustic DOFs in C)

more realistic configuration as typically only structural response measurements (modes) would be made on a coupled system. Then, the acoustic modes for the B component could be obtained from a model. It was found that 20 C component modes and 5 B component modes provide good results. Curiously, increasing the number of modes in both components to 40 in C and 30 in B introduced inaccuracies. Typically in CMS problems there is monotonic convergence with increasing modes, which indicates that this process on this type of system is not robust.

16.5 Conclusions and Future Work

Traditional CMS methods were applied to a new type of system modeling, acoustoelastic systems. CMS was effective to couple together a structure and acoustic model, and the resulting coupled system response matched predictions from a coupled system direction frequency response simulation. As expected, the number and type of modes used in the component basis affects the accuracy of results. Next, the Transmission Simulator method was used to decouple a coupled system, subtracting off the acoustic component from an acoustoelastic system leaving the *in vacuo* structural response. Frequency response comparisons of a plate extracted from the coupled system compare reasonably well with response from direct simulation of a plate, but errors do exist and the results were quite sensitive to the choice of modes in the basis. Additionally, as is common for subtraction CMS problems several of the de-coupled system modes from the Transmission Simulator process are erroneous, either highly complex or degenerate with duplicate frequencies and near zero amplitude.

Further study is needed to determine, in general, what type of acoustic modes are required to sufficiently span the necessary space for this class of problems. Other acoustic domain boundary conditions at the wetted surface will be considered, for example impedance boundary conditions that lie between pressure release and rigid boundaries. Future work will focus on how to apply this technique to the ultimate goal: removing acoustic coupling effects from structural response measurements. This will require better understanding of how few structure and acoustic DOF are needed to provide component modes. Using simulation to determine coupled system modes would also be useful to validate these methods, provided the coupled system mode results are reliable.

References

1. Dowell, E.H., Gorman III, G.F., Smith, D.A.: Acoustoelasticity: general theory, acoustic natural modes and forced response to sinusoidal excitation, including comparisons with experiment. J. Sound Vib. **52**(4), 519–542 (1977)
2. Schultz, R., Pacini, B.R.: "Mitigation of Structural-Acoustic Mode Coupling in a Modal Test of a Hollow Structure". In: IMAC-XXXV International Modal Analysis Conference, Garden Grove, CA (2017)
3. de Klerk, D., Rixen, D.J., Voormeeren, S.N.: General framework for dynamic substructuring: history, review,and classification of techniques. AIAA J. **46**(5), 1169–1180 (2008)
4. Roettgen, D.R.: "Experimental Dynamic Substructuring Using Nonlinear Modal Joint Models," Doctoral Dissertation, University of Wisconsin – Madison, Madison, WI (2016)
5. Mayes, R.L., Arviso, M.: "Design studies for the transmission simulator method of experimental dynamic substructuring." In: International Seminar on Modal Analysis (ISMA2010), Lueven, Belgium (2010)
6. Hambrick, S.A., Sung, S.H., Nefske, D.J.: Engineering Vibroacoustic Analysis: Methods and Applications. Wiley, Chichester, West Sussex, United Kingdom (2016)
7. Pretlove, A.J.: Free vibrations of a rectangular panel backed by a closed rectangular cavity. J. Sound Vib. **2**(3), 197–209 (1965)
8. Sierra Structural Dyanmics Development Team. "Sierra Structural Dynamics – User's Notes," Sandia National Laboratories SAND Report (2015)
9. Blackstock, D.T.: Fundamentals of Physical Acoustics. John Wiley & Sons, Austin, TX (2000)
10. Pierce, A.D.: Acoustics, An Introduction to Its Physical Principles and Applications, 1981. Acoustical Society of America, Melville, NY (1994)

Chapter 17
Interface Reduction in Component Mode Synthesis of Bladed Disks by Orthogonal-Polynomial Series

Luigi Carassale, Andrea Bessone, and Andrea Cavicchi

Abstract The component mode synthesis based on the Craig-Bampton method has two strong limitations that appear when the number of the interface degrees of freedom is large. First, the reduced-order model obtained is overweighed by many unnecessary degrees of freedom. Second, the reduction step may become extremely time consuming. Several interface reduction techniques addressed successfully the former problem, while the latter remains open. In this paper we tackle this latter problem through a simple interface-reduction technique based on an a-priory choice of the interface modes. An efficient representation of the interface displacement field is achieved adopting a set of orthogonal basis functions determined by the interface geometry.

Keywords CMS · Craig-Bampton method · Interface reduction · Rotor blades

17.1 Introduction

The main drawback of the Craig-Bampton (CB) method [1, 2] when applied to the modelling of rotor blades is in the very large number of constraint modes due to the very refined mesh that is typically employed in these applications [3]. The large number of constraint modes has two serious consequences: (1) the order of the Reduced-Order Models (ROM) obtained by the CB method is much larger than the order of a ROM derived directly from the full structure (e.g. using its global normal modes) due to the presence of numerous interface degrees of freedom (DOF); (2) the calculation of the constraint modes is very time consuming as it involves the solution of a number of static analyses equal to the number of interface DOFs (usually in the order of a few thousands for a typical model having 10^5–10^6 internal DOFs).

Several authors proposed methods to reduce the number of interface DOFs formulating numerous numerical approaches called, collectively, Interface Reduction (IR) techniques. In some approaches the interface DOFs are reduced through a secondary modal analysis [4–7], while other authors proposed the extraction of an orthogonal basis from the normal modes of a full or partial model of the structure [8–10]. Both these approaches solve brilliantly the issue of the ROM large dimension, but require the time-consuming computation of the constraint modes or the modal analysis of the full model. In this paper we try to solve this latter problem by representing the interface DOFs through a set of a-priori basis functions, whose shape is determined by the interface geometry. These functions are derived using an ortogonalization algorithm formulated using the Finite Element (FE) method. The method is applied to the reduction and assemblage of a rotor blade composed by two substructures.

17.2 Interface Reduction in Craig-Bampton Method

According to the CB method, the DOFs of a substructure s are divided into a set of internal DOFs $\mathbf{q}_B^{(s)}$ and a set of interface DOFs $\mathbf{q}_\Gamma^{(s)}$. The full displacement vector is then reconstructed by the expression [2]:

L. Carassale (✉)
Department of Mechanical, Energy, Management and Transportation Engineering, University of Genova, Genoa, Italy
e-mail: luigi.carassale@unige.it

A. Bessone · A. Cavicchi
Ansaldo Energia, Genoa, Italy
e-mail: andrea.bessone@ansaldoenergia.com; andrea.cavicchi@ansaldoenergia.com

A. Linderholt et al. (eds.), *Dynamics of Coupled Structures, Volume 4*, Conference Proceedings of the Society for Experimental Mechanics Series, https://doi.org/10.1007/978-3-319-74654-8_17

$$\mathbf{q}^{(s)} = \begin{bmatrix} \mathbf{q}_B^{(s)} \\ \mathbf{q}_\Gamma^{(s)} \end{bmatrix} = \begin{bmatrix} \mathbf{\Phi}^{(s)} & \mathbf{\Psi}^{(s)} \\ \mathbf{0} & \mathbf{I} \end{bmatrix} \begin{bmatrix} \mathbf{p}^{(s)} \\ \mathbf{q}_\Gamma^{(s)} \end{bmatrix} \tag{17.1}$$

where $\mathbf{I}$ and $\mathbf{0}$ are, respectively, the identity and the zero matrix (of suitable size); the matrix $\mathbf{\Phi}^{(s)}$ contains the first n_s fixed-interface normal modes of the substructure s; the matrix $\mathbf{\Psi}^{(s)}$ contains the constraint modes representing the displacement of the internal DOFs due to a unit displacement of the interface DOFs.

Several researchers proposed techniques to reduce the number of interface DOFs [7–10]. These techniques follow different strategies and principles, however, most of them can be casted into a common framework. The interface DOFs of the substructure s are approximated through a linear combination of vectors $\boldsymbol{\theta}_k = \mathbb{R}^{N_\Gamma}$ $(k = 1, \ldots, m)$ as:

$$\mathbf{q}_\Gamma^{(s)} = \mathbf{\Theta}^{(s)} \mathbf{z}^{(s)} \tag{17.2}$$

where $\mathbf{\Theta}^{(s)}$ is a matrix of size $N_\Gamma \times m$ containing the vectors $\boldsymbol{\theta}_k$ referred to as interface modes, and $\mathbf{z}^{(s)} \in \mathbb{R}^m$ is a vector of generalized coordinates. The difference between different reduction techniques is essentially in the choice of $\mathbf{\Theta}^{(s)}$.

The Orthogonal Polynomial Series (OPS) method assumes that the interface connecting the two substructures a and b can be described by a region $\mathcal{D}$ belonging to a 2D surface regular enough to be parametrized by a curvilinear reference system. Then, it also assumes that all the interface nodes of both the substructures lay on $\mathcal{D}$, so that the jth node of the interface for substructure s can be identified by the coordinates $x_j^{(s)}$ and $y_j^{(s)}$. Finally, the interface modes for the substructure s are obtained by sampling a set of scalar function $f_k : \mathcal{D} \to \mathbb{R}$ $(k = 1, \ldots, m)$:

$$\Theta_{jk}^{(s)} = f_k \left(x_j^{(s)}, y_j^{(s)} \right) \quad \begin{pmatrix} j = 1, \ldots N_\Gamma^{(s)} \\ k = 1, \ldots m \end{pmatrix} \tag{17.3}$$

where $\Theta_{jk}^{(s)}$ is the j,k-entry of the reduction matrix $\mathbf{\Theta}^{(s)}$.

In order to obtain a fast convergence of the interface displacement representation, it is best to select the functions f_k as mutually orthogonal, so that each interface mode added to the representation delivers fresh information without overwriting the existing one. To impose this condition, it is necessary to establish a metrics. This is done by assuming that $f_k \in L^2(\mathcal{D})$ and defining the inner product:

$$\langle f, g \rangle = \int_{\mathcal{D}} f(x, y)\, g(x, y)\, \mathrm{d}x\mathrm{d}y \tag{17.4}$$

with $f, g \in L^2(\mathcal{D})$. To construct the basis function, we start from the set of polynomial functions in $\mathcal{D}$ and apply the Gram-Schmidt orthogonalization procedure leading to a set of orthogonal polynomials. To compute explicitly the basis functions f_k, it is necessary to implement numerically the inner product defined by Eq. (17.4). To this purpose, we use the FE method applied to a 2D mesh $\mathcal{D}$. For convenience, we may construct this 2D mesh using the same nodes and topology of the 3D structural mesh emerging on $\mathcal{D}$.

The OPS method is applied to the reduction of a rotor blade whose airfoil (substructure a) and root (substructure b) are modelled separately (Fig. 17.1). The airfoil has $N^{(a)} \approx 150{,}000$ DOFs and the root $N^{(b)} \approx 300{,}000$ DOFs. The interface has compatible mesh on the two substructures with $N_\Gamma \approx 2000$.

Figure 17.2 shows the relative error in the calculation of the natural frequency as a function of the number of interface modes retained in the interface reduction. The first 10 modes are used for the radial component of the displacement, modes 11 to 20 are for the tangential component and modes 21 to 30 are for the axial component. The convergence of the Characteristic Constraint (CC) Mode method [4, 5] is extremely fast for all the considered natural frequencies, and also the convergence velocity of the Local Interface Reduction (LIR) method [7] is excellent. As far as the OPS is concerned, we note that the convergence velocity is slower than the other methods, but is still very good.

The advantage of OPS compared to other methods is in the very important saving of the computational time required for model reduction.

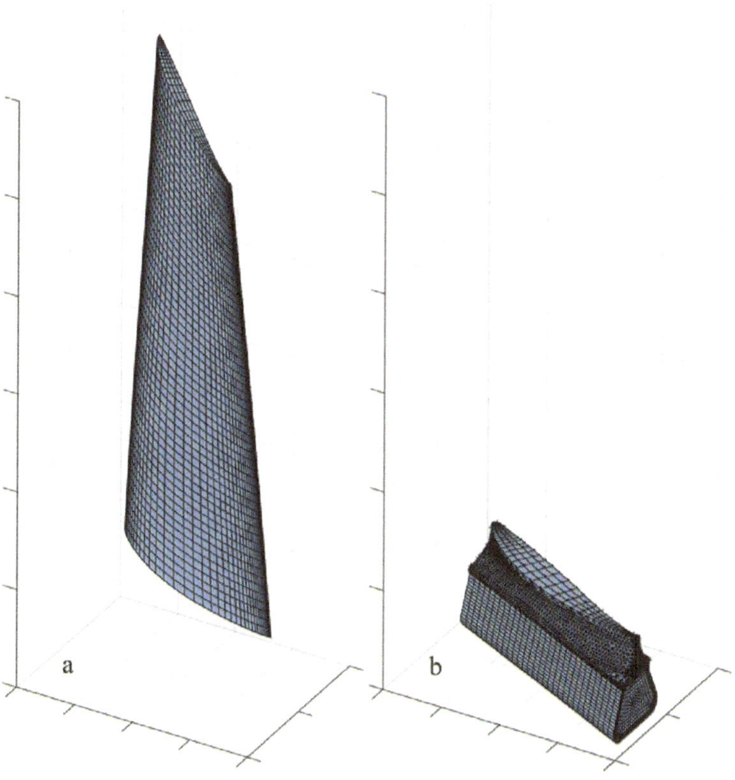

Fig. 17.1 Mesh of airfoil (**a**) and root (**b**)

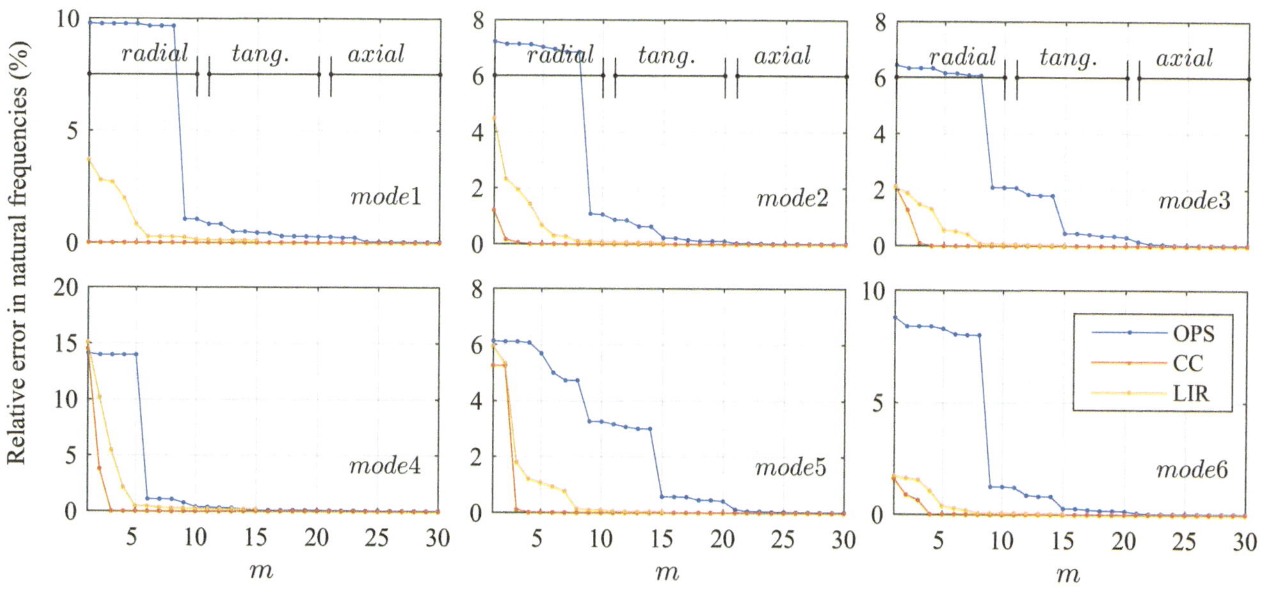

Fig. 17.2 Error in the natural frequency with respect to the classical CB method as a function of the IR order. The separation among modes with radial, axial and tangential direction is pertinent only for the OPS method

References

1. de Klerk, D., Rixen, D.J., Voormeeren, S.N.: General framework for dynamic substructuring: history, review, and classification of techniques. AIAA J. **46**(5), 1169–1181 (2008)
2. Craig, R.R., Bampton, M.C.C.: Coupling of substructures for dynamic analyses. AIAA J. **6**, 1313–1319 (1968)
3. Bladh, R., Castanier, M.P., Pierre, C.: Component-mode-based reduced order modeling techniques for mistuned bladed disks – Part I: theoretical models. J. Eng. Gas Turb. Power. ASME. **123**, 89–99 (2001)
4. Castanier, M.P., Tan, Y.-C., Pierre, C.: Characteristic constraint modes for component mode synthesis. AIAA J. **39**, 1182–1187 (2001)
5. Brahmi, K., Bouhaddi, N., Fillod, R.: Reduction of junction degrees of freedom in certain methods of dynamic substructure synthesis. In: Proc. 13th International Modal Analysis Conference, February 13–16, 1995, Nashville, Tennessee, USA, pp. 1763–1769 (1995)
6. Tran, D.-M.: Component mode synthesis methods using interface modes. Application to structures with cyclic symmetry. Comput. Struct. **79**, 209–222 (2001)
7. Hong, S.-K., Epureanu, B.I., Castanier, M.P.: Next-generation parametric reduced-order models. Mech. Sys. Sig. Proc. **37**, 403–421 (2013)
8. Balmès, E.: Use of generalized interface degrees of freedom in component mode synthesis. In: Proc. 14th International Modal Analysis Conference, February 12–15, 1996, Dearborn, Michigan, USA, pp. 204–210 (1996)
9. Aoyama, Y., Yagawa, G.: Component mode synthesis for large-scale structural eigenanalysis. Comput. Struct. **79**, 605–615 (2001)
10. Donders, S., Pluymers, B., Ragnarsson, P., Hadjit, R., Desmet, W.: The wave-based substructuring approach for the efficient description of interface dynamics in substructuring. J. Sound Vib. **329**, 1062–1080 (2010)

Chapter 18
Frequency Based Substructuring with the Virtual Point Transformation, Flexible Interface Modes and a Transmission Simulator

E. A. Pasma, M. V. van der Seijs, S. W. B. Klaassen, and M. W. van der Kooij

Abstract In this contribution, Frequency Based Substructuring is applied to a well-known benchmark structure with special attention to correct modelling of the interface problem. Various approaches with the Virtual Point Transformation are used to incorporate the effect of interface flexibility that is present in the clamped connections. An extension of the rigid displacement basis with flexible interface displacement modes is proposed and evaluated. Furthermore, a practical approach using a Transmission Simulator is used in order to improve the quality of the interface description.

Keywords Experimental modelling · Virtual point transformation · Interface reduction · Dynamic substructuring · Transmission simulator

18.1 Introduction

Dynamic Substructuring (DS) has proven to be an effective technique to gain insight into the behaviour of a system by coupling models of its sub-components or substructures. DS uses clear interfaces which allows for separation and optimization on a component level, which in practice is beneficial for systems composed by components from different suppliers or departments. Although DS is domain-independent, it is most often applied in the frequency domain, because of the ease of using FRF's to express dynamics and the possibility to use experimentally obtained data. Frequency based substructuring (FBS) is now one of the more popular methods which slowly gets adopted by industry.

Despite the mathematics behind DS being straightforward (see for example [1]) it is still challenging to implement DS techniques in practice. One of the main challenges presents itself in the coupling of the different models, especially when one of the models used is obtained experimentally. Experimental models are gaining popularity due to an increasing interest in higher frequency ranges (>1 kHz) where numerical models oftentimes cannot provide reliable results. Trends such as rapid prototyping, modular design and the transition from combustion engines to electric drive lines also fuel the interest in experimental modelling [2].

Hybrid modelling techniques give engineers more freedom in selecting the most appropriate method for a specific case. One can either mix numerical and experimental data to obtain improved component models (using for example modal updating), or one can describe a complete system using an assembly of both experimental and numerical component models. A key challenge in the latter is to properly describe the interfaces that connect the component models in a effective and unambiguous manner. A recent method to describe the interface of an (experimentally obtained) model is the Virtual Point Transformation, as introduced in [3]. The Virtual Point Transformation reduces the measured dynamics at the interfaces to 6-DoF nodes. These virtual nodes are inherently compatible with nodes in FE models and allow to couple both translational and rotational dynamics.

The goal of this paper is to explore multiple ways to model the interface problem. A comparison is made between the standard Virtual Point Transformation, an extension with flexible interface displacement modes (IDMs) and an approach using the Transmission Simulator [4]. The Virtual Point Transformation is extended in various ways to account for additional

E. A. Pasma (✉) · M. V. van der Seijs · M. W. van der Kooij
VIBES.technology, Delft, The Netherlands

Department of Precision and Microsystems Engineering, Delft University of Technology, Delft, The Netherlands
e-mail: epasma@vibestechnology.com

S. W. B. Klaassen
Faculty of Mechanical Engineering, Institute of Applied Mechanics, Technische Universität München, Garching, Germany

Department of Precision and Microsystems Engineering, Delft University of Technology, Delft, The Netherlands

A. Linderholt et al. (eds.), *Dynamics of Coupled Structures, Volume 4*, Conference Proceedings of the Society for Experimental Mechanics Series, https://doi.org/10.1007/978-3-319-74654-8_18

local flexibility not captured by a single 6-DOF node. The usage and effectiveness of these techniques is compared and validated using a well-known benchmark structure [5]. In Sect. 18.2 the theory of the standard Virtual Point Transformation is covered and an extension to flexible IDMs is proposed. Section 18.3 discusses the application of the different interface modelling methods, and describes the test setup, the methods and the validation results.

18.2 Theory

The Virtual Point Transformation [3] is a method to create experimental 'super-elements' from impact or shaker measurements on a structure, much like super-elements from FE models. The main idea is to choose a point on or near a physical interface of a substructure that one wants to make compatible with other (experimental or numerical) substructures to couple. To do so, the dynamics of such point are said to be described by a fixed number of DoFs m, which for instance represent the XYZ translations and rotations about the XYZ axes in the global coordinate frame. The standard form of the virtual point transformation indeed proposes $m = 6$ DoFs per virtual point, suitable to describe the rigid-body displacements and loads at the interface point by virtual translations & rotations and virtual forces & moments. In an extended form, m can be larger than six to also accommodate for non-rigid behaviour of the interface, as will be described further on.

18.2.1 Virtual Point Transformation

All $n_u > m$ measured displacements $\mathbf{u}$ around an interface can be transformed to the virtual point by means of a kinematic relation between the measured DoFs $\mathbf{u}$ and the virtual DoFs, here denoted by $\mathbf{q}$. This relation is governed by the so-called Interface Displacement Mode (IDM) matrix $\mathbf{R_u}$. A similar relation $\mathbf{R_f}$ can be set up for the $n_f > m$ forces $\mathbf{f}$ and virtual point forces/moments, denoted by $\mathbf{m}$. This gives rise to the following transformations [2, 3]:

$$\text{Displacements}: \quad \mathbf{u} = \mathbf{R_u q} \quad \Longrightarrow \quad \mathbf{q} = (\mathbf{R_u})^+ \mathbf{u} \qquad \mathbf{R_u} \in \mathbb{R}^{n_u \times m} \tag{18.1a}$$

$$\text{Forces}: \quad \mathbf{m} = \mathbf{R_f}^T \mathbf{f} \quad \Longrightarrow \quad \mathbf{f} = \left(\mathbf{R_f}^T\right)^+ \mathbf{m} \qquad \mathbf{R_f} \in \mathbb{R}^{n_f \times m} \tag{18.1b}$$

The inverted IDM matrices $(\mathbf{R_u})^+$ and $(\mathbf{R_f})^+$ are the actual transformation matrices that convert a $n_u \times n_f$ measured FRF matrix $\mathbf{Y}(\omega)$ to a $m \times m$ virtual point FRF matrix $\mathbf{Y_{qm}}(\omega)$.

$$\text{Measured FRFs}: \quad \mathbf{u} = \mathbf{Yf} \tag{18.2a}$$

$$\text{Virtual point FRFs}: \quad \mathbf{q} = (\mathbf{R_u})^+ \mathbf{Y} \left(\mathbf{R_f}^T\right)^+ \mathbf{m} \quad \Longrightarrow \quad \mathbf{q} = \mathbf{Y_{qm} m} \tag{18.2b}$$

The experimental super-element is in fact governed by this $\mathbf{Y_{qm}}(\omega)$ matrix, as it pertains perfectly collocated force and displacement DoF with correct driving-point behaviour. Also note that, as the two IDM matrices $\mathbf{R_u}$ and $\mathbf{R_f}$ may be different, it is not necessary to excite the structure at the same points as the sensors are located.

18.2.2 Rigid IDMs

The essence of the virtual point transformation lies in the construction of the IDM matrices $\mathbf{R}$ that determine how n measured DoFs around an interface can be expressed by m virtual point DoFs. For the rows of $\mathbf{R}$, one must know the location and orientation of all measured DoF relative to the location and orientation of the virtual point. The columns of $\mathbf{R}$ consequently define which displacement modes are used to express the displacements of physical DoFs around the interface.

The following kinematic relation can be established between the virtual point DoFs $\mathbf{q}^v$ and the rigid interface displacements $\mathbf{u}^k$ of a tri-axial sensor k, already rotated to the global coordinate frame XYZ. Vector $\mathbf{r}^k$ holds the position of the sensor relative to the location of the virtual point.

$$\begin{bmatrix} u_X^k \\ u_Y^k \\ u_Z^k \end{bmatrix} = \left[\begin{array}{ccc|ccc} 1 & 0 & 0 & 0 & r_Z^k & -r_Y^k \\ 0 & 1 & 0 & -r_Z^k & 0 & r_X^k \\ 0 & 0 & 1 & r_Y^k & -r_X^k & 0 \end{array}\right] \mathbf{q}^v + \begin{bmatrix} \mu_X^k \\ \mu_Y^k \\ \mu_Z^k \end{bmatrix} \tag{18.3}$$

This relation states how much displacement is measured on the channels of sensor k for a given displacement and rotation of the corresponding virtual point v. Any residual displacement ends up in $\boldsymbol{\mu}$, which in this form also includes displacement due to flexibility of the interface region. Relation (18.3) can be established for each sensor surrounding the virtual point in order to add more rows to the system of equations. This renders the matrix full rank, such that it can be inverted according to (18.1a). Relations for the force transformations are set up in a similar way, which is more thoroughly described in [2, 3].

18.2.3 Flexible IDMs

In order to accommodate for flexibility of the interface, additional modes must be added to the IDM matrices $\mathbf{R}$, namely as the $(6+n)$th columns next to the rigid IDM columns. A 'top-down' approach to do so, is to inspect the measured interface dynamics, for instance the left singular vectors $\mathbf{U}$ of the decomposition $\mathbf{U}\boldsymbol{\Sigma}\mathbf{V}^T = \mathbf{Y}$, and use these to define one or more vectors orthogonal to the space already spanned by the six rigid vectors of $\mathbf{R}$. However, as effective as this may be for a single substructure, one can never guarantee that these modes are present in the substructure to couple. Therefore it is suggested to use a 'bottom-up' definition of flexible interface behaviour, which doesn't require a priori knowledge of the dynamics of the component at hand, and is therefore generally applicable. The following extended IDM matrix adds 6 flexible modes, namely three extension or 'strain' modes in XYZ and three torsion modes about the XYZ axes:

$$\begin{bmatrix} u_X^k \\ u_Y^k \\ u_Z^k \end{bmatrix} = \left[\begin{array}{ccc|ccc|ccc|ccc} 1 & 0 & 0 & 0 & r_Z^k & -r_Y^k & r_X^k & 0 & 0 & 0 & r_Y^k r_Z^k & -r_Z^k r_Y^k \\ 0 & 1 & 0 & -r_Z^k & 0 & r_X^k & 0 & r_Y^k & 0 & -r_X^k r_Z^k & 0 & r_Z^k r_X^k \\ 0 & 0 & 1 & r_Y^k & -r_X^k & 0 & \underbrace{0 \;\; 0 \;\; r_Z^k}_{\text{Extension}} & & & \underbrace{r_X^k r_Y^k \;\; -r_Y^k r_X^k \;\; 0}_{\text{Torsion}} & & \end{array}\right] \begin{bmatrix} \mathbf{q}_{\text{tr.}}^v \\ \mathbf{q}_{\text{rot.}}^v \\ \mathbf{q}_{\text{ext.}}^v \\ \mathbf{q}_{\text{tor.}}^v \end{bmatrix} + \begin{bmatrix} \mu_X^k \\ \mu_Y^k \\ \mu_Z^k \end{bmatrix} \tag{18.4}$$

As can be observed, flexible IDMs can be defined with similar ease as rigid IDMs, using the elements in vector $\mathbf{r}^k$ as the only ingredients. Even more modes can be added using a similar reasoning to achieve compound flexibility modes, such as illustrated by the shapes in Fig. 18.1. However, it is important to always preserve a full-rank IDM matrix $\mathbf{R}$, which implies that adding additional modes is only effective if there is sufficient row rank in the measurement channels $\mathbf{u}$. In practice, this means that one needs to 'cherry-pick' the right IDM modes for the particular coupling case and generally avoid an abundance of interface modes in the coupling conditions. This will be illustrated in the following application examples.

18.3 Application

This section demonstrates practical applications of frequency based substructuring (FBS) using the virtual point transformation with six degrees of freedom, the virtual point transformation extended with flexible modes and making use of a transmission simulator [4]. For the test-cases, we will make use of structures that have been previously used for experimental substructuring and source characterisation method validation [5]. In summary, there are three structures: substructure A, substructure B and the transmission simulator TS; see Fig. 18.2. In this paper, focus lies on comparing different strategies for modelling the interface. Therefore, the following test cases are used:

Case 1: The virtual point transformation with six rigid interface modes (Fig. 18.3);
Case 2: The virtual point transformation extended with a flexible interface mode ("extension Z" in Fig. 18.1) (Fig. 18.4);
Case 3: A transmission simulator and a FBS decoupling step to remove the dynamics of the TS substructure (Fig. 18.5).

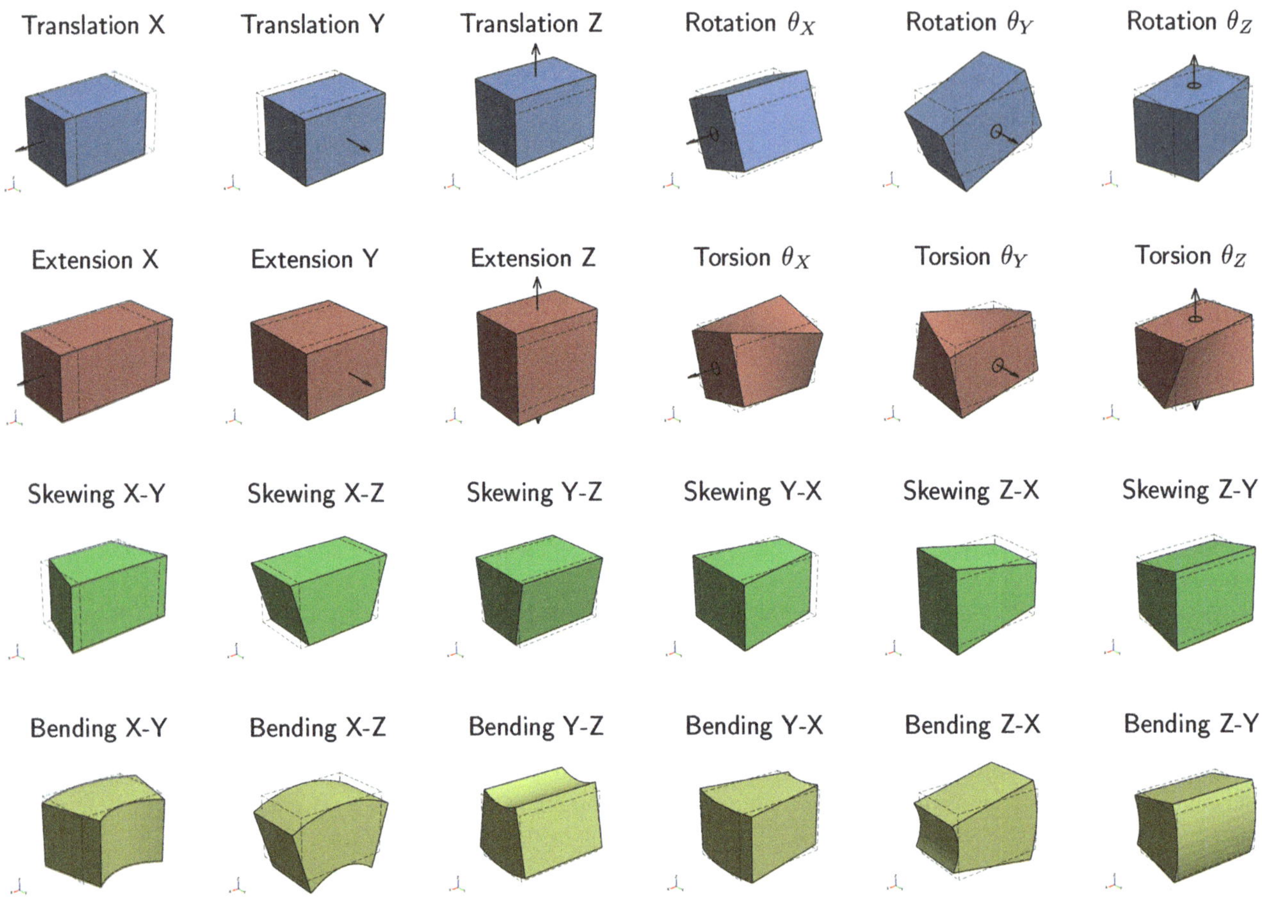

Fig. 18.1 The *interface displacement mode cookbook*: 6 rigid displacement modes (blue), 6 simple extension and torsion modes (red), 12 compound modes for skewing (green) and bending (yellow) behaviour

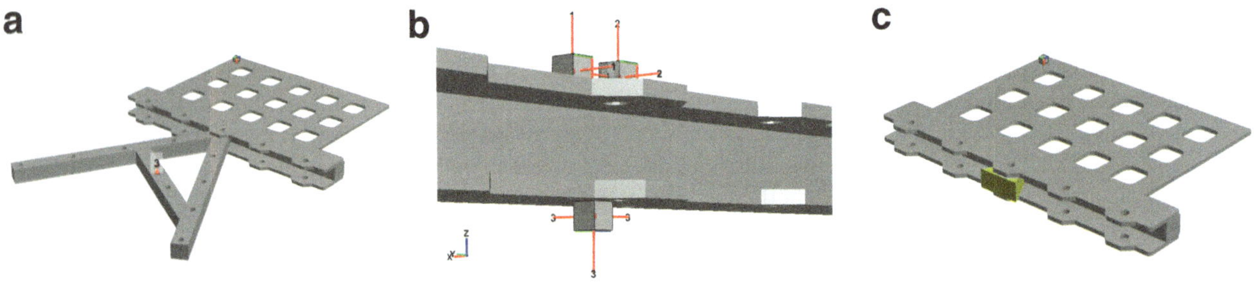

Fig. 18.2 The coupled structure AB with the validation with the excitation on structure A and the response sensor on structure B is shown in (**a**); the impact and response locations around the coupling point, the virtual point lies right between the two flanges (**b**); structure B with the transmission simulator (yellow) between the two flanges (**c**)

The admittance data is obtained numerically by synthesising the Frequency Response Functions using modal superposition. For the sake of simplicity, it is chosen to use collocated excitation and response locations. Note that this is not a requirement for the application of the virtual point transformation in practice; rather it is chosen here to eliminate sources of uncertainty. The locations and orientations of the excitations and responses are kept the same for every case, see Fig. 18.2b. The quality of the interface models is evaluated for the three test cases using criteria for sensor and impact consistency [3]. Here, only the quality indicators for structure B are shown, as this substructure has the more challenging interface. The comparison for substructure coupling is covered in Sect. 18.4.

18.3.1 Case 1: The Virtual Point Transformation with Six Rigid Interface Modes

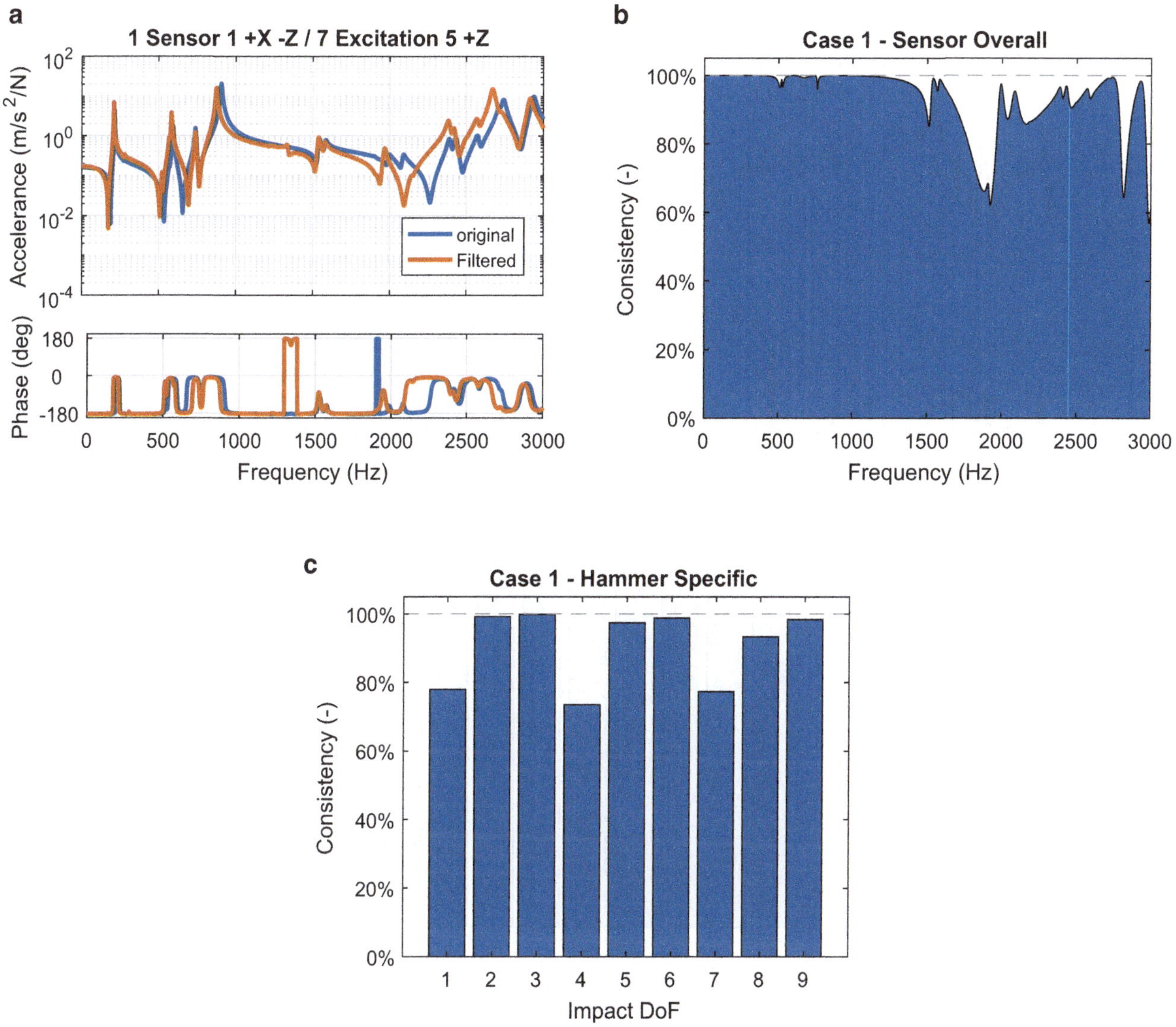

Fig. 18.3 Modelling the interface using the virtual point transformation with six rigid interface modes. (**a**) A comparison of a typical FRF and a filtered FRF; (**b**) the overall consistency of the sensors; (**c**) the consistency of the impacts

18.3.2 Case 2: The Virtual Point Transformation Extended with a Flexible Extension Mode

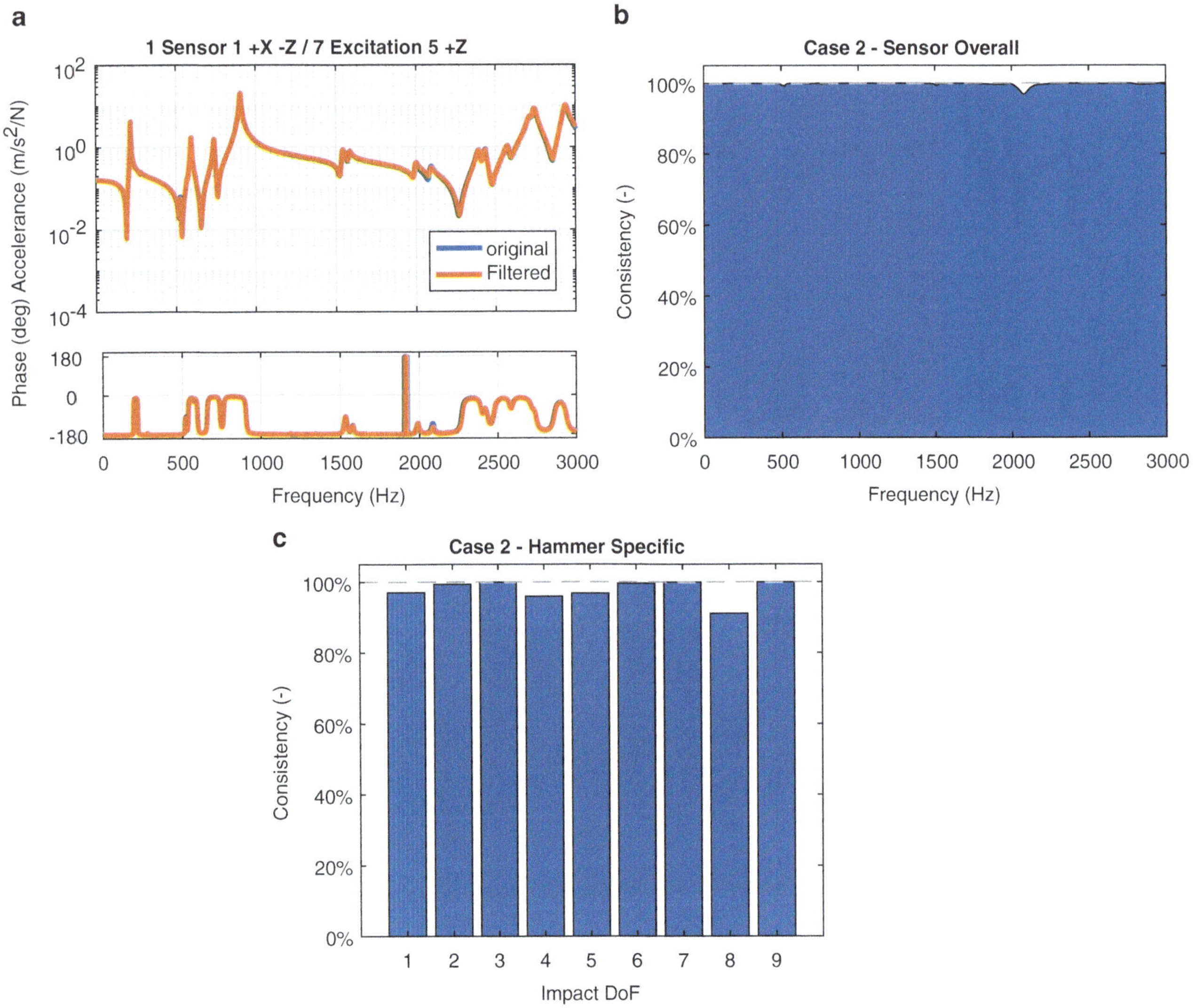

Fig. 18.4 Modelling the interface using the virtual point transformation with six rigid interface modes extended with an extension mode (Z-direction). (**a**) A comparison of a typical FRF and a filtered FRF; (**b**) the overall consistency of the sensors; (**c**) the consistency of the impacts

18.3.3 Case 3: Transmission Simulator

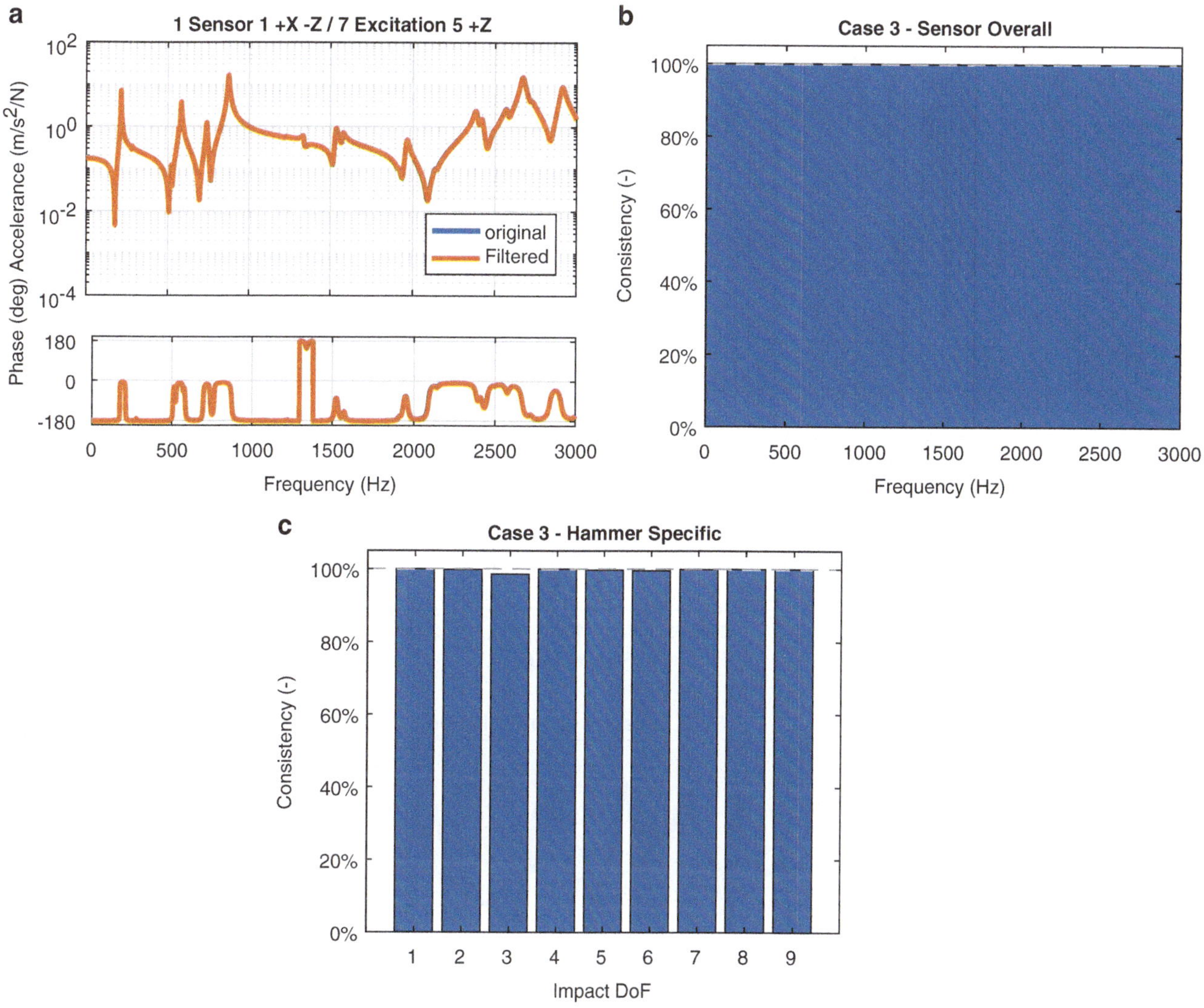

Fig. 18.5 Modelling the interface using the virtual point transformation on the substructure B with a transmission simulator. (**a**) A comparison of a typical FRF and a filtered FRF; (**b**) the overall consistency of the sensors; (**c**) the consistency of the impacts

18.4 Validation

Now all structures have been experimentally modelled, the admittance for a coupled system $\mathbf{Y}^{\mathrm{AB}}$ can be derived. The FRF matrices $\mathbf{Y}^{\mathrm{A}}$, $\mathbf{Y}^{\mathrm{B}}$ and (for case 3) $\mathbf{Y}^{\mathrm{TS}}$ are put in block-diagonal form and coupled using the LM-FBS algorithm [1, 6], making use of an appropriate Boolean localisation matrix $\mathbf{B}$ (not discussed here) to match the shared interface degrees of freedom:

$$\mathbf{Y}_{\mathrm{coupled}} = \mathbf{Y} - \mathbf{Y}\mathbf{B}^{T}\left(\mathbf{B}\mathbf{Y}\mathbf{B}^{T}\right)^{-1}\mathbf{B}\mathbf{Y} \qquad \mathbf{Y} \triangleq \begin{bmatrix} \mathbf{Y}^{\mathrm{A}} & \mathbf{0} & \mathbf{0} \\ \mathbf{0} & \mathbf{Y}^{\mathrm{B}} & \mathbf{0} \\ \mathbf{0} & \mathbf{0} & -\mathbf{Y}^{\mathrm{TS}} \end{bmatrix} \tag{18.5}$$

As validation, so called transfer dynamics will be considered, hence from an excitation point on structure A to a response point on structure B, see Fig. 18.2. Results of the substructured FRFs of AB are shown in Fig. 18.6. In the left Fig. 18.6a, a

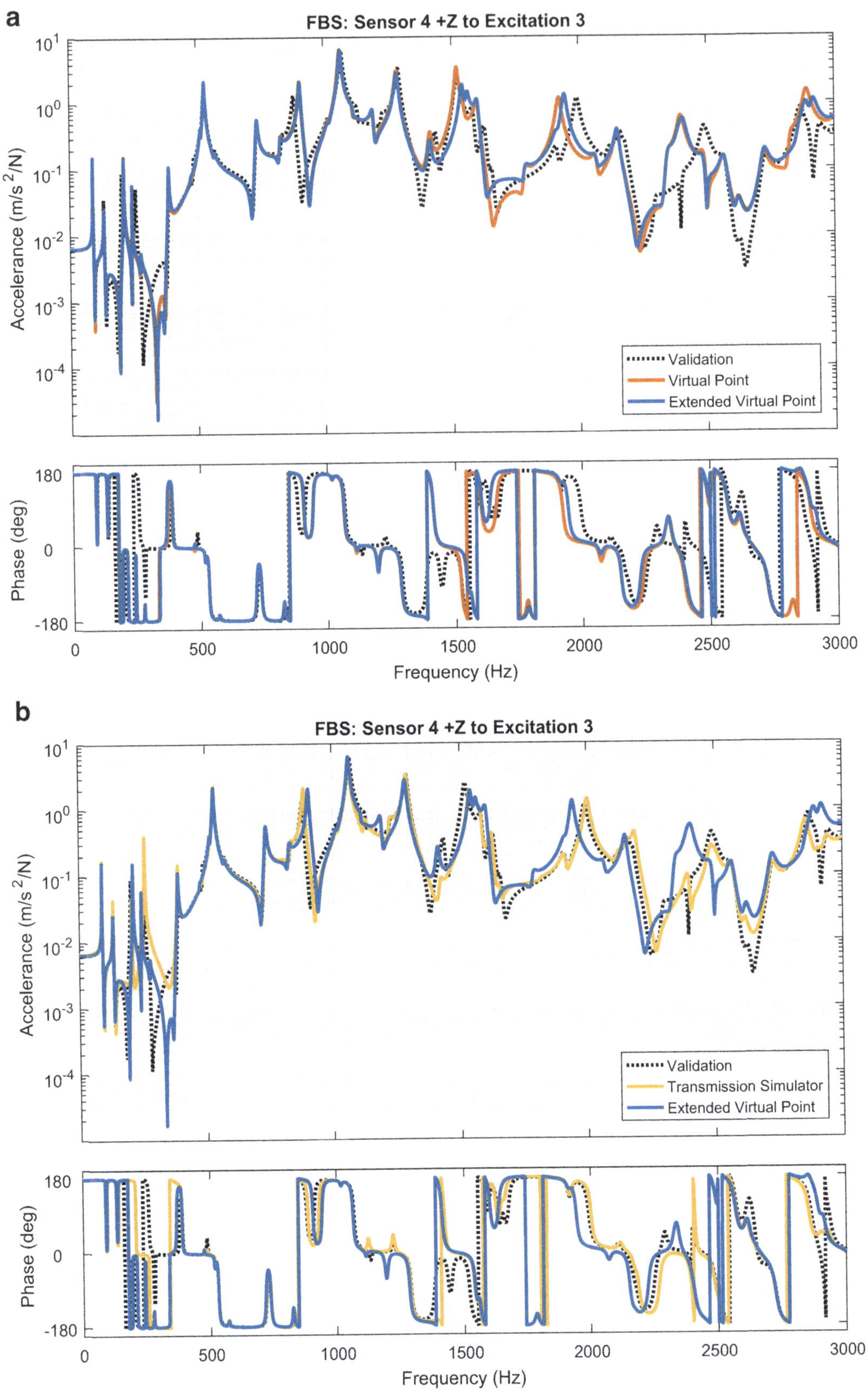

Fig. 18.6 The transfer FRFs of the coupled structures A and B compared with a validation FRF. (**a**) The virtual point FRF model with 6 degrees of freedom, compared with the 7 DoF extended virtual point FRF model; (**b**) The 7 DoF extended virtual point FRF model compared with the coupled A and B using a transmission simulator TS

transfer FRF with the phase of the coupled structures of case 1 and case 2 are shown for an internal excitation on structure A to the response sensor in Z-direction on structure B. It can be observed that both interface modelling strategies result in FRFs that match the validation FRF fairly well. Around 300, 2000 and 2500 Hz it can be observed that the coupled structures are less stiff than the validation structure. The effect of the flexible extension mode in case 3 can be observed around 1500 Hz, where the coupling is slightly improved. In Fig. 18.6b the transfer FRFs for the case 2 and case 3 are shown. Apart from the coupling around 300 Hz, the transmission simulator out-performs the coupling of case 2.

18.5 Conclusions

Three strategies to model the interfaces of connecting structures for the application of frequency based substructuring are shown. It has been shown that extending the 'classic' virtual point with flexible modes boosts the consistency of the separate structures, but only slightly improves the coupling for the presented case. The transmission simulator seems to be the most suitable approach for this case. This can be explained by the fact that the dynamics around the coupling point are close to the coupled configuration. Because of its low weight and high stiffness, the transmission simulator only marginally effects the 'global' dynamics of structure B, which is again beneficial for this coupling case. Furthermore, it has to be noted that in the validation model the coupling is probably too stiff compared to reality, as all coinciding DoFs of structure A and B are coupled in the numerical assembly. Therefore, this can be considered as a (upper) limit case of interface coupling for validation purposes.

References

1. de Klerk, D., Rixen, D.J., Voormeeren, S.N.: General framework for dynamic substructuring: history, review and classifcation of techniques. AIAA J. **46**(8), 1169–1181 (2008). http://dx.doi.org/10.2514/1.33274
2. van der Seijs, M.V.: Experimental dynamic substructuring: analysis and design strategies for vehicle development. Ph.D. thesis, Delft University of Technology (2016). http://dx.doi.org/10.4233/uuid:28b31294-8d53-49eb-b108-284b63edf670
3. van der Seijs, M.V., van den Bosch, D.D., Rixen, D.J., de Klerk, D.: An improved methodology for the virtual point transformation of measured frequency response functions in dynamic substructuring. In: Papadrakakis, M., Papadopoulos, V., Plevris, V. (eds.) 4th ECCOMAS Thematic Conference on Computational Methods in Structural Dynamics and Earthquake Engineering (COMPDYN), Kos Island, pp. 4334–4347 (2013). http://dx.doi.org/10.7712/120113.4816.C1539
4. Mayes, R.: Tutorial on experimental dynamic substructuring using the transmission simulator method. In: Topics in Experimental Dynamics Substructuring and Wind Turbine Dynamics. Proceedings of the 30th IMAC, a Conference on Structural Dynamics, vol. 2, pp. 1–9. Springer (2012). http://dx.doi.org/10.1007/978-1-4614-2422-2_1
5. van der Seijs, M.V., Pasma, E.A., van den Bosch, D.D., Wernsen, M.W.M.: A benchmark structure of validation of experimental substructuring, transfer path analysis and source characterisation techniques. In: Dynamics of Coupled Structures, Proceedings of the 35th IMAC, a Conference and Exhibition on Structural Dynamics, vol. 1, chapter 6, pp. 73–82, Springer, New York (2017)
6. de Klerk, D., Rixen, D., de Jong, J.: The frequency based substructuring method reformulated according to the dual domain decomposition method. In: Proceedings of the XXIV International Modal Analysis Conference (IMAC), St. Louis. Society for Experimental Mechanics, Bethel (2006)

The manufacturer's authorised representative in the EU is Springer Nature Customer Service Centre GmbH, Europaplatz 3, 69115 Heidelberg, Germany. If you have any concerns regarding our products, please contact ProductSafety@springernature.com

Printed and bound by CPI Group (UK) Ltd, Croydon, CR0 4YY
05/10/2026
02228168-0001